New Approaches in Spectral Decomposition

Recent Titles in This Series

(*Continued in the back of this publication*)

CONTEMPORARY MATHEMATICS

128

New Approaches in Spectral Decomposition

Ridgley Lange
Shengwang Wang

American Mathematical Society
Providence, Rhode Island

1991 *Mathematics Subject Classification*. Primary 47A10, 47A13, 47A15, 47A55, 47B20, 47B40; Secondary 46E10, 46E35, 46F05, 46J15, 46M20, 46M40, 47A53, 47A60, 47B10, 47B15, 47B66, 47F05.

Library of Congress Cataloging-in-Publication Data

Lange, Ridgley, 1942–

New approaches in spectral decomposition/Ridgley Lange and Shengwang Wang.

p. cm.—(Contemporary mathematics, ISSN 0271-4132; v. 128)

Includes bibliographical references and index.

ISBN 0-8218-5139-X

1. Operator theory. 2. Spectral theory (Mathematics) 3. Decomposition (Mathematics) I. Wang, Shengwang, 1932– . II. Title. III. Series: Contemporary mathematics (American Mathematical Society); v. 128.

QA329.L36 1992   92-6183

515′.724—dc20   CIP

Printed in the United States of America.

The paper used in this book is acid-free and falls within the guidelines established to ensure permanence and durability. ♾

This volume was printed directly from author-prepared copy.

10 9 8 7 6 5 4 3 2 1   97 96 95 94 93 92

TABLE OF CONTENTS

PREFACE

In this expository volume on some recent developments in the various theories of spectral decomposition, we attempt to bring to a general mathematics audience a readable, reasonably advanced account of the techniques used in the theory and applications of decomposable operators and related classes of operators.

The first chapter on the fundamental properties of decomposable operators begins with a characterization theorem for these operators; this material is drawn mostly from several recent papers by the authors. The remainder of Chapter I is devoted to the study of various subclassses of decomposable operators. For example, in Section 3 we give criteria for strongly decomposable and quasi-strongly decomposable operators as well as those which are decomposable relative to the identity. In the last section of the chapter we study the relationships among those classes already mentioned and many others (ten altogether) which have been introduced in the literature. We show that these ten classes are linearly ordered by inclusion and that most of them are distinct.

In Chapter II, Section 1, we present several types of results on perturbation of a decomposable operator by a commuting operator. In the next section we specialize from Banach to Hilbert space and we prove that certain compact perturbations of a normal operator having spectrum in a Jordan curve are decomposable. This result hinges on the estimate of the growth of the local resolvent of the perturbed operator. Many results in the last section on quasisimilarity follow from the earlier perturbation theorems.

Chapter III introduces a class of operators which we call "weakly decomposable relative to the identity." We begin with the basic properties of these operators; for example, they have the important single-valued extension property. The next section discusses the multiplication operator associated with one weakly decomposable relative to the identity. By example, we show that, although weakly decomposable operators in the present sense are quasidecomposable, they need not be decomposable. In fact, unlike the situation in Chapter I, this new class is not comparable to other classes. We also generalize to our new class some old results of Colojoara and Foias for maximal hyperinvariant subspace chains of a given A-spectral operator.

Our principal aim in Chapter III comes in Section 5, where we prove that any linear transformation intertwining a decomposable operator and one weakly decomposable relative to the identity is automatically bounded (given supplementary conditions). (Operator A intertwines T and S if AT = SA). The following section of the chapter gives several applications of the automatic continuity theorem, and in the final section of the chapter we give further perspectives on this notion.

Chapter IV deals with some recent applications of subdecomposable operators, i.e. those which are restrictions of decomposable operators. The principal general result here is that if the restriction of an "unconditional- ly" decomposable operator has spectrum with nonempty interior, then such a restriction has nontrivial invariant subspaces.

Next we prove that hyponormal operators are subdecomposable, more precisely, subscalar. Those hyponormal operators with spectra having empty interior may also have nontrivial invariant subspaces, but the proof is different from the case above since the scalar extension of a hyponormal operator is not clearly unconditionally decomposable.

Our final chapter is an expository survey of some research that has been done on the spectral decomposition of commuting systems of (bounded) operators in Banach space. This direction of study, initiated by E. Albrecht in the early 1970's, is based on J. L. Taylor's notion of joint spectrum. We include this material to give a certain completeness to our discussion.

Our point of view in writing the proofs in this book has been to make them accessible to the nonspecialist and to bring out the rich interplay between current operator theory and other branches of classical analysis. Thus the reader with a sound, rudimentary acquaintance with functional analysis (Banach space), function theory, harmonic analysis, elementary topology and homological algebra should be able to follow the arguments. To this degree, the volume is self-contained. But, of course, no work of mathematics can be truly "self-contained," so we have given the details of all proofs except those which would take the discussion clearly out of the essential line of development. In those few cases we cite an outside reference.

When the authors first prepared this volume, the second named author was visiting Central Michigan University and the University of Pittsburgh. He would like to express his thanks to Professors R. Fleming, W. Deskins, S. Hastings, K. S. Lau and H. Cohen for their kind invitations. He is also indebted to the committee of the National Natural Science Foundation of China for their valuable support.

The Authors

Chapter I - FOUNDATIONS

1. Basic Concepts.

Our basic idea is the following. Let T be a bounded linear operator on the complex Banach space X. Then T has a "spectral decomposition" if for each finite open cover $\{G_1, \ldots G_n\}$ of the complex plane there correspond n invariant subspaces $M_1, M_2, \ldots, M_n$ that span X and each spectrum $\sigma(T|M_i)$ is contained in G_i. Thus a spectral decomposition of T consists of a decomposition of X and a corresponding decomposition of $\sigma(T)$ each part of which is localized by some G_i. In the sequel this necessarily vague notion will be modified, generalized and specialized, but for now we note that it generalizes normal operators (Hilbert space) and the Dunford spectral operators. Whereas these latter types produce direct sum decompositions, let us remark that in our present case the sum $M_1 + M_2 + \ldots + M_n$ need not be direct.

1.1.1 Definition. Let T be a bounded linear operator on the Banach space X: $T \in L(X)$. We say T has the _spectral decomposition property_ (SDP) if, for every finite open cover $\{G_i: 1 \leq i \leq n\}$ of the plane **C**, there corresponds a system $\{M_i\}$ of T-invariant subspaces such that

(1.1) $$X = M_1 + M_2 + \ldots + M_n$$

(1.2) $$\sigma(T|M_i) \subset G_i \ (1 \leq i \leq n)$$

One variant of this notion is the _decomposable_ operator, where in (1.1) the spaces M_i must be "spectral maximal," i.e. each M contains every invariant subspace Y for which $\sigma(T|Y) \subset \sigma(T|M)$. In the next section we give nine criteria for an operator to be decomposable, while Section 3 studies characterizations of some subclasses of decomposable operators, including those of strongly decomposable, decomposable relative to the identity and quasi-strongly decomposable operators. Section 4 is devoted to the hierarchy of special classes of spectral decomposition.

2. Decomposable Operators

To make the statement of Theorem 1.2.1 intelligible, we recall the following notions and notations. Let T be a bounded linear operator on the arbitrary Banach space X. For F closed in **C** define

(2.1)
$$X(T,F) = \{x \in X: \ x = (\lambda - T)f(\lambda) \text{ for some analytic } f: \mathbf{C}\backslash F \to X\};$$

for $y \in X$ define
$$\gamma(y,T) = \cap\{F: \ y \in X(T, F)\};$$

T has the <u>single valued extension property</u> (SVEP) if, for any open $D \subset \mathbf{C}$, $f = 0$ is the only analytic $f: D \to X$ satisfying $(\lambda - T)f(\lambda) = 0$. In this case, for each $x \in X$ there is a maximal analytic $f_x: D_x \to X$ satisfying $x = (\lambda - T)f_x(\lambda)$ on D_x [24. p. 1]. Put $\sigma(x,T) = \mathbf{C}\backslash D_x$. Then $\sigma(x,T) = \gamma(x,T)$ and is called the <u>local spectrum</u> at x.

A stronger property than SVEP is (β); T has property (β) if for every sequence $f_n: D \to X$ such that $(\lambda - T)f_n(\lambda) \to 0$ uniformly on compact sets in D it follows that $f_n \to 0$ uniformly on compact sets in D. It is an easy exercise to show that $(\beta) \Rightarrow$ SVEP and (β) implies that the manifolds (2.1) are norm-closed if F is closed. More generally, if T has SVEP such that (2.1) is closed if F is, then we say T has property (κ).

Other notations are as follows: T^* is the adjoint of T; $\sigma(T)$, $\rho(T)$ are the spectrum, resolvent set of T, resp.; $R(\lambda;T)$ is the resolvent operator of T at λ; if Y is T-invariant, then $T|Y$ is its restriction and T/Y is the coinduced operator on X/Y (i.e. define T/Y at coset $x + Y$ by $(T/Y)(x + Y) = Tx + Y$); $Y^\perp$ denotes the annihilator of Y in X^*. Finally for a set A in a topological space we denote the closure of A by A^-.

1.2.1. Theorem. Let $T \in L(X)$. Then the following assertions are equivalent.

(i) T is decomposable.

(ii) for every pair of open discs G, H, with $G^- \subset H$, there exist T-invariant subspaces X_G, X_H such that

$$X = X_G + X_H$$
$$\sigma(T|X_H) \subset H \text{ and } \sigma(T|X_G) \subset \mathbf{C}\backslash G;$$

(iii) for every pair G, H of open discs with $G^- \subset H$, there exist T-invariant subspaces Y, Z such that

(a) $\sigma(T|Y) \subset \mathbf{C}\backslash G$ and $\sigma(T/Y) \subset H$,

(b) $\sigma(T|Z) \subset H$ and $\sigma(T/Z) \subset \mathbf{C}\backslash G$;

(iv) both T and T* have property (β);

(v) T has property (β) and T* has property (κ);

(vi) T has property (β) and T* has SVEP such that $X^*(T^*, \mathbf{C}\backslash D)$ is closed for each open disc D;

(vii) there exists a T-invariant subspace Y such that $T|Y$ and T/Y are decomposable;

(viii) there is a T-invariant subspace Y such that $T|Y$ and $T^*|Y^\perp$ are decomposable;

(ix) for every open cover {G, H} of **C** there is a linear map $P: X \to X$ such that

(a) for $x \in X$, $\gamma(Px,T) \subset G^-$ and $\gamma(x - Px, T) \subset H^-$,

(b) for closed $F \subset G\backslash H^-$ and $x \in X(T,F)^-$ we have $Px = x$,

(c) for closed $K \subset H\backslash G^-$ and $x \in X(T,K)^-$ we have $Px = 0$.

(x) for every cover {G,H} of **C** where G is an open disc and H is the complement of a closed disc, there is a linear transformation $P: X \to X$ satisfying (ix, a-c).

The proof of Theorem 1.2.1. will be organized into three cycles:

(1) (i) ⇒ [(ii) or (iii)] ⇒ (iv) ⇒ (v) ⇒ (vi) ⇒ (i);

(2) (i) ⇒ (vii) ⇒ (viii) ⇒ (i);

(3) (i) ⇒ (ix) ⇒ (x) ⇒ (i).

In order to prove cycle (1) we must develop some properties of pairs of Banach spaces of analytic functions. These results were proved by Bishop [20] for the case of reflexive Banach spaces. Our aim here is to extend them to the general case. Our notations follow [20].

Let U_1 and U_2 be Cauchy domains in **C** such that U_2 is bounded and U_1 is unbounded and such that $U_1 = \mathbf{C}\backslash U_2^-$. We call (U_1, U_2) a <u>complementary simple pair</u>. Let W_1 be the set of analytic $f : U_1 \to X$ vanishing at $z = \infty$, and let W_2 be the set of analytic $g : U_2 \to X^*$. The seminorms

$$\|f\|_{K_1} = \max\{\|f(z)\|: z \in K_1\}, \quad K_1 \text{ compact in } U_1,$$

$$\|g\|_{K_2} = \max\{\|g(z)\|: z \in K_2\}, \quad K_2 \text{ compact in } U_2,$$

generate locally convex topologies on W_1, W_2 respectively. For $i = 1, 2$, let V_i denote the set of vectors in W_i extending continuously to U_i^-. Then let the respective norms on V_i be the usual sup-norm over U_i (or equivalently, U_i^-). This makes V_1 and V_2 Banach spaces. For $x \in X$, $\lambda \in U_2$ define an <u>elementary</u> vector $\alpha(x, \lambda, \mu)$ in V_1 by

$$\alpha(x, \lambda, \mu) = (\mu - \lambda)^{-1}x \qquad (\mu \in U_1)$$

Let $V \subset V_1$ be the closed span of elementary vectors. Moreover, the formula

$$\langle f, g \rangle = (2\pi i)^{-1}\int_\Gamma \langle f(\lambda), g(\lambda)\rangle\, d\lambda \qquad (f \in V_1, g \in V_2) \tag{2.2}$$

defines a bilinear form on $V_1 \times V_2$ where Γ is the mutual boundary of U_1 and U_2.

1.2.2 Lemma. With U_i, V_i, W_i ($i = 1, 2$) and V defined as above, there exists a linear subspace Y in W_2 such that

(i) Y is isometrically isomorphic to V^*;

(ii) $V_2 \subset Y$;

(iii) the imbeddings $V_2 \to Y$ and $Y \to W_2$ are continuous;

(iv) the bilinear form (2.2) on $V \oplus V_2$ extends to $V \oplus Y$ via the identification (i).

Proof. (i) Fix $\lambda \in U_2$, $\phi \in V^*$. Then $\langle\alpha(x, \lambda, \cdot), \phi\rangle$ $(x \in X)$ defines a linear functional on X. If we put $\langle x, g(\lambda)\rangle = \langle\alpha(x, \lambda, \cdot), \phi\rangle$, then $\lambda \to \langle x, g(\lambda)\rangle$ is analytic on U_2, hence $g : \lambda \to g(\lambda)$ is analytic. Let Y be the set of all such g. Clearly the correspondence $\phi \to g$ is one-to-one. Then, defining $\|g\| = \|\phi\|$, we make V^* isometrically isomorphic to Y.
(ii) To prove $V_2 \subset Y$ let $g \in V_2$, and let $f = \alpha(x, \lambda, \cdot)$. By (2.2)

$$\langle f, g\rangle = (2\pi i)^{-1}\int_\Gamma (\mu - \lambda)^{-1}\langle x, g(\mu)\rangle\, d\mu = \langle x, g(\lambda)\rangle.$$

Thus $V_2 \subset V^* = Y$ by (i). For (iii) let $\|g_n\| \to 0$ in V_2. Then the calculation

$$\|g_n\|_Y = \sup\{|\langle f, g_n\rangle| : \|f\| = 1, f \in V\}$$

$$= (2\pi i)^{-1} \sup\{|\int_\Gamma \langle f(\mu), g_n(\mu)\rangle d\mu| : f \in V, \|f\| = 1\}$$

$$\leq (2\pi i)^{-1} \int_\Gamma |g_n(\mu)||d\mu|$$

$$\leq (2\pi i)^{-1} M\|g_n\| \qquad (M > 0 \text{ independent of } n)$$

proves that $V_2 \to Y$ is continuous. Next let $\|g_n\|_Y \to 0$, let $K \subset U_2$ be compact and let $x \in X$ with $\|x\| \leq 1$. Hence

$$|\langle x, g_n(\lambda)\rangle| = |\langle \alpha(x, \lambda, \cdot), g_n\rangle| \leq \delta^{-1}\|x\|\,\|g_n\|_Y \to 0$$

uniformly for $\lambda \in K$, where $\delta = \mathrm{dist}(K, U_1)$. Hence $g_n \to 0$ in W_2. Assertion (iv) follows from the proof of (ii).

From the definition of V it is not hard to see that the map τ: $V \to X$ given by $\tau f = \lim_{\mu \to \infty} \mu f(\mu)$ is well-defined and surjective: $\tau\alpha(x, \lambda, \cdot) = x$ for $x \in X$. Moreover, its adjoint τ^*: $X^* \to V^*$ is given by $(\tau^* x^*)(\lambda) = x^*$, for $x^* \in X^*$, $\lambda \in U_2$ as shown by the following calculation: if $f \in V$ then

$$\langle \tau f, x^*\rangle = \lim_{\mu\to\infty} \langle \mu f(\mu), x^*\rangle = \lim_{\mu\to\infty} \frac{\mu}{2\pi i} \int_\Gamma \frac{\langle f(\lambda), x^*\rangle}{\mu - \lambda} d\lambda$$

$$= \frac{1}{2\pi i} \int_\Gamma \langle f(\lambda), x^*\rangle \, d\lambda = \langle f, x^*\rangle$$

We also define $H \in L(V)$ by

$$(2.3) \qquad (Hf)(\mu) = \tau f - (\mu - T) f(\mu) \qquad (f \in V, \mu \in U_1)$$

To find the formula for H^*, first recall that if $f = \alpha(x, \lambda, \cdot)$ ($x \in X$) and $g \in V^*$, then $\langle f, g\rangle = \langle x, g(\lambda)\rangle$; thus for $\mu \in U_1$,

$$\begin{aligned}(Hf)(\mu) &= x - (\mu - T)(\mu - \lambda)^{-1} x \\ &= \alpha(Tx - \lambda x, \lambda, \mu)\end{aligned}$$

or $Hf = \alpha(Tx - \lambda x, \lambda, \cdot)$. Hence for $\lambda \in U_2$, $x \in X$,

$$\begin{aligned}\langle x, (H^*g)(\lambda)\rangle &= \langle f, H^*g\rangle = \langle Hf, g\rangle \\ &= \langle \alpha(Tx - \lambda x, \lambda, \cdot), g\rangle \\ &= \langle x, (T^* - \lambda) g(\lambda)\rangle,\end{aligned}$$

i.e. $(H^*g)(\lambda) = (T^* - \lambda) g(\lambda)$, for $\lambda \in U_2$.

On $V \oplus V$ define the operator K by $K(f_1, f_2) = (f_1, Hf_1 - f_2)$. Straightforward calculations show $K^2 = I$ and $K^*(g_1, g_2) = (H^*g_2+g_1, g_2)$. We shall also consider a family of norms on $V \oplus V$ defined as follows. For $\eta > 0$ let

(2.4) $$\|(f_1,f_2)\|_\eta = (\eta\|f_1\|^2 + \|f_2\|^2)^{1/2} \quad (f_i \in V)$$

Obviously this norm is equivalent to the original one on $V \oplus V$ and its dual on $V^* \oplus V^*$ is

$$\|(g_1,g_2)\|_\eta = (\eta^{-1}\|g_1\|^2 + \|g_2\|^2)^{1/2} \quad (g_i \in V^*).$$

Now let M and N be Banach spaces with norms denoted by $\|\cdot\|$, and let $A: M \to N$ be a continuous surjection and renorm N by

$$\|g\|_1 = \inf\{\|f\| : g = Af\} \quad (g \in N).$$

Then $\|\cdot\|_1$ is clearly equivalent to $\|\cdot\|$ on N and its "dual" norm on N^* satisfies

$$\|g^*\|_1 = \|A^*g^*\| \quad (g^* \in N^*).$$

Dually, if $A: M \to N$ is a homeomorphism into N, then the norm on M defined by $\|f\|_2 = \|Af\|$ is equivalent to the original one and

$$\|f^*\|_2 = \inf\{\|v\| : v \in N^*, f^* = A^*v\}$$

is the dual of $\| . \|_2$ on M.

Now, returning to X, X^*, etc., above, we define for $\eta > 0$ norms on X, X^* as follows. Let $x \in X$, $x^* \in X^*$, and put

(2.5) $$\|x\|_\eta = \inf\{(\eta\|f\|^2 + \|Hf\|^2)^{1/2}: \ x = \tau f, f \in V\}$$

$$\|x^*\|_\eta = \inf\{(\|g\|^2 + \eta^{-1}\|H^*g - \tau^*x^*\|^2)^{1/2}: g \in V^*\}$$

1.2.3. Lemma. The norms (2.5) are dual to each other and each is equivalent to the original one on X, X^*.

Proof. It is convenient to consider the following diagram of maps.

$$\begin{array}{ccc} & & X \\ & & \downarrow \alpha \\ & \beta & \\ V \oplus V & \rightarrow & X \oplus V \end{array}$$

where α is the injection $x \rightarrow (x,0)$ and β is the surjection $(f_1,f_2) \rightarrow (\tau f_1,f_2)$. With the norm $\|\cdot\|_\eta$ given by (2.4) on $V \oplus V$, these maps induce a norm $\|.\|_1$ on X equivalent to the original one according to the remarks above. Now by (2.4) and (2.5) we have

$$\|x\|_1 = \|(x, 0)\|_\eta \qquad \text{(induced by } \beta,\eta)$$

$$= \inf\{\|(f,0)\|_\eta : x = \tau f\}$$

$$= \inf\{\|K(f,0)\|_\eta : x = \tau f\}$$

$$= \inf\{\|(f,Hf)\|_\eta : x = \tau f\}$$

$$= \inf\ \{(\eta\|f\|^2 + \|Hf\|^2)^{1/2} : x = \tau f\} = \|x\|_\eta .$$

A similar calculation shows that the dual norm $\|x^*\|_1$ on X^* satisfies $\|x^*\|_1 = \|x^*\|_\eta$, and this completes the proof.

We now reach the crucial step. Define submanifolds of X, X^* as follows:

$$N = \{x \in X: \text{ for } \varepsilon > 0 \text{ there is } f_\varepsilon \in V \text{ with } x = \tau f_\varepsilon,\ \|Hf_\varepsilon\| < \varepsilon\},$$

$$M = \{x^* \in X^*: \ \tau^* X^* = H^* g \text{ for some } g \in V^*\}.$$

1.2.4 Lemma. The subspace N consists of all x such that $\|x\|_\eta \rightarrow 0$ (as $\eta \rightarrow 0$) and M consists of all $x^* \in X^*$ with $\{\|x^*\|_\eta : \eta > 0\}$ bounded; moreover $N^\perp$ is the weak* closure of M.

Proof. Let $S = \{y \in X: \|y\|_\eta \rightarrow 0 \text{ as } \eta \rightarrow 0\}$. For $x \in N$ and $\eta > 0$, choose $f \in V$ with $x = \tau f$ and $\|Hf\| < \varepsilon$. Then $\lim_{\eta\rightarrow 0}\|x_\eta\| \leq \varepsilon$, and it follows that $N \subset S$. To prove $S \subset N$, reverse this argument.

Next, let $Q = \{x^* \in X^* : \|x^*\|_\eta \text{ bounded }\}$. Let $u \in M$ and $g \in V^*$ be such that $\tau^* u = H^* g$. For $\eta > 0$,

$$\|u\|_\eta \leq \{\eta^{-1}\|\tau^* u - H^* g\|^2 + \|g\|^2\}^{1/2} = \|g\|.$$

Thus $M \subset Q$. Now let $u \in Q$, i.e. let $\sup \|u\|_\eta < r$ for some $r > 0$. For n

$=1,2,\ldots$, choose $g_n \in V^*$ such that

$$\eta\|\tau^*u - H^*g_n\|^2 + \|g_n\|^2 < r^2.$$

In particular, $\{g_n\}$ is bounded and hence there is a subnet $g_m \to g_0$ (weak*) for some $g_0 \in V^*$. Then $H^*g_m \to H^*g_0$ (weak*). Obviously $\tau^*u = H^*g_0$, so $u \in M$; this proves $M = Q$.

To prove that $N^\perp$ is the weak* closure M^w, note first that for $x \in N$, $x^* \in M$ we must have $|\langle x,x^*\rangle| = 0$ since $|\langle x,x^*\rangle| \leq \|x\|_\eta\|x^*\|_\eta \to 0$ $(\eta \to 0)$. Hence $M \subset N^\perp$ so $M^w \subset N^\perp$. For the opposite inclusion, let $x \notin N$. Then there exist $a > 0$ and $\eta_m \to 0$ such that $\|x\|_{\eta_m} > a$ for $m = 1, 2, \ldots$. Use the Hahn-Banach theorem to find for each m some $x_m^* \in X^*$ such that $\|x_m^*\|_\eta < 1$, $x_m^* \perp N$ and

$$|\langle x, x_m^*\rangle| > a. \tag{2.6}$$

For each m also choose $g_m \in V^*$ such that

$$\|g_m\|^2 + \eta_m^{-1}\|H^*g_m - \tau^*x_m^*\|^2 < 1$$

Then $\{(x_m^*,g_m)\}$ is bounded in $X^* \times V^*$ since $\{g_m\}$ obviously is and since

$$\begin{aligned}\|x_m^*\| &= \|\tau^*x_m^*\| \leq \|\tau^*x_m^* - H^*g_m\| + \|H^*g_m\| \\ &\leq \eta_m^{1/2}(1 - \|g_m\|^2)^{1/2} + \|H^*\|\,\|g_m\| \\ &\leq C + \|H^*\|\,\|g_m\| \quad \text{(for some } C > 0).\end{aligned}$$

It follows that there is a subnet which we also denote $\{(x_m^*, g_m)\}$ converging weak* to some (x^*,g). Then $\|\tau^*x_m^* - H^*g_m\| \leq \eta_m^{1/2} \to 0$ implies $x^* = \tau^*x^* = H^*g$, and hence $x^* \in M$. In view of (2.6) we must have

$$|\langle x, x^*\rangle| \geq a \tag{2.7}$$

This implies that the preannihilator ${}^\perp M \subset N$ because $x \notin N$ was arbitrary. By (2.7) we thus obtain $N^\perp \subset M^w$, and this completes the proof.

We shall also need the following results.

1.2.5 Lemma. If $T \in L(X)$ has SVEP and M is a T-invariant subspace, then $TM = M$ implies that T is bijective on M.

Proof. Suppose $Tx_0 = 0$ for nonzero $x_0 \in M$. By the open mapping theorem there is $k > 0$ and nonzero $x_1 \in M$ with $x_0 = Tx_1$ and $\|x_1\| < k\|x_0\|$. Inductively, choose $x_{j+1} \in M$ such that $x_j = Tx_{j+1}$ and $\|x_{j+1}\| \leq k\|x_j\|$ $(j = 0, 1, 2, \ldots)$. Then $\|x_j\| \leq k^j\|x_0\|$ for each j, and the function defined by

$$f(\lambda) = \sum_{j=0}^{\infty} x_j\lambda^j$$

is analytic on the disc $D = \{\lambda: |\lambda| < k^{-1}\}$. By construction, $f \neq 0$ and $(\lambda - T)f(\lambda) = 0$ on D. This contradicts SVEP, hence T is injective on M.

1.2.6 Lemma. Let $M \subset X$ be a closed subspace of X and suppose $f:D \to X/M$ is analytic on disc $D = \{\lambda: |\lambda| < r\}$. Then there is a concentric disc $D_1 \subset D$ and $g:D_1 \to X$ such that $f(\lambda) = g(\lambda) + M$ (= coset of $g(\lambda)$) for
$\lambda \in D_1$.

Proof. Write $f(\lambda) = \Sigma_0^\infty a_n\lambda^n$ $(a_n \in X/M)$. For $n = 0, 1, 2, \ldots$, choose $b_n \in X$ with $a_n = b_n + M$ and $\|b_n\| \leq \|a_n\| + (r/2)^{-n}$. Then the power series

$$g(\lambda) = \sum_{n=0}^{\infty} b_n\lambda^n$$

converges to an analytic function on $D_1 = \{\lambda: \lambda < r/2\}$ and satisfies the conclusion.

1.2.7. Lemma. Suppose $T^* \in L(X^*)$ has SVEP and F is closed. If $X^*(T^*,F)$ is norms closed, then it is weak* closed.

Proof. By the Krein-Smulyan theorem it suffices to prove that the norm closed unit ball S^* in $X^*(T,F)$ is weak* closed. Let $\{x_\beta^*\} \subset S^*$ be a net converging weak* to x^*. Let (U_1, U_2) be a complementary simple pair with U_1 unbounded and $F \subset U_1$. For each β the function f_β defined by

$$f_\beta(\lambda) = R(\lambda;\, T^*|X^*(T,F))x_\beta^*$$

is analytic on U_2^-. Since $U_2^- \subset \rho(T^*|X^*(T,F))$ and $\|x_\beta^*\| \leq 1$ it follows that $\{f_\beta\}$ is bounded in V_2. By Lemma 1.2.2 $\{f_\beta\}$ is also bounded in V^*, so this net has a subnet $\{f_\gamma\}$ converging weak* to $f \in V^*$. Then for $x \in X$, $\lambda \in U_2^-$,

$$\begin{aligned} \langle x, x^*\rangle &= \lim_\gamma \langle x, x_\gamma^*\rangle = \lim_\gamma \langle x, (\lambda - T)^* f_\gamma(\lambda)\rangle \\ &= \lim_\gamma \langle(\lambda - T)x, f_\gamma(\lambda)\rangle = \lim_\gamma \langle \alpha((\lambda - T)x, \lambda, \cdot), f_\gamma\rangle \\ &= \langle \alpha((\lambda - T)x, \lambda, \cdot), f\rangle = \langle x, (\lambda - T^*)f(\lambda)\rangle. \end{aligned}$$

This calculation proves $x^* = (\lambda - T^*)f(\lambda)$ on U_2^-. Hence $x^* \in S^* \cap X^*(T^*,F)$ since U_2 was arbitrary with $F \cap U_2^- = \emptyset$, so $X^*(T,F)$ is weak* closed.

The next lemma gives sufficient conditions for property (β).

1.2.8 Lemma. Let $T \in L(X)$. Then T has property (β) if either of the following holds:

(i) there exists an invariant subspace Y such that both $T|Y$ and T/Y have property (β);

(ii) for each pair of open discs G, H with $G^- \subset H$, there exists an invariant subspace Y such that

$$\sigma(T|Y) \subset C\backslash G \quad \text{and} \quad \sigma(T/Y) \subset H.$$

Proof. Let $f_n\colon D \to X$ be a sequence of analytic functions such that $(\lambda - T)f_n(\lambda) \to 0$ uniformly on compact sets in D. Let K be an open disc in D. Without loss of generality we take $K \subset K^- \subset K_0 \subset K_0^- \subset D$ to be concentric discs.

Suppose (i) holds. We can write

$$f_n(\lambda) = \sum_{k=0}^{\infty} a_{nk}\lambda^k$$

where each series is analytic and converges on K_0^-. Denote by $f_n(\lambda)^\wedge$ the coset of $f_n(\lambda)$ in X/Y. By hypothesis (i) $f_n^\wedge \to 0$ uniformly on K_0^- in X/Y and

$$f_n(\lambda)^{\hat{}} = \sum_{k=0}^{\infty} a_{nk}{}^{\wedge}\lambda^k \qquad (\lambda \in K_0^-)$$

Put $p_n = \max\{\|f_n(\lambda)^{\hat{}}\|: \lambda \in K_0^-\}$, so that $p_n \to 0$. Let r_0 be the radius of K_0. By Cauchy's inequality $\|a_{nk}{}^{\hat{}}\| \leq p_n/r_0{}^k$, i.e. $\operatorname{dist}(a_{nk}, Y) \leq p_n/r_0{}^k$.

Now choose $b_{nk} \in Y$ such that

$$\|a_{nk} - b_{nk}\| \leq 2p_n r_0{}^{-k}, \tag{2.8}$$

From the resulting inequality,

$$\|b_{nk}\| \leq \|a_{nk}\| + 2p_n r_0{}^{-k},$$

it follows that the series

$$g_n(\lambda) = \sum_{k=0}^{\infty} b_{nk}\lambda^k$$

converges and is analytic on K^-. But then by (2.8) $g_n - f_n \to 0$ on K^- and

$$(\lambda - T)g_n(\lambda) = (\lambda - T)(g_n(\lambda) - f_n(\lambda)) + (\lambda - T)f_n(\lambda) \to 0$$

and uniformly on K^-. Since $\operatorname{range}(g_n) \subset Y$, we see $g_n \to 0$ uniformly on K^-. Hence $f_n \to 0$ uniformly in K^-. By the Heine-Borel theorem, $f_n \to 0$ uniformly on compact sets in D.

Now suppose (ii) holds. Let G, H be discs concentric with D such that $K \subset G^- \subset H \subset H^- \subset D$. By hypothesis there exists a T-invariant Y such that $\sigma(T/Y) \subset H$ and $\sigma(T|Y) \subset C\backslash G$. Since clearly $(\lambda - T/Y)f_n(\lambda)^{\hat{}} \to 0$ uniformly on H, where $f_n(\lambda)^{\hat{}} = f_n(\lambda) + Y$, then on ∂H (boundary of H) $f_n{}^{\hat{}} \to 0$ uniformly. Hence $f_n{}^{\hat{}} \to 0$ uniformly on H^- by the maximum modulus principle. Also $(T/Y)f_n{}^{\hat{}} \to 0$ uniformly on H^-.

The graphs $G(T)$ and $G(T|Y)$ are closed in $X \times X$ and $Y \times Y$, respectively. Define $\tau: G(T)/G(T|Y) \to G(T/Y)$ by

$$\tau[(x,Tx) + G(T|Y)] = [x + Y, Tx + Y].$$

Since

$$\|(x,Tx) + G(T|Y)\| = \inf\{\|(x,Tx) + (y,Ty)\|: y \in Y\}$$

$$\geq \inf\{\|(x + y_1, Tx + y_2)\|: y_1, y_2 \in Y\}$$

$$= \|(x + Y, Tx + Y)\|$$

and τ is easily seen to be bijective, τ is bicontinuous by the closed graph theorem.

The map $\lambda \to (f_n(\lambda)^\wedge, (T/Y)f_n(\lambda)^\wedge)$ is analytic on H^-, hence the map

$$\phi: \lambda \to \tau^{-1}(f_n(\lambda)^\wedge, (T/Y)\, f_n(\lambda)^\wedge)$$

is also analytic on H^-. Since ϕ has values in $G(T)/G(T|Y)$, by Lemma 1.2.6 there is an analytic function (for each n) from H to G(T) of the form (h_n, Th_n) such that

$$h_n(\lambda) \oplus Th_n(\lambda) \in \tau^{-1}(f_n(\lambda)^\wedge, (T/Y)f_n(\lambda)^\wedge)$$

and we may choose h_n such that there exists $A > 0$ independent of n with

(2.9) $\sup_{\lambda \in G^-} \|(h_n(\lambda), Th_n(\lambda))\| \leq A \max_{\mu \in H^-} \|\tau^{-1}(f_n^\wedge(\mu), (T/Y)f_n^\wedge(\mu))\|.$

Then $h_n^\wedge = f_n^\wedge$ on G so $h_n(\lambda) - f_n(\lambda) \in Y$ for $\lambda \in G$. But $(f_n(\lambda)^\wedge, (T/Y)f_n(\lambda)^\wedge) \to 0$ uniformly on H^-, hence (2.9) implies $(h_n(\lambda), Th_n(\lambda)) \to 0$ uniformly on G^-. From the hypothesis $(\lambda - T)f_n(\lambda) \to 0$, we have $(\lambda - T)(f_n(\lambda) - h_n(\lambda)) \to 0$ uniformly on G^-. Since $G^- \subset \rho(T|Y)$ and $\text{range}(f_n - h_n) \subset Y$, we conclude that $h_n - f_n \to 0$, hence $f_n \to 0$, uniformly on G^-. This completes the proof.

We recall that an operator is 2-decomposable if (1.1) and (1.2) hold for covers of pairs of sets (i.e. $n = 2$).

1.2.9. Lemma. Every 2-decomposable operator is decomposable.

Proof. Let $\{H_1, H_2, \ldots, H_n\}$ be an open cover of the compact set $F \subset \sigma(T)$ and let T be 2-decomposable. It suffices to prove

$$X(T,F) \subset X(T,H_1^-) + \ldots + X(T,H_n^-),$$

for then it is clear that (1.1) holds for n summands. Let $n = 2$, and put $K = H_1^- \cap H_2^-$ and $M = X(T,K)$. We show that $\sigma(T/M)$ is disjoint from $H_1 \cap H_2$. For $\lambda \in H_1 \cap H_2$, let V be open such that $\lambda \notin V^-$ and

$\{V, H_1 \cap H_2\}$ covers C. By the remark above each $y \in X$ can be expressed $y = w_1 + w_2$ with $\sigma(w_1,T) \subset V^-$ and $\sigma(w_2,T) \subset K$. The coset $y\hat{} = y + M$ satisfies $y\hat{} = (\lambda - T/M)w_1\hat{}$. Since an easy argument shows that T/M has SVEP [31, Prop. 4.14], Lemma 1.2.5 implies that $\lambda \in \rho(T/M)$. This proves the claim $\sigma(T/M) \cap H_1 \cap H_2 = \emptyset$. Hence for $x \in X(T,F)$ the local spectrum $\sigma(x\hat{},T/M)$ is contained in the disjoint union of $F\backslash H_i$ $(i = 1,2)$. Let f be the local resolvent of $x\hat{}$. and define

$$v_i = (2\pi i)^{-1}\int_{\Gamma_i} f(\lambda)d\lambda,$$

where Γ_i is a contour in $\rho(T/M)$ surrounding $F\backslash H_i$ and having disjoint "interiors." Hence $\sigma(v_i,T/M) \subset F\backslash H_j \subset H_i$ $(i \neq j)$. For $y_i \in v_i \in X/M$, standard arguments now show that $\sigma(y_i,T) \subset \sigma(v_i, T/M) \cup K$ [31, Prop. 3.1], hence $x = y_1 + y_2 + u$ for some $u \in M$ and

$$\sigma(y_i + (1/2)u,T) \subset H_i^- \cup K = H_i^-.$$

This proves the proposition for $n = 2$, and the result for $n > 2$ follows by induction.

Proof of Theorem 1.2.1. (i) $\Rightarrow$ (ii) is clear.

(ii) $\Rightarrow$ (iv). Since $\sigma(T|X_G) \subset C\backslash G$, in order to invoke Lemma 1.2.8(ii) we need only prove $\sigma(T/X_G) \subset H$. But $X = X_G + X_H$, hence T/X_G is similar to $(T|X_H)/(X_G \cap X_H)$ and thus their spectra coincide. Because H is convex, $\sigma(T/X_G) \subset H$. Hence T has property (β) by Lemma 1.2.8(ii). As for T^*, let K be an open disc with $G^- \subset K \subset K^- \subset H$. Then there exist T-invariant X_G and X_K such that

$$(2.10) \qquad \begin{aligned} X &= X_G + X_K \\ \sigma(T|X_K) &\subset K \\ \sigma(T|X_G) &\subset C\backslash G. \end{aligned}$$

Now T having property (β) ensures that $X(T,K^-)$ and $X(T, C\backslash G)$ are closed, hence $X_K \subset X(T,K^-)$ and $X_G \subset X(T,C\backslash G)$. Thus by (2.10) X is the linear span of $X(T,K^-)$ and $X(T, C\backslash G)$. The coinduced operator $T/X(T, K^-)$ is similar to

$$[T|X(T,\mathbf{C}\backslash G)]/[X(T,\mathbf{C}\backslash G)\cap X(T,K^-)] = [T|X(T,\mathbf{C}\backslash G)]/[X(T,(\mathbf{C}\backslash G)\cap K^-)],$$

hence $\sigma(T/X(T,K^-)) \subset \mathbf{C}\backslash G$ by [31, Prop. 3.1]. Now T^* has property (β) by Lemma 1.2.8(ii).

(i) $\Rightarrow$ (iii). Let G, H be open discs with $G^- \subset H$ and let K be another open set with $G^- \subset K \subset K^- \subset H$. Put $Y = X(T,\mathbf{C}\backslash G)$ and $Z = X(T,K^-)$. Since $\{K, \mathbf{C}\backslash G^-\}$ is a cover of $\mathbf{C}$ and T is decomposable, we must have $X = Y + Z$. It then follows (as in a previous part of the proof) that

$$\sigma(T/Y) = \sigma[(T|Z)/(Z \cap Y)] \subset K^- \subset H.$$

This and the fact that $X(T,\mathbf{C}\backslash G)$ is closed imply (iii,a); (iii,b) follows in a similar way.

(iii) $\Rightarrow$ (iv). That T has property (β) follows from (a) and Lemma 1.2.8(ii). To see that T^* has property (β), let Z be as in (b). Hence by duality

$$\sigma(T^*|Z^\perp) = \sigma(T/Z) \subset \mathbf{C}\backslash G$$
$$\sigma(T^*/Z^\perp) = \sigma(T|Z) \subset H,$$

so the result follows by Lemma 1.2.8.

(iv) $\Rightarrow$ (v). Clear because $(\beta) \Rightarrow (\kappa)$.

(v) $\Rightarrow$ (vi) is clear.

(vi) $\Rightarrow$ (i). Let G be open. To prove T decomposable it suffices to find a T-invariant subspace Y with $\sigma(T|Y) \subset G^-$ and $\sigma(T/Y) \subset \mathbf{C}\backslash G$. To see this, let $\{H_1, H_2\}$ be an open cover of $\mathbf{C}$, and choose T-invariant Y with $\sigma(T|Y) \subset (H_1 \cap H_2)^-$ and $\sigma(T/Y) \cap (H_1 \cap H_2) = \emptyset$. We see easily from this that $\sigma(T/Y)$ is the disjoint union $K_1 \cup K_2$ of compact sets with $K_1 \subset H_1^-$, $K_2 \subset H_2^-$. We use the functional calculus to write $X/Y = Z_1 + Z_2$ with each Z_i invariant for T/Y. Hence $X = \phi^{-1}(Z_1) + \phi^{-1}(Z_2)$ where $\phi: X \to X/Y$ is the canonical surjection. Moreover $\phi^{-1}(Z_i)$ is a closed T-invariant subspace such that

$$\sigma(T|\phi^{-1}(Z_i)) \subset K_i \cup (H_1 \cap H_2) \subset H_i.$$

It follows that $X = X(T,H_1^-) + X(T,H_2^-)$, so T is 2-decomposable. By Prop. 1.2.9, T is decomposable.

Returning to (vi), we let $Y = X(T,G)^-$. Then it follows from property (β) that $\sigma(T|Y) \subset G^-$. To prove $\sigma(T/Y) \subset \mathbf{C}\backslash G^-$, we may suppose G is bounded and we apply the theory of complementary simple pairs. Let K be any closed neighborhood of $\lambda = \infty$ such that $\{G, \operatorname{Int} K\}$ covers $\mathbf{C}$. Let U_1, U_2, be a complementary simple pair with U_1 unbounded such that $\mathbf{C}\backslash G \subset U_1 \subset U_1^- \subset \operatorname{Int} K$. Then $\mathbf{C}\backslash K \subset U_2 \subset U_2^- \subset G$. Define M and N as in Lemma 1.2.4. We prove

$$N \subset X(T,G)^-, \tag{2.11}$$

$$M^w \subset X^*(T^*,K). \tag{2.12}$$

To prove (2.11), let $x \in N$. For $n = 1, 2, \ldots$, choose $f_n \in V$ with $Tf_n = x$ and $\|Hf_n\| < 1/n$, where H is defined by (2.3). Hence

$$\|(\lambda - T)f_n(\lambda) - x\| < 1/n \quad (\lambda \in U_1).$$

Since T has property (β), $\{f_n\}$ converges uniformly on compact sets in U_1. Put $f(\lambda) = \lim f_n(\lambda)$ $(\lambda \in U_1)$. Then $(\lambda - T)f(\lambda) = x$ for all $\lambda \in U_1$. In particular, $\sigma(x, T) \subset \mathbf{C}\backslash U_1 = U_2^- \subset G$, and (2.11) follows.

To see (2.12) observe first that $M \subset X^*(T^*,K)$. Since K is a closed neighborhood of ∞, $X^*(T^*,K) = \cap\{X^*(T^*,\mathbf{C}\backslash D)$: D an open disc in $\mathbf{C}\backslash K\}$; thus $X^*(T^*,K)$ is closed. By Lemma 1.2.7 it is weak* closed, and (2.12) follows.

Hence, from (2.11), (2.12) and Lemma 1.2.4, we infer

$$X(T,G)^\perp \subset N^\perp = M^w \subset X^*(T^*,K).$$

Now K is arbitrary with $K \supset \mathbf{C}\backslash G$, hence we obtain

$$X(T,G)^\perp \subset X^*(T^*,\mathbf{C}\backslash G) \subset X(T,G)^\perp$$

and thus $X(T,G)^\perp = X^*(T^*,\mathbf{C}\backslash G)$. By duality,

$$\sigma(T/X(T,G)^-) = \sigma(T^*|X^*(T^*,\mathbf{C}\backslash G)) \subset \mathbf{C}\backslash G$$

and T is decomposable by the previous argument. We now have shown that (i)-(vi) are equivalent.

We next prove (i) $\Rightarrow$ (vii) $\Rightarrow$ (viii) $\Rightarrow$ (i). Since the second of these implications follows easily from duality considerations, we prove (i) $\Rightarrow$ (vii) and (viii) $\Rightarrow$ (i).

(i) $\Rightarrow$ (vii). Let Y = (0). Then $T|Y$ and $T/Y = T$ are both decomposable. (viii) $\Rightarrow$ (i). Let Y be T-invariant and such that $T|Y$ and T/Y are both decomposable. Then these operators both satisfy property (β). By Lemma 1.2.8 T has property (β). Dually, $(T|Y)^* = T^*/Y^\perp$ and $(T/Y)^* = T^*|Y^\perp$ are decomposable, so again by Lemma 1.2.8, T^* has property (β). By (iv) $\Leftrightarrow$ (i), T is decomposable.

To prove (i), (ix) and (x) equivalent, since (ix) $\Rightarrow$ (x) is obvious, we finish the proof by showing (i) $\Rightarrow$ (ix) and (x) $\Rightarrow$ (i).

(i) $\Rightarrow$ (ix). Let $\{G,H\}$ be an open cover of **C**. Then $X = X(T,G^-) + X(T,H^-)$. Put $Y = X(T,G^- \cap H^-)$. By a Hamel basis argument there is a manifold $Z \subset X(T, H^-)$ such that

$$X(T,H^-) = Y \oplus Z \quad \text{(algebraic)}. \tag{2.13}$$

We further show that X has algebraic direct sum

$$X = X(T,G^-) \oplus Z \tag{2.14}$$

Let $x \in X$ be written as $x = x_1 + x_2$ with $x_1 \in X(T,G^-)$ and $x_2 \in X(T,H^-)$. By (2.13) $x_2 = y + z$ with $y \in Y$ and $z \in Z$. Since $y \in X(T,G^-)$, clearly $X = X(T,G^-) + Z$. To prove (2.14), let $u + z = 0$ with $u \in X(T,G^-)$ and $z \in Z$. Then $u = -z \in Z \subset X(T, H^-)$, so $u \in Y$. Hence (2.14) follows from (2.13).

Now let $P: X \to X$ be the projection onto $X(T, G^-)$ along Z. Then $Px = x$ if $x \in X(T,G^-)$ and $Px = 0$ if $x \in Z$. In fact, we have $PX = X(T,G^-)$ and $(I - P)X = Z$. Thus for $x \in X$

$$\gamma(Px,T) = \sigma(Px,T) \subset G^-$$
$$\gamma(x - Px, T) = \sigma(x - Px, T) \subset H^-.$$

This is (ix, a). For (ix,b) let $F \subset G\backslash H^-$ be closed. Then for $x \in X(T,F)$ we have $\sigma(x,T) \subset G$, so $Px = x$. This proves (ix,b); (ix,c) follows in a like way.

(x) $\Rightarrow$ (i). We first prove that for closed F and $\lambda_0 \in F$, $x_0 \in X$, such that $(\lambda_0 - T)x_0 \in X(T,F)$, one has $x_0 \in X(T,F)$. In fact, let f be an analytic X-valued function on $\mathbf{C}\backslash F$ satisfying

$$(\lambda - T)f(\lambda) = (\lambda_0 - T)x_0 \qquad (\lambda \notin F).$$

Then the function $h(\lambda) = (\lambda - \lambda_0)^{-1}[x_0 - f(\lambda)]$ satisfies

$$(\lambda - T)h(\lambda) = (\lambda - \lambda_0)^{-1}[(\lambda-\lambda_0)x_0 + (\lambda_0 - T)x_0 - (\lambda - T)f(\lambda)] = x_0$$

for $\lambda \in \mathbf{C} \backslash F$, which shows $x_0 \in X(T,F)$.

To prove that T has SVEP, let $f\colon D \to X$ be analytic on a connected region D such that

$$(\lambda - T)f(\lambda) = 0 \quad (\lambda \in D).$$

Let δ_1, δ_2 be discs in D with disjoint closures in D. From the previous argument, for $\lambda \in \delta_j$ ($j = 1,2$), we have $f(\lambda) \in X(T,\delta_j^-) \subset X(T,\delta_j^-)^-$. By analytic continuation

$$\text{(2.15)} \qquad f(\lambda) \in X(T,\delta_1^-)^- \cap X(T,\delta_2^-)^- \quad (\text{all } \lambda \in D).$$

Now choose by (x) an open cover $\{G, H\}$ of $\mathbf{C}$ where G is an open disc and H is the complement of a closed disc such that $\delta_1^- \subset G \backslash H^-$ and $\delta_2^- \subset H \backslash G^-$. Let $P\colon X \to X$ be as in (x) satisfying (a)-(c). It follows from (2.15) that $f(\lambda) = Pf(\lambda) = 0$ for $\lambda \in D$. This proves that T has SVEP.

We next prove that T has property (κ). Let F be closed with $\lambda \notin F$, and put $H_\lambda = \{\mu \in \mathbf{C}\colon |\mu - \lambda| > (1/2)\text{dist}(\lambda,F)\}$. The hypothesis asserts existence of $P\colon X \to X$ such that $Px = x$ for $x \in X(T,F)^-$ and $\gamma(Px,T) = \sigma(Px,T) \subset H_\lambda^-$ for all $x \in X$. Thus if $x \in X(T,F)^-$ then $\sigma(x,T) \subset H_\lambda^-$, hence $\lambda \notin \sigma(x,T)$. Since $\lambda \notin F$ was arbitrary, $x \in X(T,F)^-$ implies $\sigma(x,T) \subset F$. Hence $X(T,F)^- \subset X(T,F)$, so T has property (κ).

Finally, let $\{K, H\}$ be an open cover of $\mathbf{C}$ where H is a disc and K is the complement of a disc, and let P be as prescribed. It follows that each $x \in X$ can be written $x = Px + (I - P)x \in X(T,K^-) + X(T,H^-)$; and since each of the latter manifolds is closed by the last paragraph, it is straightforward to verify that (ii) holds where $G = \mathbf{C} \backslash K^-$. Hence T is decomposable by the first part of the proof. This completes the proof of Theorem 1.2.1.

Conditions (ix) and (x) of Theorem 1.2.1 seem to be the first purely "algebraic" criteria for decomposability.

1.2.10. Corollary. $T \in L(X)$ is decomposable if and only if T^* is.

Proof. Let T be decomposable. For each pair of open discs $G^- \subset H$ by Th. 1.2.1(iii,a) there is a T-invariant subspace Y such that

$$\sigma(T|Y) \subset \mathbf{C}\backslash G \qquad \text{and} \qquad \sigma(T/Y) \subset H.$$

Then $W = Y^{\perp\perp} \subset X^{**}$ is the second dual of Y and is invariant under T^{**}, and X^{**}/W is the second dual of X/Y. Since $\sigma(S) = \sigma(S^*)$ for any bounded S,

$$\sigma(T^{**}|W) \subset \mathbf{C}\backslash G \qquad \text{and} \qquad \sigma(T^{**}/W) \subset H.$$

By Lemma 1.2.8(ii) T^{**} has property (β); and since T^* has property (β) by Th. 1.2.1 (iv), T^* is decomposable by Th. 1.2.1(iv). The converse is proved similarly.

3. Further Criteria

1.3.1 Definition. Let $T \in L(X)$ be decomposable. Then T is <u>strongly decomposable</u> if for closed sets F_1, F_2 with the property $\sigma(T) \subset \text{Int } F_1 \cup \text{Int } F_2$ we have

$$X(T,K) = X(T, K \cap F_1) + X(T, K \cap F_2) \tag{3.1}$$

for every closed K in **C**. (We remark that this definition is equivalent to the original [11] because of Lemma 1.2.9.)

1.3.2 Theorem. Let $T \in L(X)$. Then T is strongly decomposable if and only if for each F closed in **C** and each open cover {G,H} of **C** there is a linear transformation $P: X \to X$ satisfying (x, a-c) of Th. 1.2.1 and leaving X(T,F) invariant.

Note that the condition makes sense because of Th. 1.2.1. The proof of Theorem 1.3.2 uses the following.

1.3.3 Lemma. Let T be decomposable, and let {G, H} be an open cover of **C**. Let $Y = X(T, G^- \cap H^-)$ and $Z = X(T, H\backslash G^-)$. Then for each closed F

$$X(T,F) \cap (Y \oplus Z) = [X(T,F) \cap Y] \oplus [X(T,F) \cap Z] \tag{3.2}$$

where the direct sums are algebraic.

Proof. Since the right-hand side of (3.2) is clearly contained in the left-hand side, let $x \in X(T,F) \cap (Y \oplus Z)$. Then $x = y + z$, where $y \in Y$, $z \in Z$. Since $\sigma(y,T) \subset \sigma(x, T) \cup \sigma(z,T)$, and $\sigma(y,T) \cap \sigma(z,T) = \emptyset$, we must have $\sigma(y,T) \subset \sigma(x,T) \subset F$. Hence $y \in X(T,F) \cap Y$ and by symmetry $x \in X(T,F) \cap Z$. Thus (3.2) follows.

Proof of Theorem 1.3.2. Sufficiency. By Theorem 1.2.1(ix) T is decomposable, so it suffices to prove (3.1) for $V = X(T,F)$ with F closed. Let $P: X \to X$ be the linear transformation leaving $X(T,F)$ invariant. Let
$x \in X(T,F)$. Then

$$Px \in X(T,F) \cap X(T,G^-), \quad (I-P)x \in X(T,F) \cap X(T,H^-).$$

It follows that (3.1) holds, so $T|X(T,F)$ is 2-decomposable. By the remark before the proof, T is strongly decomposable.

Necessity. Let T be strongly decomposable let $\{G,H\}$ be an open cover of $\mathbf{C}$, and let F be closed. Then by (3.1)

$$X(T,F) = [X(T,F) \cap X(T,G^-)] + [X(T,F) \cap X(T,H^-)]. \tag{3.3}$$

With the notation of Lemma 1.3.3, there is a manifold $W_0 \subset X$ such that

$$X(T,F) \cap X(T,H^-) = [X(T,F) \cap Y] \oplus [X(T,F) \cap Z] \oplus W_0 \tag{3.4}$$

by a Hamel basis argument. We claim that $W_0 \cap (Y \oplus Z) = (0)$. Let x lie in the last intersection. Then $x \in X(T,F)$, so by Lemma 1.3.3

$$x \in (X(T,F) \cap Y) \oplus (X(T,F) \cap Z),$$

hence $x = 0$ by (3.4).

By another Hamel basis argument we find a $W \subset X(T,H^-)$ such that $W_0 \subset W$ and

$$X(T,H^-) = Y \oplus Z \oplus W \tag{3.5}$$

We next prove $W_0 = W \cap X(T,F)$. Since it is obvious that W_0 is contained in $W \cap X(T,F)$, let $x \in W \cap X(T,F)$. Then by , (3.4), $x = u + w_0$ where

$$u \in [X(T,F) \cap Y] \oplus [X(T,F) \oplus Z], \quad w_0 \in W_0.$$

Then $u \in Y \oplus Z$ and $u = x - w_0 \in W$ imply that $u = 0$ by (3.5), so $x \in W_0$.

It now follows from (3.5) that

$$(3.6) \qquad X(T,F) \cap X(T,H) = [X(T,F) \cap Y] \oplus [X(T,F) \cap Z] \oplus [X(T,F) \cap W]$$

Since $Y \subset X(T,G^-)$, (3.3) and (3.6) imply

$$(3.7) \qquad X(T,F) = [X(T,F) \cap X(T,G^-)] \oplus [X(T,F) \cap Z] \oplus [X(T,F) \cap W],$$

and $X = X(T,G^-) + X(T,H^-)$ implies

$$X = X(T,G^-) \oplus Z \oplus W.$$

Let P be the projection onto $X(T,G^-)$ along $Z \oplus W$. Then $Px = x$ for $x \in X(T,G^-)$ and $Px = 0$ for $x \in Z \oplus W$. Hence for $x \in X(T,F)$ we obtain $x = g + z + w$ as in (3.7) where $g \in X(T,F) \cap X(T,G^-)$, $w \in W$ and $z \in Z$. Then $Px = Pg = g \in X(T,F)$. The properties (a), (b), (c) follow as in the proof of (i) $\Rightarrow$ (x) of Th. 1.2.1, hence Th. 1.3.2 is proved.

We now turn to the study of another operator class.

1.3.4 Definition. Let $T \in L(X)$ be decomposable. Then T is said to be <u>decomposable relative to the identity</u> if for each open cover $\{G_i\}$ of **C** there exist corresponding systems of invariant subspaces $\{Y_i\}$ and operators $\{P_i\} \subset \{T\}'$ (commutant of T) such that

$$\sigma(T|Y_i) \subset G_i, \qquad P_iX \subset Y_i \qquad \text{and} \qquad I = P_1 + \dots + P_n.$$

1.3.5 Theorem. Let $T \in L(X)$. Then the following are equivalent:

(i) T is decomposable relative to the identity;

(ii) for every open cover $\{G, H\}$ of **C** there exists $P \in L(X)$ commuting with T such that

$$(3.8) \qquad \gamma(Px,T) \subset G^- \quad \text{and} \quad \gamma(x - Px, T) \subset H^-$$

for each $x \in X$;

(iii) for each open cover $\{G_1, G_2\}$ of **C**, each $x \in X$ has decomposition $x = x_1 + x_2$ with

$$(3.9) \qquad \gamma(x_i, T) \subset G_i^-,$$

and for disjoint closed sets F_1, F_2 there is $P \in \{T\}'$ such that

(3.10) $$Px = x \text{ if } \gamma(x, T) \subset F_1, \text{ and } Px = 0 \text{ if } \gamma(x, T) \subset F_2;$$

(iv) for each cover $\{G,H\}$ of $\mathbf{C}$, where G is a disc and H is the complement of a disc, there is $P \in \{T\}'$ satisfying (3.8);

(v) if in (iii) G_1 is a disc and G_2 is complement of a disc then (3.9) and (3.10) hold.

In the proof of Th. 1.3.5 we use the following elementary fact.

1.3.6 Lemma. Let $T \in L(X)$ have SVEP, and let $P, Q \in \{T\}'$. Suppose that $F \subset \mathbf{C}$, $K \subset \mathbf{C}$, and for each $x \in X$ we have $\sigma(Px,T) \subset F$, $\sigma(Qx,T) \subset K$. Then $\sigma(PQx, T) \subset F \cap K$ for each x.

Proof of Theorem 1.3.5. (i) $\Rightarrow$ (ii) Let $\{G,H\}$ be a cover of $\mathbf{C}$. Choose P_1, $P_2 = I - P_1$ as in Definition 1.3.4. It follows that $P_1X \subset X(T,G^-)$ and $P_2X \subset X(T,H^-)$. Hence for $x \in X$ we obtain $\sigma(P_1x, T) = \gamma(P_1x, T) \subset G^-$ and $\sigma(x - P_1x, T) \subset H^-$, and thus (ii) is proved.

Since (ii) $\Rightarrow$ (iii) $\Rightarrow$ (v) and (ii) $\Rightarrow$ (iv) $\Rightarrow$ (v) are evident we finish the proof with (v) $\Rightarrow$ (i).

Under the hypothesis (v), we first prove that T is decomposable. Let $\{G,H\}$ be an open cover of $\mathbf{C}$ with G an open disc and H the complement of a closed disc. Now let G_1 be a second open disc with $G_1 \subset G_1^- \subset G$ such that $\{G_1, H\}$ covers $\mathbf{C}$, and let H_1 be the complement of a closed disc such that $\{G, H_1\}$ covers $\mathbf{C}$ and $G_1^- \cap H_1^- = \emptyset$. By hypothesis there is some $P \in L(X)$ commuting with T such that

(3.11) $$\begin{aligned} Px &= x \text{ for } \gamma(x,T) \subset G_1^- \\ Px &= 0 \text{ for } \gamma(x,T) \subset H_1^- \end{aligned}$$

Moreover, every $x \in X$ has a decomposition $x = x_1 + x_2$ with $\gamma(x_1,T) \subset G_1^-$ and $\gamma(x_2,T) \subset H^-$. By (3.11) $(I - P)x_1 = 0$, hence $(I - P)x = (I - P)x_2$. Thus $\gamma((I - P)x,T) \subset \gamma(x_2,T) \subset H^-$. In a similar way $\gamma(Px, T) \subset G^-$. Let $F \subset G\backslash H^-$ be closed. Since $\{G_1, H\}$ covers $\mathbf{C}$, $F \subset G_1\backslash H^-$. By (3.11) P is the identity on $X(T,F)$ and so also on $X((T,F)^-$ by continuity. Similarly P is zero on $X(T,K)^-$ for every closed $K \subset H\backslash G$. By Theorem

1.2.1(x) T is decomposable.

Let now $\{G,H\}$ be an open cover of $\mathbf{C}$. Without loss of generality assume G is bounded. Since $\mathbf{C}\backslash H$ is closed in G, it is compact. Let $\delta = (1/3)\mathrm{dist}(\mathbf{C}\backslash H, \mathbf{C}\backslash G)$. Then there are finitely many open discs $D_1, D_2, \ldots, D_k$ covering $\mathbf{C}\backslash H$ with radius δ and centers in $\mathbf{C}\backslash H$. Let G_i be the δ–neighborhood of D_i^-, $i = 1, 2, \ldots, k$; hence G_i is a disc (of radius 2δ) and $\cup_i G_i \subset G^-$. For each $i = 1, \ldots, k$, let $H_i = \mathbf{C}\backslash D_i^-$. It follows from the evident inclusions

$$\mathbf{C}\backslash H^- \subset \mathbf{C}\backslash H \subset \cup_i D_i = \cup_i(\mathbf{C}\backslash H_i^-)$$

that $\cap_i H_i^- \subset H^-$. Moreover, by construction each pair $\{G_i, H_i\}$ $(i = 1, \ldots, k)$ covers $\mathbf{C}$. For each $i = 1, 2, \ldots, k$, by the previous argument there exist $P_i \in \{T\}'$ such that

$$\sigma(P_i x, T) \subset G_i^-, \quad \sigma(x - P_i x, T) \subset H_i^- \qquad (x \in X).$$

Put $P = I - \Pi_i(I - P_i)$, so that $I - P = \Pi_i(I - P_i) \in \{T\}'$. By Lemma 1.3.6 applied k times

$$\sigma((I-P)x, T) \subset \cap H_i \subset H^-. \tag{3.12}$$

But since

$$P = (P_1 + P_2 + \ldots + P_n) - (P_1P_2 + \ldots + P_{n-1}P_n) + \ldots + (-1)^{n+1}P_1 \ldots P_n,$$

Lemma 1.3.6 and [24, p. 1] show that

$$\sigma(Px, T) \subset \cup_i G_i^- \subset G^- \qquad (x \in X). \tag{3.13}$$

By (3.12) and (3.13) T is 2-decomposable relative to the identity. Hence by [31, Th. 17.2] T is decomposable relative to the identity. For completeness, we sketch the proof, which proceeds by induction. Suppose Definition 1.3.4 holds for covers of C consisting of n sets $(n \geq 2)$, and let $\{G_i\}_1^{n+1}$ be an open cover of $\mathbf{C}$. Choose open sets H_1, H_2 such that $H_1^- \subset G_1$, $\{G_1, H_2\}$ covers $\mathbf{C}$, $\{G_2, \ldots, G_{n+1}, H_1\}$ covers $\mathbf{C}$, and $H_1^- \cap H_2^- = \emptyset$. Let $P, Q \in \{T\}'$ satisfy Definition 1.3.4 for the cover $\{G_1, H_2\}$. By the induction hypothesis, since $\{G_2, \ldots, G_n, H_1 \cup G_{n+1}\}$ covers $\mathbf{C}$, we can choose $Q_2, Q_3, \ldots, Q_{n+1} \in \{T\}'$ satisfying the definition for this cover. Let $P_1 = P$, $P_j = QQ_j$ $(2 \leq j \leq n+1)$. Then the reader can verify that $\{P_i\}_1^{n+1}$ satisfies Definition 1.3.4.

1.3.7 Corollary. Every operator decomposable relative to the identity is strongly decomposable.

Proof. Let T be decomposable relative to the identity, and let F be closed in **C**. It suffices to show that $T|X(T, F)$ is 2-decomposable. Let $\{G, H\}$ be an open cover of **C**, and let $P \in \{T\}'$ satisfying (3.8). Hence $X(T,F)$ is P-invariant, so if $x \in X(T,F)$ we have

$$Px \in X(T,F) \cap X(T,G^-)$$
$$(I - P)x \in X(T,F) \cap X(T,H^-).$$

This implies $X(T,F) \subset X(T,F) \cap X(T,G^-) + X(T,F) \cap X(T,H^-)$; and since the reverse inclusion is obvious, (3.1) is verified. Hence $T|X(T,F)$ is 2-decomposable, and the result follows.

1.3.8 Remarks. (1) The converse of Corollary 1.3.7 is false; in §4 we give an example that shows this.

(2) In [6] there is an example of a decomposable operator which is not strongly decomposable. Below we show that this example does satisfy the following weaker condition.

1.3.9 Definition. Let $T \in L(X)$ be decomposable. We say that T is quasi-strongly decomposable if for each open G the restriction

$$T|X(T,[G \cap \sigma(T)]^-)$$

is also decomposable.

Obviously every strongly decomposable operator is quasi-strongly decomposable. The following theorem characterizes quasi-strongly decomposable operators.

1.3.10 Lemma. If T has SVEP, then $\sigma(T) = \cup\{\sigma(x,T): x \in X\}$.

Proof. The inclusion $\sigma(T) \supset \cup_{x\in X}\sigma(x,T)$ is clear. If $\lambda \notin \sigma(x,T)$ for each $x \in X$, then each x satisfies $x = (\lambda - T)x_\lambda$ for some x_λ. Hence $\lambda - T$ is surjective, so $\lambda \notin \sigma(T)$ by Lemma 1.2.5., so the equality follows.

1.3.11 Theorem. Let $T \in L(X)$ and let $\sigma = \sigma(T)$. Then the following assertions are equivalent.

(i) T is quasi-strongly decomposable;

(ii) T is decomposable and if G, H are open and G_ε is the ε-

neighborhood of G^-, then

$$(3.14) \qquad X(T,[\sigma \cap (G \cup H)]^-) \subset X(T,[\sigma \cap G_\varepsilon]^-) + X(T,[\sigma \cap H]^-)$$

(iii) T is decomposable and if G is open then $T/X(T, [\sigma \cap G]^-)$ has SVEP and for each $x \in X$, the coset $x^\wedge$ in $X/X(T,[\sigma \cap G]^-)$ satisfies

$$(3.15) \qquad \sigma(x^\wedge,T/X(T, [\sigma \cap G]^-)) = [\sigma(x, T)\backslash(\sigma \cap G)^-]^-.$$

Proof. (i) $\Rightarrow$ (ii) Let T be quasi-strongly decomposable, and let G, G_ε, H be as in (ii). Since it is easy to see that

$$(3.16) \qquad [\sigma \cap (G \cup H)]^- = (\sigma \cap G)^- \cup (\sigma \cap H)^-,$$

choose an open cover $\{G_1, H_1\}$ of the set (3.16) such that $G \subset G_1 \subset G_\varepsilon$ and $H_1 \cap \sigma \subset H \cap \sigma$. Let $M = X(T,[\sigma \cap (G \cup H)]^-)$, so that $T_1 = T|M$ is decomposable. Then

$$M = M(T_1, G_1) + M(T_1, H_1) \subset X(T,(\sigma \cap G_\varepsilon)^-) + X(T,(\sigma \cap H)^-),$$

hence (3.14) holds.

(ii) $\Rightarrow$ (iii) Let G be open, $x \in X$ and let $M = X(T, (\sigma \cap G)^-)$. Now let G_1 be open containing $[\sigma(x,T)\backslash(\sigma \cap G)^-]^-$, and let H be an ε-neighborhood of G_1. Then

$$\sigma(x,T) \subset [G_1 \cup (G \cap \sigma)^-] \cap \sigma$$
$$= (G_1 \cap \sigma) \cup (G \cap \sigma)^- \subset (G_1 \cup G) \cap \sigma]^-.$$

By (3.14) $x = x_1 + x_2$ where $\sigma(x_1,T) \subset (H \cap \sigma)^-$ and $\sigma(x_2,T) \subset (G \cap \sigma)^-$. For $y \in X$ put $y^\wedge = y + M$. Then $\sigma(x^\wedge, T/M)$ is defined [31, Props. 4.14, 4.26] and $\sigma(x^\wedge,T/M) = \sigma(x_1^\wedge,T/M) \subset H^-$. Since G_1 and H are arbitrary with their defining properties, we obtain

$$\sigma(x^\wedge,T/M) \subset [\sigma(x,T)\backslash(G \cap \sigma)^-]^-.$$

The reverse inclusion follows from the inclusion

$$(3.17) \qquad \sigma(x,T) \subset \sigma(x^\wedge, T/M) \cup (\sigma \cap G)^-.$$

To see this let $\lambda \notin \sigma(x^\wedge,T/M)$ and $\lambda \notin (\sigma \cap G)^-$ and let D be a disc with center λ and disjoint from both $\sigma(x^\wedge, T/M)$ and $(\sigma \cap G)^-$ such that $x^\wedge = (\mu - T/M)f(\mu)$ on D. By Lemma 1.2.6 lift f to a function g analytic on a concentric disc $D_1 \subset D$ with $g(\mu)^\wedge = f(\mu)$ $(\mu \in D_1)$. For such μ define $h(\mu) = (\mu - T)g(\mu)$ and $h_1(\mu) = h(\mu) - x$. Then $h_1(\mu)^\wedge = 0$ on D_1, so $h_1(\mu) \in M$. We may then write $h_2(\mu) = (\mu - T)^{-1}h_1(\mu)$ for $\mu \in D_1$, so

$$x = h(\mu) - h_1(\mu) = (\mu - T)[g(\mu) - h_2(\mu)]$$

on D_1. In particular, $\lambda \notin \sigma(x,T)$, hence (3.17) follows and (3.15) is proved.

(iii) $\Rightarrow$ (i). Let G, H be open sets and put $M = X(T,(\sigma \cap G \cap H)^-)$. and $N = X(T,(\sigma \cap G)^-)$. Then M is closed and $\sigma(T|M) \subset H^-$. If $x \in N$, then (iii) implies that, for $x^\wedge = x + M$,

$$\begin{aligned}\sigma(x^\wedge, (T|N)/M) &\subset [(\sigma \cap G)^- \backslash (\sigma \cap G \cap H)^-]^- \\ &\subset (\sigma \cap G)^- \backslash [(\sigma \cap G)^- \cap H] \\ &\subset \mathbf{C} \backslash H.\end{aligned}$$

Now $((T|N)/M) \subset \mathbf{C}\backslash H$ by Lemma 1.3.10, and since $(T|N)|M = T|M$, $T|N$ is decomposable by [31, Th. 5.17] or [59, Th. 1]; so (i) follows.

1.3.12 Corollary. If T is quasi-strongly decomposable, then for each open G the quotient operator $T/X(T,(G \cap \sigma)^-)$ is decomposable where $\sigma = \sigma(T)$.

Proof. Let $F = (\sigma \cap G)^-$ and let $M = X(T,F)$. For K closed in $\mathbf{C}$, define $X(K)^\wedge = X(T,K \cup F)/M$. Let $x^\wedge$ be the coset $x + M$ for $x \in X$. Hence $X(K)^\wedge$ is closed in X/M and (T/M)-invariant. For $x^\wedge \in X(K)^\wedge$ clearly $\sigma(x,T) \subset K \cup F$. But by (3.15)

$$\sigma(x^\wedge,T/M) = (\sigma(x,T)\backslash F)^- \subset [(K \cup F)\backslash F]^- \subset K. \tag{3.18}$$

For an open cover $\{G_1, G_2\}$ of $\mathbf{C}$ it is evident that $X/M = X(G_1^-)^\wedge + X(G_2^-)^\wedge$. From (3.18) and Lemma 1.3.10

$$\sigma[(T/M)|X(G_i^-)^\wedge) = \cup\{\sigma(x^\wedge,T/M): \sigma(x,T)\} \subset G_i^- \cup K \subset G_i^-$$

Hence T/M is decomposable by [7, 58 or 70].

4. Special Classes and Examples

In their monograph on generalized spectral operators, the authors of [24] begin with the larger class of decomposable operators. Since the publication of [24] in 1968 a whole hierarchy of subclasses between these two (as well as others) have been defined and studied. In this section we consider the following classes of operators.

1. spectral [Dunford]
2. boundedly decomposable
3. generalized scalar [A-scalar]
4. decomposable relative to the identity
5. strongly decomposable
6. quasi-strongly decomposable
7. superdecomposable
8. decomposable
9. quasi-decomposable
10. analytically decomposable.

The program of this section is to prove that each class above is contained in its successor and that most of these nine containments are proper. Classes 4, 5, 6 and 8 have already been treated in §§2 and 3. Definitions of the others follow.

There is a generalization of (1.1) called "asymptotic" which has received some attention ([50],[55]). We say that $T \in L(X)$ has asymptotic spectral decomposition (ASD) if T satisfies (1.2) and

$$X = \vee_i M_i \qquad \text{(ASD)}$$

where i runs from 1 to n. We use this notion below.

1.4.1 Definition. Let $T \in L(X)$. Then T is

(a) <u>spectral</u> if there is a spectral measure $E: B \to L(X)$ from the σ-ring of Borel sets in **C** into the set of idempotents (projections) on X commuting with T where E is uniformly bounded on B and is σ-additive in the strong operator topology. (For details see [26]).

(b) <u>boundedly decomposable</u> with bound m if T is decomposable and there exists $m > 0$ such that for each open cover $\{G_1,G_2\}$ of **C** each $x \in X$ has decomposition $x = x_1 + x_2$ and for $j = 1,2$

$$\sigma(T,x_j) \subset G_j \quad \text{and} \quad \|x_j\| \le m\|x\|$$

(see [36]).

(c) A-scalar if A is a normal "admissible" algebra [24, p. 59] and $U: A \to L(X)$ is an algebra homomorphism into the commutant $\{T\}'$ and $U_1 = I$ and $U_{id} = T$; U is called an A-spectral function of T; finally, T is A-spectral if $T = S + Q$ where S is A-scalar and Q is a quasinilpotent commuting with the range of an A-spectral function of S.

(d) generalized scalar if T is A-scalar where $A = C^{\infty}(\mathbf{C})$ and U is continuous in the Montel tolopogy of $C^{\infty}(\mathbf{C})$; T is generalized spectral if it is C^{∞}-spectral; in this case U is called a spectral distribution.

(e) superdecomposable if T is decomposable and for $F \subset G$ with F closed, G open there exists a T-invariant subspace M such that $T|M$ is decomposable and $X(T,F) \subset M \subset X(T,G)$; (see [83]).

(f) quasi-decomposable if T has ASD and property (κ) [51].

(g) analytically decomposable if T has ASD consisting of analytically invariant subspaces [55]; (Y is "analytically invariant" if T/Y has SVEP).

1.4.2 Theorem. In the list above $(i) \Rightarrow (i+1)$ for $i = 1, 2, \dots, 9$.

Proof. (1) ⇒ (2): See Prop. 1.4.3. (2) ⇒ (3): Prop. 1.4.7. (3) ⇒ (4): Prop 1.4.8. (4) ⇒(5): Cor. 1.3.7. (5) ⇒(6) is clear. (6) ⇒ (7): Let $F \subset H \subset H^{-} \subset G$. Put $M = X(T,[H \cap \sigma(T)]^{-})$. Then M satisfies Def. 1.4.1(e). (7) ⇒ (8) and (8) ⇒ (9) are clear from the definitions.

(9) ⇒ (10): It clearly suffices to prove that $X(T,F)$ is analytically invariant for F closed. Let $M = X(T,F)$ and we may assume $F = \sigma(T|M)$. Let $f: \Omega \to X/M$ be analytic with $(\lambda - T/M)f(\lambda) = 0$ on Ω, where we suppose Ω to be a disc. We prove $f = 0$ on Ω. By Lemma 1.2.6 we "lift" f to $g: D \to X$ where D is a disc concentric with Ω and $g(\lambda) + M = \text{coset} = f(\lambda)$ on D. Thus $(\lambda - T)g(\lambda) \in M$ for $\lambda \in D$, and we show $g(\lambda) \in M$. It is enough to take $D \subset F$. Let $\lambda \in D$ and $y \in M$ such that $y = (\lambda - T)z$ for some $z \in X$. If $\mu \notin F$ then

$$(\lambda - T)z = (\lambda - \mu)z + (\mu - T)z = (\mu - T)v$$

for some $v \in M$. Hence $z = (\mu - T)[(\lambda - \mu)^{-1}(v - z)]$. This shows that $\mu - T$ is surjective on $Y = M + \mathbf{C}z$. Since Y is closed and T has SVEP, by Lemma 1.2.5 $\mu - T$ is bijective on Y, i.e. $\sigma(T|Y) \subset F = \sigma(T|M)$. By maximality of M we have $Y \subset M$, and so $z \in M$. It follows that the range of g is contained in M, hence $f = 0$ on Ω.

1.4.3 Proposition. Every spectral operator is boundedly decomposable.

Proof. Let T be spectral with spectral measure E. For the open cover $\{G_1, G_2\}$ of $\mathbb{C}$ let β_i $(i = 1,2)$ be Borel sets with $\beta_1 = \mathbb{C}\setminus\beta_2$ and $\beta_i^- \subset G_i$. For $x \in X$ put $x_i = E(\beta_i)x$. Definition 1.4.1(b) is now satisfied by taking m as the uniform bound on E.

We now sketch a characterization of boundedly decomposable operators. If T is decomposable the map ρ: $x \to \sigma(x,T)$ is called support representation for T if, whenever $\sigma(x,T) \subset G_1 \cup G_2$ (G_i open), there exist $x_1, x_2 \in X$ with $x = x_1 + x_2$ and $\sigma(x_i,T) \subset G_i$. Moreover, ρ is bounded if it satisfies Def. 1.4.1(b). This condition gives rise to a unique continuous algebra homomorphism $\rho\hat{}$: $C(\sigma(T)) \to L(X)$ satisfying the properties

(B1) $\rho\hat{}(f)x = 0$ if $\operatorname{supp} f \cap \sigma(x,T) = \emptyset$;
(B2) ran $\rho\hat{}$ commutes with $A \in L(X)$ iff $\sigma(Ax,T) \subset \sigma(x,T)$ for each $x \in X$.

(See [35,36] for details).

1.4.4 Theorem. Let $T \in L(X)$. Then T is boundedly decomposable if and only if $T = S + Q$ where S is C-scalar ($C = C(\sigma(T))$) and Q is quasinilpotent. Moreover, in this case the decomposition $T = S + Q$ is unique and the range of the C-spectral function lies in the bicommutant $\{T\}''$.

Proof. (Sketch) If T is boundedly decomposable, let $U_f = \rho\hat{}(f)$ for $f \in C$. Let $S = U_{id}$, and $Q = T - S$. It follows that $\sigma(x,S) = \sigma(x,T)$ for all $x \in X$. Hence $X(S,F) = X(T,F)$ for all closed F. Since $ST = TS$ by (B2), it follows that Q is quasinilpotent by [24, Th. 2.2.2]. Conversely if $T = S + Q$ where S is C-scalar and Q is quasinilpotent, and the map $f \to U_f$ is continuous, then it is straightforward to prove that S is boundedly decomposable. The map $x \to \sigma(x,S)$ is a bounded support representation. Since $\sigma(x,T) = \sigma(x,S)$ for all $x \in X$ and Q is a commuting quasinilpotent, it follows that $x \to \sigma(x,T)$ is a bounded support representation for T, so T is boundedly decomposable.

For uniqueness, suppose $T = S_1 + Q_1$. Then $T - S_1$ is quasinilpotent and $\sigma(x,T) = \sigma(x,S_1)$. The uniqueness of the map $\rho\hat{}$ implies $S_1 = S$. The property ran $\rho\hat{} \subset \{T\}''$ follows from (B2), since $AT = TA$

implies $\sigma(Ax,T) \subset \sigma(x,T)$. (For details see [36]).

Boundedly decomposable operators also have the following dual characterization. Recall that $T \in L(X)$ is of class $\Gamma \subset X^*$ if E is a spectral measure of T satisfying a weaker version of Def. 1.4.1(a) requiring the scalar Borel measure $\langle E(\cdot)x,x^*\rangle$ to be σ-additive for all $x^* \in \Gamma$.

1.4.5 Theorem. For $T \in L(X)$ the following are equivalent.

(i) T is boundedly decomposable;

(ii) T^* is boundedly decomposable;

(iii) T^* is prespectral of class X and $E(F)X^* = X^*(T^*,F)$ for closed F.

Proof. (ii) $\Rightarrow$ (i). By Lemma 1.2.7 $X^*(T^*,F)$ is weak*-closed for each closed F, hence by [35, Prop. 7] each $f(S)$ ($f \in C$) is weak* continuous. Thus $f(S) = (V_f)^*$ for $V_f \in L(X)$, $f \in C$. The map $f \rightarrow V_f$ is a continuous C-spectral function, hence the predual $T - V_{id}$ of $T^* - S$ is quasinilpotent and $V_{id} \in \{T\}''$. Thus T is boundedly decomposable.

(i) $\Rightarrow$ (iii). [25, Th. 5.21].

(iii) $\Rightarrow$ (ii). This follows by the argument in Prop. 1.4.3.

The next corollary is immediate from Th. 1.4.5 and the fact that on reflexive spaces the classes of spectral and prespectral operators coincide.

1.4.6 Corollary. On reflexive spaces every boundedly decomposable operator is spectral.

1.4.7 Proposition. Every boundedly decomposable operator is generalized spectral.

Proof. Let T be boundedly decomposable. By Th. 1.4.4 $T = S + Q$ where S is C-scalar and $f(S) \in \{T\}''$. Hence it suffices to prove that S is generalized scalar. But the continuous algebra homomorphism $\rho\hat{}$ is clearly continuous in the Montel topology of $C^\infty(\sigma(S))$, hence S is generalized scalar.

1.4.8 Proposition. Every A-spectral operator is decomposable relative to the identity.

Proof. Let $T = S + Q$ be A-spectral and let U be an A-spectral function commuting with Q. For an open cover $\{G,H\}$ of $\mathbf{C}$, let $\phi \in A$ with $\text{supp}\,\phi \subset G$ and $\phi = 1$ on a neighborhood of $\mathbf{C}\backslash H$. Let $U_f = P$. By hypothesis there is an analytic function $\mu \rightarrow V(\mu)$ from $\mathbf{C}\backslash\text{supp}\,\phi$ to ran U such that $(\mu-S)V(\mu) = P$, hence

$$\gamma(Px,S) \subset G \text{ and } \gamma(x - Px,S) \subset H$$

(see [24, pp. 59, 61]). Since QP = QP we have also

$$\gamma(Px,T) \subset G, \qquad \gamma(x - Px,T) \subset H,$$

hence T is decomposable relative to the identity by Th. 1.3.5(ii).

1.4.9 Example. A boundedly decomposable operator which is not spectral. In [29, p. 2097] the authors indicate that the compact diagonal operator on $\ell^1(N)$ given by $T\{e_n\} = \{n^{-1}e_n\}$ is spectral while T^* is not spectral. But T^* is boundedly decomposable by Prop. 1.4.3 and Th. 1.4.5.

1.4.10 Example. A generalized scalar operator which is not boundedly decomposable. By Theorem 1.4.4 every boundedly decomposable operator is regular in the sense of [24, p. 98]. But in [2] the author shows that the operator T defined on p. 32 below is a nonregular generalized scalar operator. Hence this T is not boundedly decomposable.

We now turn to some examples that show that implications (6) ⇒ (5), (5) ⇒ (4) and (4) ⇒ (3) all fail.

Let $\bar\partial$ denote the differential operator $\bar\partial = (1/2)(\partial/\partial x + i\partial/\partial y)$ where x = Re z and y = Im z.

Let Q_1 be the unit square 0 < Re z < 1, 0 < Im z < 1. For n > 1, and j,k = 1,2,....,n, let Q_{njk} be n^2 congruent overlapping open squares such that for each n fixed $\{Q_{njk}\}_{j,k}$ is an open cover of Q_1^-. For each triple index njk, let B_{njk} denote the Banach space of functions f on Q_{njk} having continuous derivatives $\bar\partial^m f$ (m = 1,2,...) up through order $p_{njk} = (1/6)n(n-1)(2n-1) + j + k(n-1)$ all of which extend continuously to the closures. Now define X to be the ℓ^1 direct sum of the B_{njk}, i.e.

$$X = \{(f_{njk}): f_{njk} \in B_{njk} \text{ and } \Sigma_{n\geq 1}\Sigma_{1\leq j,k\leq n} \|f_{njk}\| < \infty\},$$

and define T as multiplication by z:

$$T(f_{njk}) = (zf_{njk}). \tag{4.1}$$

1.4.11 Theorem. The operator T defined by (4.1) is decomposable relative to the identity, but T is not A-scalar for any inverse-closed algebra A. In particular, T is not generalized scalar.

Proof. Let $\{G_1, G_2,\dots,G_m\}$ be an open cover of C. Then there are corresponding functions $\phi_1, \phi_2,\dots, \phi_m$ in $C^\infty(R^2)$ such that

$$1 = \phi_1 + \phi_2 + \dots + \phi_m$$
$$0 \le \phi_q \le 1 \quad \text{and} \quad \operatorname{supp}\phi_q \subset G_q \quad (q = 1.2,\dots,m).$$

Fix this particular partition of unity; then each $f \in X$ has the unique representation

$$f = \Sigma_{1 \le q \le m}\, g_q,$$

where $g_q = \phi_q f = (\phi_q f_{njk})$. Evidently, $g_q \in X(T,G_q^-)$ for $q = 1,2,\dots,m$. If we define $P_q f = g_q$, then P_q is well-defined for each q. Clearly, $I = \Sigma P_q$ on X and $P_q X \subset X(T,G_q^-)$ since $g_q \in X(T,G_q^-)$. Hence Definition 1.3.4 is satisfied. Since it was proved in [4] that T is not generalized scalar, we are finished.

Now we describe another construction which we use for our next example (see [2]). Let $D = \{z \in \mathbf{C}: |z| < 1/2\}$ and let $\mathcal{D}^\infty(D)$ consist of those C^∞-functions with compact support. Let $B^0(D^-)$ be the set of all C-valued continuous functions on D^-, and let $B^1(D^-)$ be those $f \in B^0(D^-)$ for which there exists $g \in B^0(D^-)$ satisfying

$$\iint_D (\bar\partial\phi)(z)f(z)\,dxdy = -\iint_D \phi(z)g(z)\,dxdy$$

for all $\phi \in \mathcal{D}^\infty(D)$. Define $g = \bar\partial f$. Then $B^1(D^-)$ with norm

$$\|f\| = \sup|f| + \sup|\bar\partial f|$$

(suprema over D) is a Banach space. Now let $X_0 = B^0(D^-)$ and $X_1 = B^1(D^-)$. For $j= 0,1$, define T_j on X_j by $T_j f(z) = zf(z)$ $(f \in X_j)$. The direct sum $T = T_0 \oplus T_1$ on $X_0 \oplus X_1$ is a generalized scalar operator [2], and for closed $F \subset \mathbf{C}$ the spectral manifolds are given by

$$X(T,F) = X_0(T_0,F) \oplus X_1(T_1,F)$$

where $X_j(T_j,F) = \{f \in X_j: \text{ supp } f \subset F\}$ $(j = 0,1)$.

For $a \in X_0$ and $(f,g) \in X$, let

$$A(a)(f,g) = (a\bar{\partial}g, 0),$$
$$Q(a)(f,g) = (ag, 0).$$

For $a, b \in X_0$, it is clear that

$$Q(a)Q(b) = Q(a)A(b) = A(a)Q(b) = A(a)A(b) = 0, \tag{4.2}$$
$$Q(a) + A(b) \text{ commutes with } T. \tag{4.3}$$

From (4.2) it is evident that

$$[Q(a) + A(b)]^2 = 0. \tag{4.4}$$

1.4.12 Lemma. If $S \in L(X)$ such that $X(T,F)$ is S-invariant for all F closed, then there exists a_0, a_1, $b_0 \in X_0$ and $b_1 \in X_1$ such that for all $(f,g) \in X$

$$S(f,g) = (b_0 f, b_1 g) + [Q(a_0) + A(a_1)](f,g). \tag{4.5}$$

Proof. Write S in the matrix form

$$S = \begin{bmatrix} S_{00} & S_{01} \\ S_{10} & S_{11} \end{bmatrix}$$

where $S_{ij}: X_j \to X_i$ $(i,j = 0,1)$. Put $a_i = S_{ii}(1)$ $(i = 0,1)$. Then $a_i \in X_i$, and it is straightforward to prove that S_{ii} $(i = 0,1)$ are multiplications by a_i. Since the restriction $S_{10}|X_1$ is continuous and $X_1(T_1,F) \subset X_0(T_0,F)$ for closed F, S_{10} is also a multiplication by $b \in X_1$. Hence $S_{10}f = bf$ for all $f \in X_0$ by density of X_1 in X_0. In fact, by Lemma 3.2.8 below the embedding $X_1 \to X_0$ is compact, and hence S_{10} is a compact operator on X_0. Thus $\sigma(S_{10})$ is totally disconnected, so b is constant. In this case, $S_{10}f = bf \notin X_1$ if $f \in X_0 \backslash X_1$, contradicting $\text{ran}(S_{10}) \subset X_1$. Hence $b = 0$. By the proof of [24, Th. 6.1.15] $\text{supp}(S_{01}f) \subset \text{supp } f$ for all $f \in X_1$. For $z \in D$ the relation $u_z(\phi) = (S_{01}\phi)(z)$ defines a functional on C^∞ $(\phi \in C^\infty)$, hence u_z is a distribution of order 1 and support $\{z\}$ [49] and has the form

$$u_z(\phi) = a_0'(z)\phi(z) + a_1'(z)(\bar{\partial}f)(z) + a_2'(z)(\partial\phi)(z),$$

and it follows that $S_{01}\phi = a_0'\phi + a_1'\bar{\partial}\phi + a_2'\partial\phi$ on $C^\infty(D)$. Applying S_{01} to $1, z, \bar{z}$, resp., we find $a_0', a_1', a_2' \in X_0$. But $C^\infty(D)$ is dense in X_1, so it follows that $a_2' = 0$, Thus, for $(f,g) \in X$, we have $S(f,g) =$

$$\begin{bmatrix} a_0 & a_0' + a_1'\bar{\partial} \\ 0 & a_1 \end{bmatrix} \begin{bmatrix} f \\ g \end{bmatrix} = \begin{bmatrix} a_0 f + a_0' g + a_1'\bar{\partial} g \\ a_1 g \end{bmatrix}$$

So (4.5) follows using the definitions of $Q(a_0')$ and $A(a_1')$.

1.4.13 Lemma. Let $S \in L(X)$. Then S commutes with $T + A(1)$ if and only if there exist $a_0, a_1 \in X_0$ and analytic $b \in X_1$ such that for all $(f,g) \in X$,

$$S(f,g) = (bf, bg) + [Q(a_0) + A(a_1)](f,g). \tag{4.6}$$

Proof. Suppose S commutes with $V = T + A(1)$. Since V is decomposable, $X(V,F)$ is defined and $X(V,F) = X(T,F)$. So $X(T,F)$ is also S-invariant for all F. By Lemma 1.4.12, S satisfies (4.5). Let $U(f,g) = (b_0 f, b_1 g)$. Since V commutes with S by hypothesis and U obviously commutes with T, a simple calculation shows that U commutes with $A(1)$. For $f = 0$, $g = 1$, we obtain

$$A(1)U(0,1) = A(1)(0, b_1) = (\bar{\partial}b_1, 0),$$
$$UA(1)(0,1) = U(\bar{\partial}1, 0) = (0,0).$$

So $\bar{\partial}b_1 = 0$, i.e. b_1 is analytic on D. Using $f = 0$, $g = \bar{z}$ in an analogous way, we find $b_0 = b_1$. Put $b = b_0$ to get (4.6).

For the converse, let $U(f,g) = (bf, bg)$ where $b \in X_1$ is analytic on D. It is easily shown that U and $A(1)$ commute. If we define $S = U + Q(a_0) + A(a_1)$, then by use of (4.2) and (4.3) it follows that $SV = VS$.

1.4.14 Theorem. The operator $V = T + A(1)$ is strongly decomposable but not decomposable relative to the identity.

Proof. we observed in the proof of Lemma 1.4.13 that V and T have the same spectral manifolds: $X(V,F) = X(T,F)$. Since we also observed that T is generalized scalar, T is strongly decomposable, hence V is also strongly decomposable.

Now suppose that V is decomposable relative to the identity. Let $F = \{z \in \mathbf{C}: \operatorname{Re} z \leq 0\}$ and let $G = \{z \in \mathbf{C}: \operatorname{Re} z < 1/4\}$. Then $F \subset G$ and $D \cap (\mathbf{C} \setminus G) \neq \emptyset$. By assumption, there is $P \in L(X)$ commuting with V such that

$$PX \subset X(V, G^-) \tag{4.7}$$

$$P(f,g) = (f.g) \ \text{ for } (f,g) \in X(V,F). \tag{4.8}$$

By Lemma 1.4.13 there is analytic $b \in X_1$ such that

$$P(f,g) = (bf, bg) + [Q(a_0) + A(a_1)](f,g). \tag{4.9}$$

For $(f,g) = (0,1)$, (4.9) implies $P(0,1) = (a_0, b)$. By (4.7) supp $b \subset G^-$, so $b(z) = 0$ for $z \in D \setminus G^-$. Since b is analytic it must be zero throughout D. Then $P = Q(a_0) + A(a_1)$ is nilpotent by (4.4), and this contradicts (4.8). Hence V is not decomposable relative to the identity.

1.4.15 Remark. In the foregoing discussion, we saw that T was generalized scalar, hence it is also decomposable relative to the identity. Note that $V = T + A(1)$ where $A(1)$ is a nilpotent commuting with T. Although such perturbations preserve strong decomposability, in the case of an operator decomposable relative to the identity we can state

1.4.16 Corollary. The class of operators decomposable relative to the identity is not preserved by quasinilpotent equivalence [24].

In the following we will prove that the operator given in [6] is quasi-strongly decomposable and its adjoint is strongly decomposable. From this it follows that (6) ⇒ (5) fails.

Suuppose $\{m_p\}$ is an increasing sequence of positive numbers satisfying $m_0 = 1$ and $m_p \to \infty$. Let I be a compact interval in $\mathbf{R}$, and let D be a compact disc in $\mathbf{C}$. Put

$$C_{<m_p>}(I) = \{f \in C^\infty(I) \colon \|f\|_{<m_p>,I} = \sum_{p=0}^{\infty} \frac{1}{p!\,m_p^p} \|f^{(p)}\|_I < \infty\}$$

where $\|f^{(p)}\| = \max_{s \in I} |f^{(p)}(s)|$. Furthermore, denote by $C_{<m_p>,s}(D)$ the space of all $f \in C(D)$ which are infinitely often continuously differentiable on Int D with respect to the real variable s such that for every $p \in N$ the function $\partial^p f/\partial s^p$ has continuous extension to D (again denoted by $\partial^p f/\partial s^p$) and such that

$$\|f\|_{<m_p>,D} = \sum_{p=0}^{\infty} \frac{1}{p! m_p^p} \left\| \frac{\partial^p f}{\partial s^p} \right\|_D < \infty,$$

where

$$\|\partial^p f/\partial s^p\|_D = \max_{(s,t) \in D} \left| \frac{\partial^p f(s,t)}{\partial s^p} \right|.$$

In the following we will suppose that either $m_p = p^2$ or $m_p = \ln(e + p)$ and $I = [0,1]$.

Let $\{s_n\}$ be the set of rational numbers in $[0,1]$. For $z \in C$, $r > 0$, we write $S(z, r) = \{w \in C: |z - w| \leq r\}$. We shall now construct a sequence of mutually disjoint discs $D_n = S(s_n + ir_n, r_n)$ with $0 < r_n < 1$. Suppose $D_j = S(s_j + ir_j, r_j)$ $(j = 1,2,\ldots,n)$ have been constructed with $0 < r_j < 1$ and $D_j \cap D_k = \emptyset$ for $j \neq k$. Since s_{n+1} is not in the union of $D_1,\ldots, D_n$, there is $r_{n+1} \in (0,1)$ such that $D_{n+1} = S(s_{n+1} + ir_{n+1}, r_{n+1})$ does not intersect $D_1,\ldots, D_n$. Obviously $r_n \to 0$ as $n \to \infty$. Put $K = I \cup (\cup_{n\geq 1} D_n)$. Then K is compact. We define two Banach spaces X_1 and X_2 by

$$X_1 = \left\{ f \in C(K): \begin{array}{c} f|I \in D_{<p^2>}(I),\ f|D_n \in D_{<p^2>,s}(D_n) \text{ for } n \geq 1 \\ \dfrac{d^p}{ds^p}(f|I)(s_n) = \dfrac{\partial^p}{\partial s^p}(f|D_n)(s_n) \text{ for } 0 \leq p \leq n \\ \|f\|_1 = \|f|I\|_{<p^2>,I} + \sup_n \|f|D_n\|_{<p^2>,D_n} < \infty \end{array} \right\}$$

$$X_2 = \left\{ g \in C(K): \begin{array}{c} g|D_n \in D_{<\ln(e+p)>,s}(D_n) \text{ for } n \geq 1 \\ \|g\|_2 = \sup_n \|g|D_n\|_{<\ln(e+p)>,D_n} < +\infty \end{array} \right\}.$$

Let $Q = \{z = s + it: -1 \leq s \leq 2, -1 \leq t \leq 1\}$. Let A be the set of all functions analytic on Int Q and continuous on Q. Put $Y = X_1 \oplus X_2 \oplus A$, endowed with the norm $\|(f,g,h)\|_Y = \|f\|_1 + \|g\|_2 + \|h\|_Q$ $((f,g,h) \in Y)$, where $\|h\|_Q$ denotes sup norm on Q.

1.4.17 Lemma. The operator $S: Y \to Y$ defined by $S(f,g,h) = (\mathrm{id}\cdot f, \mathrm{id}\cdot g, \mathrm{id}\cdot h)$ is bounded on Y where $\mathrm{id}(z) = z$ for $z \in K$.

Proof. The continuity of the maps $f \to \mathrm{id}\cdot f$ and $g \to \mathrm{id}\cdot g$ follows from [6, Lemma 1.1]; the continuity of $h \to \mathrm{id}\cdot h$ is obvious.

Now let $N = \{(f,g,h) \in Y: f|I = h|I$ and $f = g$ on $K\}$ Then N is closed and clearly S-invariant. Let $X = Y/N$ and let T denote S/N on X. Our next important aim is to prove

1.4.18 Theorem. T is quasi-strongly decomposable.

To prove Th. 1.4.18 we show that (3.15) of Th. 1.3.11 holds. This will require a rather extensive analysis of the spectral properties of T. We begin with some preliminary observations.

Suppose F_0 is closed in $\mathbf{C}$ and $x = (f,g,h)^{\wedge} + N \in X$. Put $F = (\sigma(x,T)\backslash F_0)^-$ and $I_0 = F \cap I$. Let H be open with $H \supset F$. Then $H \cap R \supset I_0$ and there exist finitely many disjoint open intervals $\Delta_1,\dots, \Delta_k$ in $H \cap R$ such that

$$\cup_{1 \le j \le k} \Delta_j \supset I_0.$$

Without loss of generality we may change Δ_j so that $\Delta_j^- \subset H$ and the endpoints of Δ_j $(j = 1, 2, \dots, k)$ lie in the interior $(F_0 \cap R)^0$ or in $R\backslash(F_0 \cap R)$.

Let η be any endpoint of some Δ_j $(j = 1,2,\dots,k)$ which lies in $(F_0 \cap R)^0$, let η' be any endpoint of Δ_j lying in $R\backslash(F_0 \cap R)$. Let $\alpha > 0$, $\beta > 0$ be small enough such that for any $j = 1, 2,\dots, k$ and $J = (-\alpha, \alpha)$

$(*_1)$ $$\alpha < \beta^2/4$$

$(*_2)$ $$[\eta' - \beta, \eta' + \beta] \times J^- \subset (H\backslash F) \cap [(F_0 \cap R)^0 \times J^-]$$

$(*_3)$ $$[\eta' - \beta, \eta' + \beta] \times J^- \subset (H\backslash F)\backslash[(F_0 \cap R) \times J^-].$$

Let ξ_j (ζ_j) be the left (right) endpoint of Δ_j. Then there are four possibilities altogether:

(a) $\xi_j \in (F_0 \cap R)^0$ $\quad\zeta_j \in R\backslash(F_0 \cap R)$;

(b) $\xi_j \in (F_0 \cap R)^0$, $\quad\zeta_j \in (F_0 \cap R)^0$;

(4.10) (c) $\xi_j \in R\setminus(F_0 \cap R)$, $\quad \zeta_j \in (F_0 \cap R)^0$;

(d) $\xi_j \in R\setminus(F_0 \cap R)$, $\quad \zeta_j \in R\setminus(F_0 \cap R)$.

Since Δ_j^- is contained in $H \cap R$, we may choose α and β so small that

$$[\xi_j - \beta, \zeta_j + \beta] \times J^- \subset H \qquad (j = 1,2,\ldots,k) \tag{4.11}$$

and

$$F \cap (R \times J^-) \subset \cup_{1 \le j \le k} \{(\xi_j + \beta, \zeta_j - \beta) \times J^-. \tag{4.12}$$

1.4.19 Proposition. The following containment holds:

$$F_0 \cap \sigma(T) \supset [F \cup (F_0 \cap \sigma(T))] \cap (R \times J^-)\setminus [\cup_j (\xi_j + \beta, \zeta_j - \beta) \times J^-. \tag{4.13}$$

Proof. Relation (4.13) follows from (4.12) and

$$\begin{aligned}[F \cup (F_0 \cap \sigma(T))] \cap (R \times J^-) &= [F \cap (R \times J^-)] \cup [F_0 \cap \sigma(T) \cap (R \times J^-)] \\ &\subset [\cup_j (\xi_j + \beta, \zeta_j - \beta) \times J^-] \cup (F_0 \cap \sigma(T)).\end{aligned}$$

Let us choose $\psi \in C(I)$ satisfying $\psi(t) = 1$ on $[0, \alpha/2]$ and supp $\psi \subset J$. For the given $x = (f,g,h)^\wedge \in X$, let

$$(**_1) \qquad f^*(z) = [1 - \psi(t)][f(z) - g(z)] \qquad (t = \operatorname{Im} z).$$

Then $f^* \in X_1$ and hence $(f^*,0,0) \in Y$. Furthermore, let

$$(**_2) \qquad f_0 = f - f^* - h, \quad \text{and} \quad g_0 = g - h \text{ on } K.$$

Then $f_0 \in X_1$, $g_0 \in X_2$ and hence $(f_0, g_0, 0) \in Y$. It follows from the definition of ψ and $(**_1)$, $(**_2)$ that

$$\begin{aligned}(**_3) \qquad \operatorname{supp} f^* &\subset \{s + it: t \ge \alpha/2\} \cap \sigma(x,T) \\ &\subset \{s + it: t \ge \alpha/2\} \cap [F \cup (F_0 \cap \sigma(T))],\end{aligned}$$

$$(**_4) \qquad \operatorname{supp}(f_0 - g_0) \subset \{s + it: 0 \le t \le \alpha\} \cap \sigma(x, T).$$

1.4.20 Proposition. For the function f^* in $(**_1)$, there exist f_i^*

$\in X_i$ $(i = 1,2)$ such that $f^* = f_1{}^* + f_2{}^*$ on K and

$$(4.14) \qquad (f_1{}^*,0,0)^\frown \in X(T, H) \text{ and } (f_2{}^*,0,0)^\frown \in X(T, F_0 \cap \sigma(T)).$$

Proof. Choose $\phi \in C_{\langle p^2 \rangle, s}(R^2)$ such that $\phi(z) = 1$ in a neighborhood of $F \setminus (R \times [0,\alpha/2])$ and

$$(4.15) \qquad \operatorname{supp} \phi \subset H \cap \{z: \operatorname{Im} z > 0\}.$$

Put $f_1{}^*(z) = \phi(z)f^*(z)$ and $f_2{}^*(z) = f^*(z) - f_1{}^*(z)$ for $z \in K$. Then $f_1{}^*(z) = f^*(z)$ in a neighborhood of $\{s + it: t \geq \alpha/2\} \cap F$. Now $(**_3)$ implies that $\operatorname{supp} f_2{}^* \subset F_0 \cap \sigma(T)$, therefore

$$(f_2{}^*,0,0)^\frown \in X(T, F_0 \cap \sigma(T)),$$

and (4.15) implies that $(f_1{}^*,0,0)^\frown \in X(T,H)$; thus (4.14) is proved.

1.4.21 Proposition. For $f_0 \in X_1$ and $g_0 \in X_2$, there exist $f_i \in X_1$, $g_i \in X_2$ $(i = 1,2)$ such that $f_0 = f_1 + f_2$ and $g_0 = g_1 + g_2$ and

$$(f_1,g_1,0)^\frown \in X(T, H) \text{ and } (f_2,g_2,0)^\frown \in X(T, F_0 \cap \sigma(T)).$$

Proof The proof will be completed in several steps.

1^0. Let Δ_j be one of the intervals as defined above with respective left [right] endpoints $\xi_j[\zeta_j]$. We first consider possibility (4.10,a).

(i) For ξ_j choose $\phi \in C_{\langle p^2 \rangle}(R)$ such that $f(s) = 1$ on $[\xi + \beta/2, 2]$ and $\operatorname{supp} f \subset [\xi - \beta/2, 3]$. Put

$$(4.16) \qquad \tilde{f}_{0,j}(z) = \begin{cases} \phi_j(s)f_0(s), & \text{if } z = s \in I; \\ \{\phi_j(s)\psi(t) + \phi_j(s_n)[1 - \psi(t)]\}f_0(z), & \text{if } z \in D_n. \end{cases}$$

$$(4.17) \qquad \tilde{g}_{0,j}(z) = \begin{cases} \phi_j(s)g_0(s), & \text{if } z \in I; \\ \phi_j(s_n)g_0(z), & \text{if } z \in D_n. \end{cases}$$

It follows from $(*_1)$ that

$$(4.18) \qquad D_n \cap (R \times J^-) \subset [\xi - \beta, \xi + \beta] \times J^- \text{ whenever } s_n \in [\xi - \beta/2, \xi + \beta/2],$$

(4.19) $\quad D_n \cap ([\xi - \beta/2, \xi + \beta/2] \times J^-) = \emptyset$ whenever $s_n \notin [\xi - \beta, \xi + \beta]$,

and from (4.16), (4.18), (4.19) that

$$\tilde{f}_{0,j}(z) = \begin{cases} f_0(z) & \text{if } z = s \geq \xi_j + \beta \text{ or } z \in D_n \text{ such that } s_n \geq \xi_j + \beta; \\ 0 & \text{if } z = s \leq \xi_j - \beta \text{ or } z \in D_n \text{ such that } s_n \leq \xi_j - \beta. \end{cases} \tag{4.20}$$

By (4.17), (4.18), (4.19) we have

$$\tilde{g}_{0,j}(z) = \begin{cases} g_0(z) & \text{if } z = s \geq \xi_j + \beta \text{ or } z \in D_n \text{ such that } s_n \geq \xi_j + \beta; \\ 0 & \text{if } z = s \leq \xi_j - \beta \text{ or } z \in D_n \text{ such that } s_n \leq \xi_j - \beta. \end{cases} \tag{4.21}$$

(ii) For ζ_j, from $(*_3)$ one has

$$[\zeta_j, -\beta, \zeta_j, +\beta] \cap [I_0 \cup (F_0 \cap R)] = \emptyset,$$

and hence $(**_2)$ implies that $f_0(s) = g_0(s) = 0$ for $s \in [\zeta_j, -\beta, \zeta_j, +\beta]$. Therefore

$$\tilde{f}_{0,j}(s) = \tilde{g}_{0,j}(s) = 0 \quad \text{for } s \in [\zeta_j, -\beta, \zeta_j, +\beta].$$

Using (4.20) define

$$f_{0,j}(z) = \begin{cases} \tilde{f}_{0,j}(z) & \text{if } z = s < \zeta_j - \beta/2 \text{ or } z \in D_n \text{ such that } s_n < \zeta_j - \beta/2; \\ 0 & \text{if } z = s \geq \zeta_j - \beta/2 \text{ or } z \in D_n \text{ such that } s_n \geq \zeta_j - \beta/2. \end{cases} \tag{4.20'}$$

and using (4.21) define

$$g_{0,j}(z) = \begin{cases} \tilde{g}_{0,j}(z) & \text{if } z = s < \zeta_j - \beta/2 \text{ or } z \in D_n \text{ such that } s_n < \zeta_j - \beta/2; \\ 0 & \text{if } z = s \geq \zeta_j - \beta/2 \text{ or } z \in D_n \text{ such that } s_n \geq \zeta_j - \beta/2. \end{cases} \tag{4.21'}$$

Since D_n are disjoint from each other, it follows that $f_{0,j} \in X_1$ and $g_{0,j} \in X_2$. From the definition above, we have

$$f_{0,j}(z) = \begin{cases} f_0(z) & \text{if } z \in [\xi_j+\beta, \zeta_j-\beta] \text{ or } z \in D_n \text{ and } s_n \in [\xi_j+\beta, \zeta_j-\beta]; \\ 0 & \text{if } z = s \notin [\xi_j+\beta, \zeta_j-\beta] \text{ or } z \in D_n \text{ and } s_n \notin [\xi_j+\beta, \zeta_j-\beta]; \end{cases} \tag{4.22}$$

$$(4.23)\qquad g_{0,j}(z)=\begin{cases} g_0(z) & \text{if } z\in[\xi_j+\beta,\zeta_j-\beta] \text{ or } z\in D_n \text{ and } s_n\in[\xi_j+\beta,\zeta_j-\beta];\\ 0 & \text{f } z=s\notin[\xi_j+\beta,\zeta_j-\beta] \text{ or } z\in D_n \text{ and } s_n\notin[\xi_j+\beta,\zeta_j-\beta].\end{cases}$$

Now $(**_2)$ implies that $f_0(z) = g_0(z)$ for $\operatorname{Im} z \geq \alpha$ and hence

$$(4.24)\qquad f_{0,j}(z) = g_{0,j}(z) \quad \text{for } \operatorname{Im} z \geq \alpha,$$

and (4.22), (4.23), (4.24) imply that

$$(4.25)\qquad \operatorname{supp}(f_{0,j} - g_{0,j}) \subset [\xi_j - \beta, \zeta_j + \beta] \times J^-.$$

2^0. It is easy to see that the remaining cases in (4.10) can be handled analogously to case (a) as in step 1^0. We thus suppose that $f_{0,j}$ and $g_{0,j}$ have been defined for all four cases and $j = 1,\dots, k$.

3^0. For $z \in K$, put

$$f_1(z) = \sum_{j=1}^{k} f_{0,j}(z), \qquad f_2(z) = f_0(z) - f_1(z) \quad \text{and}$$

$$g_1(z) = \sum_{j=1}^{k} g_{0,j}(z), \qquad g_2(z) = g_0(z) - g_1(z)$$

Then $f_i \in X_1$, $g_i \in X_2$ $(i = 1,2)$, and it follows from (4.25) that

$$(4.26)\qquad \operatorname{supp}(f_1 - g_1) \subset \bigcup_{1\leq j\leq k} [\xi_j + \beta, \zeta_j - \beta] \times J^- \subset H.$$

From the definition of f_1 we have $\operatorname{supp}(f_1|I) \subset H$; thus by (4.26) $(f_1,g_1,0)^\wedge \in X(T,H)$. It follows from $(**_2)$ and Prop. 1.4.19 that

$$\begin{aligned}\operatorname{supp}(f_2 - g_2) &\subset (F \cup (F_0 \cap \sigma(T)) \cap (R \times J^-)\setminus \cup_j (\xi_j + \beta, \zeta_j - \beta) \times J^- \\ &\subset F_0 \cap \sigma(T),\end{aligned}$$

and

$$\begin{aligned}\operatorname{supp}(f_2|I) &\subset \{(F \cup (F_0 \cap \sigma(T)) \cap R\}\setminus \cup_j (\xi_j + \beta, \zeta_j - \beta) \\ &\subset F_0 \cap \sigma(T),\end{aligned}$$

and hence $(f_2,g_2,0)^\wedge \in X(T, F_0 \cap \sigma(T))$. Proposition 1.4.21 is thus proved.

1.4.22 Corollary. Let f_0, g_0 be defined by $(**_2)$. Then for every open set G in C and every open neighborhood H of $[\sigma(x,T)\backslash(G \cap \sigma(T))^-]^-$, there exist $f_i \in X_1$ and $g_i \in X_2$ $(i = 1,2)$ such that $f_0 = f_1 + f_2$ and $g_0 = g_1 + g_2$ and

$$(4.27) \qquad (f_1,g_1,0)^\wedge \in X(T, H) \quad \text{and} \quad (f_2,g_2,0)^\wedge \in X(T, (G \cap \sigma(T))^-).$$

Proof. Put $F_0 = (G \cap \sigma(T))^-$. Then the corollary follows directly from Prop. 1.4.21.

Similarly we have the following.

1.4.23 Corollary. Suppose $x = (f,g,h)^\wedge \in X(T, H_0 \cup G_0)$ where H_0 and G_0 are open sets in **C**. Then for every open set $H \supset H_0^-$, there exist $f_i \in X_1$, $g_i \in X_2$ such that

$$(4.28) \qquad f_0 = f_1 + f_2 \quad \text{and} \quad g_0 = g_1 + g_2,$$

$$(f_1,g_1,0)^\wedge \in X(T,H) \text{ and } (f_2,g_2,0)^\wedge \in X(T,G_0).$$

Proof. Set $F_0 = \sigma(x, T)\backslash H_0$ and $F = (\sigma(x,T)\backslash F_0)^-$. Then $F \subset H_0^- \subset H$, and it follows from Prop. 1.4.21 that there exist $f_i \in X_1$, $g_i \in X_2$ $(i = 1,2)$ satisfying (4.27) and hence (4.28).

Using corollary 1.4.23, we can prove Theorem 1.4.18.

<u>Proof of Theorem 1.4.18.</u> Let $\tilde{T} = T/X(T,(G \cap \sigma(T))^-)$, and for $x \in X$ let $\tilde{x}$ be the coset $x + X(T,(G \cap \sigma(T))^-$. Let H be open with $H \supset [\sigma(x,T)\backslash(G \cap \sigma(T))^-]$. Then $\{G,H\}$ covers $\sigma(x,T)$. Moreover, x has the decomposition $x = x_1 + x_2$ where $x_1 \in X(T,H)$ and $x_2 \in X(T, (G \cap \sigma(T))^-)$, and if $x = (f,g,h)^\wedge$ then

$$x_1 = (f_1^*,0,0)^\wedge + (f_1,g_1,0)^\wedge, \qquad x_2 = (f_2^*,0,0)^\wedge + (f_2,g_2,0)^\wedge.$$

Since $x_2 \in X(T,(G \cap \sigma(T))^-)$, we have $\tilde{x} = \tilde{x}_1$, hence

$$\sigma(\tilde{x}, \tilde{T}) = \sigma(\tilde{x}_1,\tilde{T}) \subset H.$$

Also, since $H \supset [\sigma(x,T)\backslash(G \cap \sigma(T))^-]^-$ is arbitrary, we obtain

$$\sigma(\tilde{x},\tilde{T}) \subset [\sigma(x,T)\backslash(G \cap \sigma(T))^-]^-.$$

The reverse inclusion always holds, hence

$$\sigma(\tilde{x},\tilde{T}) = [\sigma(x,T)\backslash(G \cap \sigma(T))^-]^-,$$

and T is quasi-strongly decomposable by Th. 1.3.11.

To prove that T^* is strongly decomposable (Th. 1.4.29) we need more results.

1.4.24. Proposition. The function f_1^* in Proposition 1.4.21 satisfies the inequality,

$$\|f_1^*\| \leq \|\phi\|_1 \|f^*\|_1,$$

where ϕ is given in the proof of (4.14).

Proof. For every positive integer p,

$$\frac{1}{p!p^{2p}}\left\|\frac{\partial^p}{\partial s^p}(\phi f^*)\right\| \leq \frac{1}{p!p^{2p}}\sum_{k+j=p}\binom{p}{k}\left\|\frac{\partial^k}{\partial s^k}\phi\right\|_{D_n}\left\|\frac{\partial^j}{\partial s^j}f^*\right\|_{D_n}$$

$$\leq \sum_{k+j=p}\frac{1}{k!k^{2k}}\left\|\frac{\partial^k}{\partial s^k}\phi\right\|_{D_n}\frac{1}{j!j^{2j}}\left\|\frac{\partial^j}{\partial s^j}f^*\right\|_{D_n}.$$

Since $f^*|I = 0$, it follows from the definition of the norm in X_1 that

$$\|f_1^*\|_1 = \|\phi f^*\|_1 = \sup_n \sum_{p=0}^{\infty}\frac{1}{p!p^{2p}}\left\|\frac{\partial^p}{\partial s^p}(\phi f^*)\right\|_{D_n} \leq$$

$$\sup_n \sum_{p=0}^{\infty}\left\{\sum_{k+j=p}\frac{1}{k!k^{2k}}\left\|\frac{\partial^k}{\partial s^k}\phi\right\|_{D_n}\frac{1}{j!j^{2j}}\left\|\frac{\partial^j}{\partial s^j}f^*\right\|_{D_n}\right\}$$

$$\leq \|\phi\|_1 \|f^*\|_1,$$

so the proof is complete.

1.4.25 Proposition. Functions f_1 and g_1 of Prop. 1.4.21 satisfy the following inequalities, where the ϕ_j are defined in the proof of Prop. 1.4.21.

$$\|f_1\|_1 \le \|f_0\|_1 \sum \|\phi_j\|_{\langle p^2\rangle, I}; \tag{4.29}$$

$$\|g_1\|_2 \le \|g_0\|_2. \tag{4.30}$$

Proof. Let Δ_j be a fixed interval whose endpoints satisfy (4.10,a). For $\tilde{f}_{0,j}$

$$\|\tilde{f}_{0,j}|I\|_{\langle p^2\rangle, I} \le \sum_{p=0}^{\infty} \frac{1}{p!p^{2p}} \|(\phi_j f_0)^{(p)}\|_I \le \|\phi_j\|_{\langle p^2\rangle, I} \|f_0\|_{\langle p^2\rangle, I}. \tag{4.31}$$

Since $|\psi(t)| \le 1$, from (4.31) one has

$$\|\tilde{f}_{0,j}|D_n\|_{\langle p^2\rangle, D_n} = \sum_{p=0}^{\infty} \frac{1}{p!p^{2p}} \sup_{z\in D_n} \left| \frac{\partial^p}{\partial s^p} \left\{ \phi_j(s)\psi(t) + \phi(s_n)[1-\psi(t)] \right\} f_0(z) \right| \tag{4.32}$$

$$= \sum_{p=0}^{\infty} \frac{1}{p!p^{2p}} \sup_{z\in D_n} \left| \frac{\partial^p}{\partial s^p} \phi_j(s)\psi(t) f_0(z) \right|$$

$$\le \|\phi_j\|_{\langle p^2\rangle, I} \|f_0\|_{\langle p^2\rangle, D_n}.$$

So (4.31), (4.32) imply

$$\|\tilde{f}_{0,j}\|_1 \le \|\phi_j\|_{\langle p^2\rangle, I} \|f_0\|_1.$$

From (4.20'), for function $f_{0,j}$, one has $\|f_{0,j}\|_1 \le \|\tilde{f}_{0,j}\|_1$, and hence

$$\|f_{0,j}\|_1 \le \|f_j\|_{\langle p^2\rangle, I} \|f_0\|_1,$$

so (4.29) is proved.

We now prove (4.30). For function $\tilde{g}_{0,j}$, notice that $|\phi_j(s)| \le 1$, so from (4.17) one has

$$\|\tilde{g}_{0,j}\|_K \le \|g_0\|_{K,}$$

and since $\phi(s_n)$ is constant on D_n, one also has

$$\|\tilde{g}_{0,j}|D_n\|_{\lhd\Gamma(e+p)\rhd, D_n} \le \|g_0|D_n\|_{\lhd\Gamma(e+p)\rhd, D_n}$$

and hence $\|\tilde{g}_{0,j}\|_2 \le \|\tilde{g}_0\|_2$. For function $g_{0,j}$, it is easy to see from (4.21') that $\|g_{0,j}\|_2 \le \|\tilde{g}_{0,j}\|_2$. Therefore $\|g_{0,j}\|_2 \le \|g_0\|_2$. Since the supports of $g_{0,j}$ and $g_{0,m}$ are disjoint for $j \neq m$, (4.30) follows and the proof is complete.

1.4.26 Proposition. Suppose the sequence $\{h^{(n)}\}$ converges to $h^{(0)}$ in A. Then $\{h^{(n)}|K\}$ converges to $h^{(0)}|K$ in the norm topology of X_1 and X_2.

Proof. We prove only that $\{h^{(n)}|K\}$ converges to $h^{(0)}|K$ in X_2; the proof for the case of X_1 is similar. Let Γ be an admissible contour in $[-1,2] \times [-1,2]$ containing K in its interior. Then for $z \in K$ and $n \ge 0$, one has by Cauchy's formula

$$(4.33) \quad \frac{\partial^p}{\partial x^p} h^{(n)}(z) = \frac{d^p}{dz^p} h^{(n)}(z) = \frac{p!}{2\pi i} \int_\Gamma \frac{h^{(n)}(\zeta)}{(\zeta - z)^{p+1}} d\zeta,$$

hence

$$(4.34) \quad \left| \frac{\partial^p}{\partial x^p} h^{(n)}(z) \right| \le \frac{p!|\Gamma|}{2\pi\delta^{p+1}} M,$$

where $|\Gamma|$ is the length of Γ, $\delta = \text{dist}(\Gamma, K)$ and $|h^{(n)}(\zeta)| \le M$ for $n \ge 0$. For every given $\varepsilon > 0$, (4.34) implies that there is p_0 such that

$$\sum_{p=p_0}^{\infty} \frac{1}{p![\ln(e+p)]^p} \left\| \frac{\partial^p}{\partial s^p} [h^{(n)} - h^{(0)}] \right\|_K \le \frac{|\Gamma|M}{\pi\delta} \sum_{p=p_0}^{\infty} \frac{1}{[\delta \ln(e+p)]^p} < \varepsilon.$$

From (4.33) we have for n sufficiently large

$$\sum_{p=0}^{p_0-1} \frac{1}{p![\ln(e+p)]^p} \left\| \frac{\partial^p}{\partial s^p} \left[h^{(n)} - h^{(0)} \right] \right\|_K < \varepsilon,$$

and hence $\{h^{(n)}|K\}$ converges to $\{h^{(0)}|K\}$ in the norm topology of X_2.

1.4.27 Proposition. For every pair of open sets H_0 and G_0 and for every open $H \supset H_0^-$, the operator T satisfies

$$X(T, H_0 \cup G_0)^- \subset X(T,H)^- + X(T,G_0)^-.$$

Proof. Let $(f,g,h)^\wedge \in X(T, H_0 \cup G_0)^-$. Then there exists a sequence $\{(f^{(n)},g^{(n)},h^{(n)})\}$ in $X(T, H_0 \cup G_0)$ such that $(f^{(n)},g^{(n)},h^{(n)})^\wedge \to (f,g,h)^\wedge$. Without loss of generality, we may suppose that

$$\sum_{n=1}^{\infty} \|(f^{(n+1)},g^{(n+1)},h^{(n+1)})^\wedge - (f^{(n)},g^{(n)},h^{(n)})^\wedge\| < \infty$$

We may also choose the elements from their cosets to satisfy

$$\|(f^{(n+1)},g^{(n++1)},h^{(n+1)}) -(f^{(n)},g^{(n)},h^{(n)})\|$$

$$\leq 2\|(f^{(n++1)},g^{(n++1)},h^{(n++1)})^\wedge - (f^{(n)},g^{(n)},h^{(n)})^\wedge\|.$$

Then $\{f^{(n)}\}$, $\{g^{(n)}\}$ and $\{h^{(n)}\}$ converge to some elements $f^{(0)}$, $g^{(0)}$, $h^{(0)}$ in X_1, X_2 and A, respectively. Clearly we have $(f^{(0)}, g^{(0)}, h^{(0)}) \in (f,g,h)^\wedge$. For the element $(f^{(0)},g^{(0)},h^{(0)})$, set

$$f^{(0)*}(z) = [1 - \psi(t)][f^{(0)}(z) - g^{(0)}(z)];$$
$$f_0^{(0)}(z) = f^{(0)}(z) - f^{(0)*}(z) - h^{(0)}(z);$$
$$g_0^{(0)}(z) = g^{(0)}(z) - h^{(0)}(z).$$

Then

$$(f,g,h)^\wedge = (f^{(0)},g^{(0)},h^{(0)})^\wedge = (f_0^{(0)},g_0^{(0)},h^{(0)})^\wedge + (f^{(0)*},0,0)^\wedge.$$

For $(f^{(n)},g^{(n)},h^{(n)})$, set

$$f^{(n)*}(z) = [1 - \psi(t)][f^{(n)}(z) - g^{(n)}(z)];$$
$$f_0^{(n)}(z) = f^{(n)}(z) - f^{(n)*}(z) - h^{(n)}(z);$$
$$g_0^{(n)}(z) = g^{(n)}(z) - h^{(n)}(z).$$

Then $\{f^{(n)*}\}$ converges to f^* in X_1. Applying Prop. 1.4.20 to $f^{(n)*}$, we have $f_i^{(n)*} \in X_1$ $(i = 1,2)$ such that

$$f^{(n)*} = f_1^{(n)*} + f_2^{(n)*}, \tag{4.35}$$

$$(f_1^{(n)*},0,0)^{\wedge} \in X(T,H) \text{ and } (f_2^{(n)*},0,0)^{\wedge} \in X(T,G_0) \tag{4.36}$$

By Proposition 1.4.24

$$\|f_1^{(n)*} - f_1^{(m)*}\|_1 \le \|\phi\|_1 \|f^{(n)} - f^{(m)}\|_1$$

and hence $\{f_1^{(n)*}\}$ converges to an element $f_1^{(0)*}$ in X_1. Consequently, $\{f_2^{(n)*}\}$ also con- verges to an element $f_2^{(0)*}$ in X_1; (4.35) implies that

$$f^{(0)*} = f_1^{(0)*} + f_2^{(0)*}. \tag{4.37}$$

From (4.36) one has

$$(f_1^{(0)*},0,0)^{\wedge} \in X(T,H)^{-} \text{ and } (f_2^{(0)*},0,0)^{\wedge} \in X(T,G_0)^{-}. \tag{4.38}$$

As for the sequences $\{f_0^{(n)}\}$ and $\{g_0^{(n)}\}$, by Prop. 1.4.25 they converge to $f_0^{(0)}$, $g_0^{(0)}$ in X_1, X_2, resp. Applying Prop. 1.4.21 to $f_0^{(n)}$, $g_0^{(n)}$, we obtain $f_i^{(n)} \in X_1$, $g_i^{(n)} \in X_2$ $(i = 1,2)$ such that

$$f_0^{(n)} = f_1^{(n)} + f_2^{(n)}, \quad g_0^{(n)} = g_1^{(n)} + g_2^{(n)},$$
$$(f_1^{(n)},g_1^{(n)},0)^{\wedge} \in X(T,H), \quad (f_2^{(n)},g_2^{(n)},0)^{\wedge} \in X(T,G_0).$$

Proposition 1.4.25 implies that

$$\|f_1^{(n)} - f_1^{(m)}\|_1 \le \|f_0^{(n)} - f_0^{(m)}\|_1 \sum \|\phi_j\|_{<p^2>,l}$$

and hence $\{f_1^{(n)}\}$ converges to an element $f_1^{(0)}$ in X_1, consequently, $\{f_2^{(n)}\}$ also converges to an element $f_2^{(0)}$ in X_1. Furthermore, we have

$$f_0^{(0)} = f_1^{(0)} + f_2^{(0)}. \tag{4.39}$$

From Prop. 1.4.25 again one has also

$$\|g_1^{(n)} - g_1^{(m)}\|_2 \le \|g_0^{(n)} - g_0^{(m)}\|_2.$$

Hence $\{g_1^{(n)}\}$ and $\{g_2^{(n)}\}$ converge to some elements $g_1^{(0)}$, $g_2^{(0)}$ in X_2, resp. Clearly,

$$g_0^{(0)} = g_1^{(0)} + g_2^{(0)}, \tag{4.40}$$

and this implies that

$$(f_1^{(0)},g_1^{(0)},0)^\frown \in X(T,H)^-, \quad (f_2^{(0)},g_2^{(0)},0)^\frown \in X(T, G_0)^-. \tag{4.41}$$

It follows from (4.38), (4.39), (4.40) and (4.41) that

$$\begin{aligned}(f,g,h)^\frown &= (f_0^{(0)},g_0^{(0)},0)^\frown + (f^{(0)*},0,0)^\frown \\ &= (f_1^{(0)},g_1^{(0)},0)^\frown + (f_2^{(0)},g_2^{(0)},0)^\frown + (f_1^{(0)*},0,0)^\frown + (f_2^{(0)*},0,0)^\frown \\ &\in X(T,H)^- + X(T,G_0)^-,\end{aligned}$$

and the proposition is proved.

1.4.28 Theorem. If $T \in L(X)$, then T^* is strongly decomposable if and only if

(i) T is decomposable;

(ii) for every pair of open sets H_0, G_0 and open set $H \supset H_0^-$, one has

$$X(T, H_0 \cup G_0)^- \subset X(T,H)^- + X(T,G_0)^-. \tag{4.42}$$

Proof. "Only if": In particular, T^* is decomposable, and so is T by Cor. 1.2.10. Let H_0, G_0 be open sets and let $H \supset H_0^-$. Put $W = X(T, H_0 \cup G_0)^-$ and let G be open such that $\{H,G\}$ covers $H_0^- \cup G_0^-$ and $G^- \cap H_0^- = \emptyset$. Hence $\{H,G\}$ covers $\sigma(T|W)$. Since T^* is strongly decomposable, for every closed F by [11] $T^*/X^*(T^*,F)$ is decomposable. Set $F = \mathbf{C}\setminus(H_0 \cup G_0)$. Then $T|W$ is decomposable by Cor. 1.2.10. Thus the following decomposition holds:

$$W = W(T|W, H)^- + W(T|W, G)^-.$$

It is easily seen that $W(T|W,H)^- \subset X(T,H)^-$, and since it follows from [31, Th. 15.13] that

$$W(T|W,G)^- = X(T,G \cap (H_0 \cup G_0))^- = X(T, G_0 \cap G)^- = X(T,G_0)^-,$$

(4.42) is proved.

"If": Since T is decomposable, so is T^* and $X^*(T^*,F)$ is closed for each closed F. To prove that T^* is strongly decomposable, we have only to show that $T^*|X^*(T^*,F)$ is decomposable. Let $G_0 = \mathbf{C}\setminus F$ and let H be an arbitrary open set. Put $\hat{X} = X/X(T,G_0)^-$ and $\hat{T} = T/X(T,G_0)^-$. It follows from the hypothesis (4.42) that every $x \in X(T, H_0 \cup G_0)^-$ has a representation

$$x = x_0 + x_H \quad \text{with } x_0 \in X(T,G_0)^- \text{ and } x_H \in X(T,H)^-,$$

where $H \supset H_0$ is arbitrary, open. Consequently, $\hat{x} = \hat{x}_H$ and since $\hat{T}$ has SVEP,

$$\sigma(\hat{x}, \hat{T}) = \sigma(\hat{x}_H, \hat{T}) \subset \sigma(x_H, T) \subset H^-.$$

Since $\hat{Z} = X(T, H_0 \cup G_0)^-/X(T,G_0)^-$ is easily shown to be analytically invariant under $\hat{T}$, we have $\sigma(\hat{x}, \hat{T}|\hat{Z}) = \sigma(\hat{x}, \hat{T}) \subset H^-$. As $\sigma(\hat{T}|\hat{Z})$ is the union of all the local spectra (Lemma 1.3.10), we have

$$(\hat{T}|\hat{Z}) = \cup\{\sigma(\hat{x}, \hat{T}|\hat{Z}): \hat{x} \in \hat{Z}\} = \cup\{\sigma(\hat{x},\hat{T}): \hat{x} \in \hat{Z}\} \subset H^-.$$

Since $H \supset H_0$ is arbitrary, we have

$$\sigma(\hat{T}|\hat{Z}) \subset H_0^-. \tag{4.43}$$

Now let $\{H_1,H_2\}$ be an open cover of $\mathbf{C}$. Since T is decomposable, we have

$$\begin{aligned} X &= X(T,H_1)^- + X(T,H_2)^- \\ &= X(T, H_1 \cup G_0)^- + X(T, H_2 \cup G_0)^-, \end{aligned}$$

and there corresponds the following decomposition of the quotient space:

$$X/X(T,G_0)^- = X(T,H_1 \cup G_0)^-/X(T,G_0)^- + X(T,H_2 \cup G_0)^-/X(T,G_0)^-. \tag{4.44}$$

It follows from (4.43) applied to $H_0 = H_i$ ($i = 1,2$) that

$$(4.45) \qquad \sigma(\hat{T}|[X(T,H_i \cup G_0)^-/X(T,G_0)^-]) \subset H_i \quad (i=1,2).$$

Hence $\hat{T}$ is decomposable by (4.44) and (4.45). It follows from

$$\hat{T}^* = [T/X(T, G_0)^-]^* = T^*|X^*(T^*, F)$$

that $T^*|X^*(T^*,F)$ is decomposable by Cor. 1.2.10.

We can now prove our desired result.

1.4.29 Theorem. There exist a complex Banach space X and a quasi-strongly decomposable but not strongly decomposable operator $T \in L(X)$ such that T^* is strongly decomposable.

Proof. Let T be the operator in Theorem 1.4.18. Then T is quasi-strongly decomposable and satisfies (4.42) by Prop. 1.4.27. Hence T^* is strongly decomposable by Th. 1.4.28.

It is an open question whether the decomposable, quasi-strongly decomposable and superdecomposable operators form distinct classes. However, Albrecht [3] proved that the decomposable operators are proper in the quasidecomposable ones. It is also unknown whether analytically decomposable operators properly contain the quasidecomposable ones. On the other hand, the former class has a property not known to be shared by any of the classes 1-9, namely,

1.4.30 Proposition. If $T \in L(X)$ is analytically decomposable, then

(i) T has SVEP;

(ii) $T|(T^mX)^-$ is also analytically decomposable for each $m \geq 1$.

Proof. (i). It suffices to prove that $\{0\}$ is analytically invariant. Let G_1, G_2 be open discs in $\mathbf{C}$ such that $G_1 \cap G_2 = \emptyset$. By analytic decomposability, there are analytically invariant subspaces X_1, X_2 with $\sigma(T|X_j) \subset G_j$ for $j = 1,2$. It follows from the convexity of the G that $\sigma(T|X_1 \cap X_2) \subset G_1 \cap G_2 = \emptyset$. Since analytic invariance is stable under intersection, we have $X_1 \cap X_2 = \{0\}$ is analytically invariant.

(ii). We prove first that, for analytically invariant Y, the space $Z = (T^mY)^-$ is also. Let $f: D \to X$ be analytic such that

$$(4.46) \qquad (\lambda - T)f(\lambda) \in Z$$

for λ in the open set D in $\mathbf{C}$. Since $Z \subset Y$ and Y is analytically invariant, one has $f(\lambda) \in Y$, Hence

$$(4.47) \qquad \lambda f(\lambda) = (\lambda - T)f(\lambda) + Tf(\lambda) \in Z + TY \subset (TY)^-.$$

Therefore, $f(\lambda) \in (TY)^-$ for $\lambda \in D$. Now assume that $f(\lambda) \in (T^kY)^-$ for some $1 \le k < m$. Then (4.47) becomes

$$\lambda f(\lambda) = (\lambda - T)f(\lambda) + Tf(\lambda) \in Z + T(T^kY)^- \subset (T^{k+1}Y)^-.$$

Hence $f(\lambda) \in (T^{k+1}Y)^-$ for $\lambda \in D$. Setting $k+1 = m$, we have $f(\lambda) \in Z$ provided (4.46) holds.

Let $Z = (T^mX)^-$ and let $\{G_j\}$ be a finite open cover of $\mathbf{C}$. There are analytically invariant subspaces X_j of T $(j = 1,\dots,n)$ such that

$$X = \vee_{1 \le j \le n} X_j \qquad \text{and} \qquad \sigma(T|X_j) \subset G_j.$$

Therefore,

$$(4.48) \qquad Z = (T^mX)^- = \vee_{1 \le j \le n} (T^mX_j)^- = \vee_{1 \le j \le n} Z_j$$

where $Z_j = (T^mX_j)^-$. By the above Z, Z_j are analytically invariant under T, so Z_j is under Z. Moreover, $\sigma(T|Z_j) \subset \sigma(T|X_j) \subset G_j$ for each j. This, together with (4.48) implies that $T|Z$ is analytically decomposable.

Another problem which has received much attention is the duality question. For $T \in L(X)$ we shall say that T has a "symmetric" duality theory if T lies in some specified class iff T^* does. The duality theories of decomposable and boundedly decomposable operators are thus symmetric (Cor. 1.2.10 and Th. 1.4.5), while spectral and strongly decomposable operators do not have symmetric duality theories. For the other classes the question is open.

In the remainder of this section we give some results dealing with duality. As noted above, ASD need not imply decomposability, but if the admissible invariant subspaces in the ASD are sufficiently restricted, then ASD may imply "decomposable"; this restricted class of subspaces is called "strongly analytic."

1.4.31 Definition. Let M be T-invariant with $T \in L(X)$. Then M is said to be strongly analytic (for T) if T/M has property (β).

1.4.32 Theorem. If $T \in L(X)$ has ASD consisting of strongly analytic subspaces, i.e. the invariant subspaces in (ASD) can be chosen to be strongly analytic, then T is decomposable and T^* is super- decomposable.

The proof requires the following lemma.

1.4.33 Lemma. Let $T \in L(X)$ and suppose $X = M \vee N$, where N and M are T-invariant and M is strongly analytic. Then

(a) $\sigma(T/M) \subset \sigma(T|N)$;

(b) if T has SVEP, then $X(T,E) \subset M$ for each E with $E \cap \sigma(T|N) = \emptyset$.

Proof. Fix $x \in X$. Then there are sequences $\{y_n\} \subset M$, $\{z_n\} \subset N$ such that

$$x = \lim_n (y_n + z_n).$$

Put $f_n(\lambda) = R(\lambda, T|N)y_n$ for $\lambda \in \rho(T|N)$. It follows that in the quotient space X/M, where $\hat{x} = x + M$, etc.,

$$\hat{x} = \lim_n (\lambda - T/M)\hat{f}_n(\lambda) \qquad (\lambda \in \rho(T|N)).$$

Since T/M has property (β) by Def. 1.4.31, $\hat{f}_n$ converges to an analytic function uniformly on compact subsets of $\rho(T|N)$. Now T/M also has SVEP, so $\lambda - T/M$ is bijective for $\lambda \in \rho(T|N)$ by Lemma 1.2.5. Thus $\rho(T|N) \subset \rho(T/M)$, and (a) follows.

(b) Let T have SVEP, and let $E \cap \sigma(T|N) = \emptyset$. Let $\sigma(x,T) \subset E$. Then $\sigma(\hat{x},T/M) \subset \sigma(x,T) \subset E$, so by (a)

$$\sigma(\hat{x}, T/M) \subset \sigma(T/M) \cap E \subset \sigma(T|N) \cap E = \emptyset.$$

Thus $\hat{x} = 0$ or $x \in M$, hence (b) follows.

Proof of Th. 1.4.32. Let G, H be open with $G^- \subset H$. Then by hypothesis there exist strongly analytic subspaces M, N with $\sigma(T|M) \subset H$ and $\sigma(T|N) \subset C\backslash G$. By Lemma 1.4.33(a), $\sigma(T/M) \subset C\backslash G$. Theorem 1.2.1(iii) implies that T is decomposable.

Hence T^* is decomposable by Cor. 1.2.10. Let $F \subset G$ with F closed and G open. Put $K = C\backslash G$, $H = C\backslash F$. Then $\{G,H\}$ is a cover of C, so we can find M, N strongly analytic for T such that $\sigma(T|M) \subset H$ and $\sigma(T|N) \subset G$. Hence $M \subset X(T,H)^-$ so

$$X^*(T^*,F) = X(T,H)^\perp \subset M^\perp.$$

Lemma 1.4.33(a) implies $\sigma(T/M) \subset \sigma(T|N) \subset G$. But $(T/M)^* = T^*|M^\perp$, hence $\sigma(T^*|M^\perp) \subset G$, or $M^\perp \subset X^*(T^*,G)$. It remains to prove that $T^*|M^\perp$ is decomposable. But the latter restriction and its predual T/M both have property (β), hence $T^*|M^\perp$ is decomposable by Th. 1.2.1(iv).

The hypotheses of the last theorem do not seem to guarantee the superdecomposability of T as well as that of T^*. However, we have

1.4.34 Theorem. Let T^* [or T] be quasi-strongly decomposable. Then both T and T^* are superdecomposable.

Proof. Since T^* is clearly decomposable, T is decomposable by Cor. 1.2.10. Let $F_1 \subset G_1$ with F_1 closed, G_1 open. Let H be open with $\mathbf{C}\setminus G_1 \subset H \subset H^- \subset \mathbf{C}\setminus F_1$. Then

$$\begin{aligned}\sigma(T)\setminus G_1 &\subset \sigma(T)\cap H \subset (\sigma(T)\cap H)^- \\ &\subset \sigma(T)\cap H^- \subset \sigma(T)\setminus F_1,\end{aligned}$$

hence

$$X(T,F_1) \subset X(T,\sigma(T)\setminus[\sigma(T)\cap H]) \subset X(T,G_1).$$

Put $M = X(T,\sigma(T)\setminus[\sigma(T)\cap H]^-)$. Then $M^- \subset X(T,G_1)$ and $M^\perp = X^*(T^*,[\sigma(T)\cap H]^-)$. By Cor. 1.3.11, $T^*/M^\perp$ is decomposable, hence since $(T|M^-)^* = T^*/M^\perp$ the restriction $T|M^-$ is decomposable by Cor. 1.2.10. Hence T is superdecomposable. The proof for the case of T is similar.

Chapter II - PERTURBATION THEORY

1. Introduction

We say (as usual) that the operator T is a perturbation of S by L if T = S + L and L is considered "small" relative to S. Typical classes of small operators include compact and quasinilpotent ones, but sometimes in our present context the perturbing operator L will come from a small subclass of the original S. The two principal types of questions we deal with in this chapter are as follows.

(A) If T = S + L with L small, then does T share special properties with S?

(B) Given T, then does T = S + L with L small and S "simpler" than T?

A classical theorem of type (B) is Dunford's decomposition of a spectral operator T = S + L into scalar-type part S and quasinilpotent part L. In §2 of the chapter we discuss problems of type (A) on arbitrary Banach spaces as well as a generalization of Dunford's theorem due to R. Evans. One result of type (A) in §3 is specialized to Hilbert space. In §4 we study the relation of perturbation to other important concepts, especially quasisimilarity.

2. General Banach Spaces.

We begin with a definition of a restrictive SDP.

2.2.1 Definition. Let $T \in L(X)$. We say T has convex SDP if for each open cover $\{G_i\}$ consisting of convex sets there are T-invariant subspaces M_i such that $X = M_1 + \dots + M_n$ and $\sigma(T|M_i) \subset G_i$ (each i).

2.2.2 Theorem. If T has convex SDP, then T has SVEP and X(T,F) is closed for F closed and convex.

Proof. Let $f: \Omega \to X$ be analytic on a connected region such that

$$(2.1) \qquad (\lambda - T)f(\lambda) = 0 \quad \text{for } \lambda \in \Omega.$$

Let $G = \{\lambda : |\lambda - \lambda_0| < r\}$ be an open disc in Ω satisfying $G^- \subset \Omega$, and let

$$G_1 = \{\lambda : \operatorname{Re} \lambda < \operatorname{Re} \lambda_0 + \varepsilon\}$$

$$G_2 = \{\lambda\colon \mathrm{Re}\,\lambda > \mathrm{Re}\,\lambda_0 - \varepsilon\}$$

be halfplanes such that $G\backslash G_i^- \neq \emptyset$ $(i = 1,2)$ for $\varepsilon > 0$ small enough. By the hypothesis "convex SDP" there exist T-invariant subspaces M_1, M_2 with $X = M_1 + M_2$ and $\sigma(T|M_i) \subset G_i$ $(i = 1, 2)$. Then there exist analytic functions f_1 and f_2 defined on G such that

$$f(\lambda) = f_1(\lambda) + f_2(\lambda), \quad f_i(\lambda) \in M_i \quad (i = 1, 2; \text{ for } \lambda \in G).$$

By (2.1)

$$(\lambda - T)f_1(\lambda) = (T - \lambda)f_2(\lambda) \in M_1 \cap M_2, \tag{2.2}$$

hence for $\lambda \in G\backslash G_1^-$ the inverse of $\lambda - T$ is defined on $M_1 \cap M_2$ and M_1 and (2.2) implies

$$f_1(\lambda) = (\lambda - T|M_1 \cap M_2)^{-1}(T - \lambda)f_2(\lambda) \qquad (\lambda \in G\backslash G_1^-).$$

Thus $f_1(\lambda) \in M_1 \cap M_2$ for all $\lambda \in \Omega$ by analytic continuation, and simiarly $f_2(\lambda) \in M_1 \cap M_2$ for all λ. If $|\mathrm{Re}\,\lambda - \mathrm{Re}\,\lambda_0| > \varepsilon$ for $\lambda \in \Omega$, then $\lambda \in \rho(T|M_1 \cap M_2)$, hence $f(\lambda) = 0$ for such λ by (2.1). By analytic continuation $f = 0$ on Ω. This proves that T has SVEP.

Next let F be a closed convex set, and let G_ε be the open ε-neighborhood of F. Then G_ε is convex for each $\varepsilon > 0$. Now choose n other open convex sets $G_1, \ldots, G_n$ such that $\{G_\varepsilon, G_1, G_2, \ldots, G_n\}$ is an open cover of $\sigma(T)$ and $F \cap G_j^- = \emptyset$ for $j = 1, 2, \ldots, n$. Let $\{M_\varepsilon, M_1, M_2, \ldots, M_n\}$ be a system of T-invariant subspaces satisfying Definition 2.2.1. Hence x in X has a decomposition

$$x = x_\varepsilon + \Sigma_j\, x_j \quad (x_\varepsilon \in M_\varepsilon,\ x_j \in M_j).$$

Now let $\Gamma \subset G_\varepsilon$ be a Jordan curve with F in its interior and each G_j in its exterior. Then $x \in X(T,F)$ can be written

$$x = (2\pi i)^{-1}\int_\Gamma x_\varepsilon(\lambda)d\lambda + (2\pi i)^{-1}\Sigma_j \int_\Gamma x_j(\lambda)d\lambda = x_\varepsilon \in M_\varepsilon,$$

where $y(\cdot)$ is the local resolvent of $y \in X$. This proves that $X(T,F) \subset$

M_ε for every $\varepsilon > 0$, hence $X(T,F) \subset \cap\{M_\varepsilon: \varepsilon > 0\}$. If x lies in the latter intersection, then $\sigma(x,T) \subset G_\varepsilon$ for $\varepsilon > 0$, hence $x \in X(T,F)$. This proves that $X(T,F)$ is closed.

2.2.3 Definition. (a) Let T and S be commuting operators on X. If $T|Y$ has convex SDP for each Y spectral maximal for S, we say that T has the <u>strong convex SDP relative to</u> S.

(b) We shall say T is <u>regularly decomposable</u> relative to the identity if in Def. 1.3.4 we may choose each $P_i \in \{T\}''$ $(i = 1,2,\ldots,n)$, i.e. P_i is in the double commutant of T.

2.2.4 Theorem. Let S and T be commuting operators on X. If one of the following conditions is satisfied, then T [resp. T*] has the strong convex SDP relative to S [resp. S*].

(i) T is spectral;

(ii) T is boundedly decomposable;

(iii) T is A-scalar and S commutes with some A-spectral function;

(iv) T is regularly decomposable relative to the identity.

Proof. We prove the theorem in case (iv), since (i)-(iii) are special cases.

Let $\{G_i\}$ be a convex open cover of $\sigma(T)$. Since T is regularly decomposable relative to the identity, by definition there exist T-invariant subspaces M_i and bounded operators $P_i \in \{T\}''$ such that

$$(2.3) \qquad I = \Sigma_i P_i, \quad P_iX \subset M_i, \text{ and } \sigma(T|M_i) \subset G_i \ (i = 1,2,\ldots,n).$$

Let Y be a spectral maximal space of S. Then Y is invariant under T and P_i and by (2.3)

$$I|Y = \Sigma_i P_i|Y, \quad P_iY \subset Y \cap M_i \quad \text{and} \quad \sigma(T|Y \cap M_i) \subset G_i,$$

hence $T|Y$ has the convex SDP. By Definition 2.2.3(a) T has the strong convex SDP relative to S. The case for adjoints T^*, S^* follows from the first part of the proof and the fact that decomposability relative to the identity is invariant under adjunction [31, Cor 17.12].

2.2.5 Theorem. Let T and S commute on X. If S is decomposable and T has the strong convex SDP relative to S, then $T^* + S^*$ has property (β).

Proof. Let $f_n: \Omega \to X^*$ be a sequence of analytic functions such that

(2.4) $$\|(\lambda - T^* - S^*)f_n(\lambda)\| \to 0 \quad (n \to \infty)$$

uniformly on every compact set in Ω. Fix $\lambda_0 \in \Omega$ and let G_1, G_2 be open discs centered at λ_0 with radii r, 2r resp. and with $G_2^- \subset \Omega$. Let $\{\eta_j\}_1^n$ and $\{\delta_k\}_1^m$ be covers of $\sigma(T)$ and $\sigma(S)$ respectively by open discs, where we further suppose that the radius of each disc $\eta_j + \delta_k$ is less than r/2. (Recall that for sets η, δ the sum $\eta + \delta = \{\lambda + \mu : \lambda \in \eta, \mu \in \delta\}$; it is easy to see that $\eta_j + \delta_k$ is a disc.) There are two possibilities:

(a) $(\eta_j + \delta_k)^- \cap G_1^- = \emptyset$,

(b) if $(\eta_j + \delta_k)^- \cap G_1 \neq \emptyset$, then $(\eta_j + \delta_k)^- \subset G_2$.

Let Y_k be spectral maximal spaces of S with

(2.5) $$X = \Sigma\, Y_k \text{ and } \sigma(S|Y_k) \subset \delta_k \quad (\text{each } k).$$

Since S clearly commutes with $R(\lambda;T)$, Y_k is invariant under T and $R(\lambda;T)$. It follows that $\sigma(T|Y_k) \subset \sigma(T)$ for all k, hence $\{\eta_j\}$ is also a cover of $\sigma(T|Y_k)$ (all k). Since $T|Y_k$ has convex SDP, let $M_j \subset Y_k$ be T-invariant with $\Sigma\, M_j = Y_k$ and $\sigma(T|M_j) \subset \eta_j$. Put $X_j = X(T,\eta_j^-)$. Then clearly $M_j \subset X_j$, and

(2.6) $$Y_k = \Sigma\, M_j \subset \Sigma_j\, Y_k \cap X_j \quad (k=1, \ldots, m).$$

The convexity of η_j and δ_k imply that

$$\sigma(T|X_j \cap Y_k) \subset \eta_j^- \text{ and } \sigma(S|X_j \cap Y_k) \subset \delta_k,$$

so that $\sigma(T+S|X_j \cap Y_k) \subset (\eta_j + \delta_k)^-$. The last inclusion follows by a spectral radius argument after translation by centers of η_j and δ_k (see [45, p.45]). From (2.5) and (2.6) we have $X = \Sigma_{j,k}\, X_j \cap Y_k$, hence there exists $C > 0$ such that for $x \in X$ there are $x_{jk} \in X_j \cap Y_k$ with $x = \Sigma\, x_{jk}$ (all j, k) and

(2.7) $$\Sigma_{jk}\, \|x_{jk}\| \leq C\|x\|.$$

For ease of notation put $Z_{jk} = X_j \cap Y_k$ and suppose (b) is satisfied.For

$\lambda \in G_2 \backslash (\eta_j + \delta_k)^-$ and corresponding x_{jk} one has

$$(2.8) \qquad |\langle x_{jk}, f_n(\lambda)\rangle| = \langle(\lambda - T - S)([\lambda - T - S]|Z_{jk})^{-1}x_{jk}, f_n(\lambda)\rangle|$$

$$= |\langle([\lambda - T - S]|Z_{jk})^{-1}x_{jk}, (\lambda - T^* - S^*)f_n(\lambda)\rangle|$$

$$\leq C_{jk}\|x_{jk}\| \, \|(\lambda - T^* - S^*)f_n(\lambda)\|$$

where $C_{jk} = \sup\{\|(\lambda - T - S|Z_{jk})^{-1}\| : \lambda \in G_2 \backslash (\eta_j + \delta_k)^-\}$. By the maximum modulus principle (2.8) is valid for all $\lambda \in G_1^-$. If (a) holds then $\lambda \in G_1^-$ implies in a similar way that

$$(2.9) \qquad |\langle x_{jk}, f_n(\lambda)\rangle| \leq \tilde{C}_{jk}\|x_{jk}\| \, \|(\lambda - T^* - S^*)f_n(\lambda)\|$$

where $\tilde{C}_{jk}$ is the supremum above over G_1^-. Hence (2.7)-(2.9) imply existence of $K > 0$ satisfying

$$|\langle x, f_n(\lambda)\rangle| \leq K\|x\| \, \|(\lambda - T^* - S^*)f_n(\lambda)\|$$

for all x. Hence $f_n \to 0$ uniformly on G_1^-, so $T^* + S^*$ has property (β) by the Heine-Borel theorem.

2.2.6 Theorem. Suppose T and S commute on X. If S is decomposable and T [resp. T^*] has the strong convex SDP relative to S [resp S^*], then $T + S$ is decomposable.

Proof. By Th. 2.2.5 $T^* + S^*$ has property (β). But S^* being decomposable shows again that $T^{**} + S^{**}$ has property (β). Since $T + S$ inherits this property under restriction, the result follows from Th. 1.2.1.

2.2.7 Corollary. Let $T \in L(X)$. Then T is decomposable iff T and T^* both have convex SDP.

Proof. Since necessity is obvious, observe that X [X^*] is the only nonzero spectral maximal space of $S = 0$ [$S^* = 0$]. Hence T is decomposable by Th. 2.2.6.

The following corollary is immediate from Ths. 2.2.4 and 2.2.6.

2.2.8 Corollary. If S is decomposable and T is regularly decomposable relative to the identity such that $TS = ST$, then $T + S$

is decomposable.

2.2.9 Theorem. If $S, T \in L(X)$ commute and satisfy one of the conditions (i)-(iv) of Th. 2.2.4, then $T + S$ is strongly decomposable.

Proof. Without loss of generality assume that T and S are both regularly decomposable relative to the identity. By Ths. 2.2.4 and 2.2.6, $T+S$ is decomposable. Let $W = X(T+S, F)$ for F closed. Then W is invariant under S and T.

We first prove that $T|W$ is decomposable. For $G \subset \mathbf{C}$ open, by regularity of T choose $P \in \{T\}''$ such that $Px = x$ for all $x \in Y = X(T,G)$. Since S, T commute, so do $(S + T)|Y$ and $(\lambda - T|Y)^{-1}P$. Thus $\sigma(T|W \cap Y) \subset G^-$. By Def. 1.3.4 we may then write $\sigma(T|W \cap Y_i) \subset G_i^-$ for $Y_i = X(T, G_i^-)$. But since $I|W = \Sigma P_j|W$, it follows that $T|W$ is decomposable relative to the identity.

If, now, Z is a closed spectral manifold of $T|W$, by the regularity of S we may show in a similar way as above that $S|Z$ is decomposable. In particular, $S|W$ has the strong convex SDP relative to $T|W$. In the same fashion we infer that $(S|W)^*$ has the strong convex SDP relative to $(T|W)^*$. By Th. 2.2.6 $(T+S)|W$ is decomposable, and this completes the proof.

2.2.10 Corollary. If $T, S \in L(X)$ commute, T is a regularly decomposable operator relative to the identity and S is compact, then $T + S$ is strongly decomposable.

Proof. It is easy to show by the functional calculus that the compact S is also a regular decomposable operator relative to the identity (see, e.g., [24]). Thus the result follows by Th. 2.2.9.

2.2.11 Remark. In [89] the author defined T to be "strongly decomposable relative to S" if $T|Y$ and T/Y are both decomposable for each Y spectral maximal for S. Under this hypothesis and the assumption that S is decomposable, he proves that $T + S$ is decomposable using a more complicated method than the one above. (For treatment of the multivariate case of these results see Chap. 5.4.)

In order to prove perturbation results for ASD operators, we need some preliminaries.

2.2.12 Lemma. (i) Let $\{T_\alpha\}$ be a net of commuting operators on X converging uniformly to S. If each T_α has SVEP, then S does also. (ii) If T has SVEP and Q is a commuting quasinilpotent, then $T + Q$ has SVEP.

Proof. Let $f: D \to X$ be analytic with $(\lambda - S)f(\lambda) = 0$ on disc D centered at $\lambda = 0$ and of radius $r > 0$. It suffices to prove $f(0) = 0$.

Let $D_1 \subset D_2 \subset D$ be concentric open discs of radii $r/3$, $2r/3$. Let $Q_\alpha = S - T_\alpha$ such that $\|Q_\alpha\| < r/6$, then fix such Q_α. From the identity

$$0 = (\lambda - S)f(\lambda) = (\lambda - T_\alpha - Q_\alpha)f(\lambda)$$

it follows that for $|\mu - \lambda| > r/4$

$$(\mu - T_\alpha)(\mu - \lambda + Q_\alpha)^{-1}f(\lambda) = f(\lambda); \tag{2.10}$$

in particular, $\sigma(f(0), T) \subset D_1$. Now let $|\mu| < r/3$. Then

$$\begin{aligned} f(0) &= (2\pi i)^{-1}\int_\Gamma \lambda^{-1}f(\lambda)d\lambda \\ &= (\mu - T_\alpha)[(2\pi i)^{-1}\int_\Gamma \lambda^{-1}(\mu - \lambda + Q_\alpha)^{-1}f(\lambda)d\lambda] \\ &= (\mu - T_\alpha)\phi(\mu) \end{aligned}$$

where $\Gamma = \partial D_2$ and ϕ is defined by

$$\phi(\mu) = (2\pi i)^{-1}\int_\Gamma \lambda^{-1}(\mu - \lambda - Q_\alpha)^{-1}f(\lambda)d\lambda \quad (\mu \in D_1)$$

and is hence analytic on D_1. Thus $\sigma(f(0), T) \subset \mathbf{C}\backslash D_1$, which together with the inclusion $\sigma(f(0), T) \subset D_1$ proves $f(0) = 0$. It follows in a routine way that $f(\lambda) = 0$ for $\lambda \in D$, and (i) is proved. Assertion (ii) follows in the same way if we note that $\lambda - Q$ is invertible for all $\lambda \neq 0$.

2.2.13 Lemma. If $T_i \in L(X_i)$ are analytically decomposable ($i = 1,2$), then so is $T_1 \oplus T_2$. Moreover, $T_1|EX_1$ is analytically decomposable whenever $E \in L(X_1)$ is a projection commuting with T_1.

Proof. The class of analytically invariant subspaces is stable under (finite) direct sums and projection. To see this, let M be analytically invariant for T and let E be a commuting projection. Then EX is easily seen to be analytically invariant for T (if T has SVEP). Hence the equality $EM = M \cap EX$ will prove that EM is analytically invariant for $T|EX$ since analytic invariance is preserved under intersection. Now $EM \subset M \cap EX$ is obvious; for the reverse inclusion let $m \in M \cap EX$ with $m = Ex$ $(x \in X)$. Then $Em = Ex = m \in EM$, so EM is analytically invariant for T and $T|EX$. For the case of direct sums, let M_i be analytically invariant under T_i $(i = 1,2)$. Then $(T_1 \oplus T_2)/(M_1 \oplus M_2) = (T_1/M_1) \oplus (T_2/M_2)$; and since a direct sum has SVEP if and only if each summand does, $M_1 \oplus M_2$ is analytically invariant for $T_1 \oplus T_2$. The conclusion on analytic decomposability of $T_1 \oplus T_2$ and $T|EX_1$ now follows by straightforward application of Def. 1.4.1(g).

2.2.14 Theorem. Let T be analytically decomposable, and let S be a commuting scalar-type spectral operator on X. Then TS and $T + S$ are analytically decomposable.

Proof. The result for the sum follows from that for the product. For suppose that TS is analytically decomposable under the hypotheses. Then for $\lambda \in \rho(-S)$ the identity

$$T + S = (\lambda + S)[I + (T - \lambda)R(\lambda;-S)]$$

shows that $T + S$ is analytically decomposable because both spectral and analytically decomposable operators are stable under translation and inversion.

To prove the result for the product TS, let S be scalar-type spectral, let E be its spectral measure and let $\varepsilon > 0$. We can find a Borel partition $\alpha = \{\beta_j\}$ of $\sigma(S)$ and $\lambda_j \in \beta_j$ (each j) such that

$$\|S - \Sigma_j \lambda_j E(\beta_j)\| < \varepsilon \|T\|^{-1}$$

Putting $E_j = E(\beta_j)$ we have

$$\|TS - \Sigma_j \lambda_j TE_j\| < \varepsilon. \tag{2.11}$$

If we put $T_j = T|E_jX$, then for α as above the operator $U_\alpha = \oplus \lambda_j T_j$ is analytically decomposable by Lemma 2.2.13.

Let M be analytically invariant for T and put $M_j = E_jM$. By the proof of Lemma 2.2.13, M is analytically invariant for U_α, i.e. U_α/M has SVEP. Lemma 2.2.12(i) and (2.11) imply that M is analytically invariant for TS.

Finally let $\{G_i\}$ be a finite open cover of $\sigma(TS)$ and let $\{H_i\}$ with $H_i^- \subset G_i$ be another cover of $\sigma(TS)$ which also covers $\sigma(U_\alpha)$ for suitable α. Choose a spanning set $\{M_i\}$ of analytically invariant subspaces for U_α with $\sigma(U_\alpha|M_i) \subset G_i$ (each i); moreover, we may choose U_α with each $\lambda_j \neq 0$. Then, again as in the proof of Prop. 2.2.13, M is analytically invariant for $T = \oplus_j(\lambda_j^{-1}\lambda_j E_j T)$ and hence for TS by the last paragraph. By convergence in the Hausdorff metric and (2.11) we can choose α so that $\sigma(TS|M_i) \subset G_i$ (each i), and thus the proof is complete.

In order to consider spectral perturbations of ASD operators, we need the following lemma.

2.2.15 Lemma. If T is quasidecomposable and Q is a quasinilpotent commuting with T, then T + Q is analytically decomposable.

Proof. For each closed F the spectral manifold $M = X(T,F)$ is Q-invariant and T/M has SVEP [31, Prop 4.14]. But then $(T + Q)/M = T/M + Q/M$ also has SVEP by Lemma 2.2.12(ii), i.e. M is analytically invariant for T + Q. The conclusion now follows easily using the equality $\sigma(T+Q|M) = \sigma(T|M)$.

2.2.16 Corollary. Let T be quasidecomposable and let S be a commuting spectral operator. Then T+S is analytically decomposable.

Proof. Since $S = A + Q$, where A is scalar-type, Q is quasinilpotent and both commute with T, we have $T + S = (T + Q) + A$, hence the conclusion follows from Th. 1.2.4, Th. 2.2.14 and Lemma 2.2.15.

3. Hilbert Space

Now let X be a Hilbert space H. To state our results we must recall some background information from various theories of Hilbert space operators. Let K(H) be the ideal of compact operators on H. For $L \in K(H)$ write $L' = (L^*L)^{1/2}$ for the positive part of the polar decomposition. For $1 \leq p < \infty$ we say $L \in K_p(H)$ (Schatten p-class) if the sequence of eigenvalues of L', arranged in descending order with multiplicities counted, lies in ℓ^p.

We begin with a perturbation theorem of Radjabalipour and Radjavi [79]. We use the following notation in the statement of Theorem 2.3.1. Let $A \in L(H)$ and suppose $\sigma(A)$ is contained in a Jordan curve J. We write i(J) to denote the bounded component of $\mathbf{C}\backslash J$ (its "interior") and define

$$\gamma(t) = \gamma(t;A) = \max\{\|(\lambda - A)^{-1}\|: \lambda \in i(J),\ \mathrm{dist}(\lambda,J) \geq t\}.$$

For $B \in K(H)$ let $\mu_n = \mu_n(B)$ be the nth eigenvalue of $(B^*B)^{1/2}$ arranged in descending order counting multiplicities. Put

$$\nu(t) = \nu(t;B) = \max\{n : \mu_n > t^{-1}\}$$

2.3.1 Theorem. Let be a C^2 Jordan curve and $G = i(J)$ be its interior. Let $A \in L(H)$ with $\sigma(A) \subset J$ and let $B \in K(H)$. Let $T = A + B$. If either (a) or (b) holds:

(a) $\int_0^\varepsilon \ln \gamma(t)dt < +\infty$ (some $\varepsilon > 0$) and $B \in K_p(H)$ $(1 < p < \infty)$;

(b) $\gamma(t) = O(t^{-1})$ and $\Sigma\, n^{-1}\mu_n < +\infty$;

then the following are equivalent:

(i) T is strongly decomposable;
(ii) T is decomposable;
(iii) T has property (β);
(iv) T has SVEP;
(v) $\sigma(T)$ does not contain $G = i(J)$.

We shall use several lemmas and a result on resolvent growth to prove Th. 2.3.1, and for these we need more notation.

For a finite-rank operator B on H define

$$D_0(B) = \prod_i (1 - \lambda_i(B)),$$

where $\lambda_i(B)$ runs over all eigenvalues of B. If $x, y \in H$ are unit vectors and $\sigma(B)$ does not contain 1, define

$$D_1(B) = D_0(B)\langle(I - B)^{-1}x, y\rangle.$$

Lemmas 2.3.2-2.3.4 are classical; their proofs may be found in the cited references.

2.3.2 Lemma [28, Ch. XI, §§9,10]. If $B \in L(H)$ has finite rank, then

$$\ln|D_k(B)| \leq C[k + \Sigma_i \ln(1 + \mu_i(B))] \quad (k = 0, 1), \tag{3.1}$$

where $C > 0$ is independent of B, k.

2.3.3 Lemma (Weyl's inequality) [44, p. 27]. For $B \in K_p(H)$ and $A \in L(H)$ and $i = 1, 2, \ldots,$

$$\mu_i(AB) \le \|A\|\mu_i(B) \quad \text{and} \quad \mu_i(BA) \le \|A\|\mu_i(B). \tag{3.2}$$

2.3.4 Lemma. Let ϕ be a conformal mapping from the open unit disc onto $G = i(J)$. Then there exist positive constants k_1 and k_2 satisfying (for $|\lambda| < 1$)

$$k_1(1 - |\lambda|) \le \operatorname{dist}(\phi(\lambda), J) \le k_2(1 - |\lambda|). \tag{3.3}$$

2.3.5 Definition. Let g be analytic on the unit disc, and for $0 < r < 1$, let

$$N(r) = \frac{1}{\pi}\int_0^{2\pi} \ln^+|g(re^{i\theta})|\, d\theta$$

define the <u>Nevanlinna characteristic</u> function of g, where $\ln^+\eta = \max\{\ln \eta, 0\}$.

Our next theorem is a fundamental inequality on growth of local resolvents. By "radius" of the Jordan curve J we mean $\max\{\operatorname{dist}(\lambda, J) : \lambda \in i(J)\}$.

2.3.6 Theorem. Let J be a C^2 Jordan curve with $G = i(J)$. Let $A \in L(H)$ with $\sigma(A) \cap G = \emptyset$ and let $B \in K(H)$ such that the spectrum of $T = A + B$ does not contain G. Let $h: \Omega \to H$ be analytic such that $\|h(\lambda)\| \le 1$ for $\lambda \in \Omega$. If $(\lambda - T)^{-1}h(\lambda)$ has analytic continuation f to $G \cap \Omega$, then there exist $C > 0$ and $K > 0$ such that

$$\ln\|f(\lambda)\| \le \frac{1}{\operatorname{dist}(\lambda,J)}\left\{1 + \ln \gamma(K\operatorname{dist}(\lambda,J)) + \int_0^{C\gamma(K\operatorname{dist}(\lambda,J))} s^{-1}\nu(s)\, ds\right\}. \tag{3.4}$$

Proof. Let $0 < t <$ radius J, let $R = (3/2)\gamma(t)$ and put $m = \nu(R)$. Write $(B^*B)^{1/2} = D_m + F_m$ where D_m is finite-rank hermitian with eigenvalues $0, \mu_1(B), \ldots, \mu_m(B)$ and $\|F_m\| \le 1/R$. Next set $T = A + B = A + B_m + E_m$ where $U(B^*B)^{1/2}$ is the polar decomposition of B and $B_m = UD_m$ and $E_m = UF_m$. Without loss of generality we may assume that the boundary of $G \cap \Omega$ is a C^2 Jordan curve. Let ϕ be a conformal map from the closed unit disc onto $(G \cap \Omega)^-$. For the unit vector $y \in H$

define $\tilde{f}(\lambda) = \langle f(\lambda), y\rangle$ for λ in the domain of f. Now for $\lambda \in (G \cap \Omega)\backslash\sigma(T)$ write

$$\begin{aligned}\tilde{f}(\lambda) &= \langle f(\lambda), y\rangle = \langle(\lambda - T)^{-1}h(\lambda), y\rangle \\ &= \langle(I - C_\lambda)^{-1}[I - (\lambda - A)^{-1}E_m](I - A)^{-1}h(\lambda), y\rangle\end{aligned}$$

where

$$C_\lambda = [I - (\lambda - A)^{-1}E_m]^{-1}(\lambda - A)^{-1}B_m.$$

Next define analytic functions for $\lambda \in G \cap \Omega$ by

$$\begin{aligned}\delta_0(\lambda) &= D_0(C_\lambda)/(\phi^{-1}(\lambda))^p; \\ \delta_1(\lambda) &= \delta_0(\lambda)\tilde{f}(\lambda)\end{aligned}$$

where p is the order of the zero of $D_0(C_\lambda)$ at $\lambda = \phi(0)$. Since δ_0 clearly has analytic continuation to all of G, δ_0 has analytic continuation to $G \cap \Omega$. Hence for $\lambda \in G \cap \Omega$ and $t = \text{dist}(\lambda, J)$ one has

$$\begin{aligned}(3.5) \qquad &\|[I - (\lambda - A)^{-1}E_m]^{-1}(\lambda - A)^{-1}\| \\ &\leq \sum_{n=0}^{\infty} \|(\lambda - A)^{-1}E_m\|^n \|(\lambda - A)^{-1}\| \\ &\leq 3\gamma(t).\end{aligned}$$

Since $\|h(\lambda)\| \leq 1$ for $\lambda \in G \cap \Omega$, it follows from (3.1) and (3.2) that if $1 \notin \sigma(C_\lambda)$ and j = 0 or 1, then

$$\begin{aligned}\ln^+|\delta_j(\lambda)| &\leq \Gamma_1\Big\{j + \sum_{i=0}^{\infty} \ln(1 + \mu_i(C_\lambda))\Big\} + \alpha_j \\ &\leq \Gamma_1\Big\{j + \sum_{i=0}^{\infty} \ln(1 + 3\gamma(t)\mu_i(B_m))\Big\} + \alpha_j \\ &\leq \Gamma_1\Big\{j + \int_0^R \ln\Big[1 + \frac{3\gamma(t)}{s}\Big] d\nu(s)\Big\} + \alpha_j \\ &\leq \Gamma_1\Big\{j + \nu(R)\ln 3 + \int_0^R \frac{2R\nu(s)}{s^2 + 2Rs} ds\Big\} + \alpha_j\end{aligned}$$

$$\leq \Gamma_1\left\{ j + \tilde{v}(R)\ln 3 + \int_0^R s^{-1}v(s)\,ds \right\} + \alpha_j$$

$$\leq \Gamma_1\left\{ j + \eta \int_0^{3R/2} s^{-1}v(s)\,ds \right\} + \alpha_j$$

$$\leq \Gamma_2\left\{ j + \int_0^{3R/2} s^{-1}v(s)\,ds \right\} + \alpha_j$$

where $\alpha_0 = 0, \alpha_1 = \ln 2R$, $\eta = \ln 3/(\ln 3 - \ln 2)$ and $\Gamma_1 > 0$ and $\Gamma_2 = \eta\Gamma_1$. The last calculation remains valid for all $\lambda \in G \cap \Omega$, so from Lemma 2.3.4 and the fact that γ is nonincreasing in its argument we infer

$$\text{(3.6)} \qquad \ln^+|\delta_0(\lambda)| \leq \Gamma_3\left\{ \int_0^{\frac{9}{4}\gamma(k_1(1-|\phi^{-1}(\lambda)|))} s^{-1}v(s)\,ds \right\}$$

and

$$\text{(3.7)} \qquad \ln^+|\delta_1(\lambda)| \leq$$

$$\ln(\gamma(k_1(1-|\phi^{-1}(\lambda)|))) + \Gamma_3\left\{ 1 + \int_0^{\frac{9}{4}\gamma(k_1(1-|\phi^{-1}(\lambda)|))} s^{-1}v(s)\,ds \right\}$$

where $\lambda \in G \cap \Omega$ and $\Gamma_3 > 0$. Because $G \cap \sigma(T)$ is countable, we may choose $0 < r < 1$ such that $\phi(re^{i\theta}) \in \rho(T)$ for $0 \leq \theta < 2\pi$. Applying Jensen's formula to $\delta_0(\phi(w))$ $(|w| < 1)$ yields

$$\ln^+|\delta_0(\phi(0))| = \frac{1}{2\pi}\int_0^{2\pi} \ln^+|\delta_0(\phi(re^{i\theta}))|d\theta \tag{3.8}$$
$$- \frac{1}{2\pi}\int_0^{2\pi} \ln^+|1/\delta_0(\phi(re^{i\theta}))|d\theta$$
$$- \sum_k \ln\frac{r}{|a_k|}$$

where $a_1, a_2, \ldots$ are the zeros of $\delta_0(\phi(w))$ in $|w| < r$. Now $\delta_1(\phi(w)) = g(w)\delta_0(\phi(w))$ where $g(w) = \tilde{f}(\phi(w))$, hence (3.8) implies

$$N(r) \le \frac{1}{2\pi}\int_0^{2\pi} \ln^+|\delta_1(\phi(re^{i\theta}))|d\theta + \frac{1}{2\pi}\int_0^{2\pi} \ln^+|1/\delta_0(\phi(re^{i\theta}))|d\theta$$
$$\le \frac{1}{2\pi}\int_0^{2\pi} \ln^+|\delta_1(\phi(re^{i\theta}))|d\theta + \frac{1}{2\pi}\int_0^{2\pi} \ln^+|\delta_0(\phi(re^{i\theta}))|d\theta - \ln|\delta_0(\phi(0))|$$

where N is the Nevanlinna characteristic of g, so it follows from (3.6) and (3.7) by integration that

$$N(r) \le \Gamma_4\left\{1 + \int_0^{\frac{9}{4}\gamma(k_1(1-r))} s^{-1}v(s)\,ds\right\} + \ln\gamma(k_1(1-r)) \tag{3.9}$$

for some $\Gamma_4 > 0$. Since $\{r \in (0, 1): \phi(re^{i\theta}) \in \rho(T), \text{ all } \theta\}$ is dense in $[0,1]$, (3.9) holds for all $0 < r < 1$. We finally apply (3.3) three times, (3.4) and (3.9) to obtain

$$\ln|\tilde{f}(\lambda)| \le \frac{C}{\text{dist}(\lambda,J)}\left\{1 + \ln\gamma(K\text{dist}(1,J)) + \int_0^{C\gamma(K\text{dist}(\lambda,J))} s^{-1}v(s)\,ds\right\}$$

where $\lambda \in G \cap \Omega$, $C > 0$ and $K > 0$. But $\tilde{f}(\lambda) = \langle f(\lambda), y\rangle$ for an arbitrary unit vector $y \in H$, hence (3.9) follows and the proof is complete.

To simplify the sequel we introduce the notation

$$(3.10) \qquad e(t) = \frac{C}{t}\left\{1 + \ln\gamma(Kt) + \int_0^{C\gamma(Kt)} s^{-1}v(s)\,ds\right\}.$$

2.3.7 Lemma [69]. Let J be a C^2 Jordan curve and let $M:(0,\varepsilon) \to (0,+\infty)$ $(\varepsilon > 0)$ be a nonincreasing function such that

$$(3.11) \qquad \int_0^{\varepsilon} \ln\ln M(t)\,dt < +\infty$$

Then for each $a \in J$ there are two piecewise smooth arcs M_a, L_a crossing J at a and functions h_a, g_a analytic at $\lambda \neq a$ such that

(i) $|h_a(\lambda)| \leq M(\mathrm{dist}(\lambda,J))^{-1}$ $(\lambda \in M_a)$,

$|g_a(\lambda)| \leq M(\mathrm{dist}(\lambda,J))^{-1}$ $(\lambda \in L_a)$;

(ii) g_a (resp. h_a) is bounded on the side of M_a (resp. L_a) opposite that of L_a (resp. M_a).

2.3.8 Lemma. Let A, B, T, f,... be as in Th. 2.3.6, and suppose either (a) or (b) of Th. 2.3.1 holds. Then there exists M satisfying (3.11) such that

$$(3.12) \qquad \|f(\lambda)\| \leq M(\mathrm{dist}(\lambda, J)) \qquad (\lambda \in i(J)).$$

Proof. Put $M(t) = \exp(e(t))$ where $e(t)$ is given by (3.10), and suppose first that (a) holds. Since

$$\int_0^{\infty} s^{-p}dv(s) = \sum_{j=1}^{\infty} \mu_j^p < +\infty,$$

$\lim s^{-p}v(s)$ exists as $s \to \infty$. Hence $v(s) \leq \Gamma s^p$ for some $\Gamma > 0$. It follows from (a) and (3.10) that

$$\int_0^{\varepsilon} \ln\ln M(t)dt = \int_0^{\varepsilon} \ln e(t)dt < +\infty$$

This proves (3.11) in case (a) and (3.12) follows from (3.11) by (3.5). In case (b) the fact that $\Sigma n^{-1}\mu_n$ converges implies

$$\int_{\alpha}^{\infty} \frac{d(\ln v(t))}{tv(t)} < +\infty,$$

where $\alpha = 1/\mu_1$, hence $\int_\alpha^\infty t^{-2}\ln v(t)\,dt$ converges. It follows that

$$\int_0^\varepsilon \ln\left[\int_0^{c/t} s^{-1}v(s)ds\right]dt \le \int_0^\varepsilon \ln(\alpha^{-1}v(ct^{-1}))\,dt = c\int_{c/\varepsilon}^\infty t^{-2}\ln\frac{v(t)}{\alpha}dt < \infty$$

whenever $c/\varepsilon > \alpha$. Thus M satisfies (3.11) in this case too.

2.3.9 Lemma. Let $T \in L(H)$, and let $G \subset C$ be an arbitrary open set. If for every sequence $f_n : G \to H$ of analytic functions for which $(\lambda - T)f_n(\lambda) \to 0$ uniformly on compact sets in G there is a subsequence $\{f_{n_k}\}$ which is uniformly bounded on compact sets in G, then T has property (β).

Proof. If T does not have property (β), then there is a sequence of H-valued analytic functions $\{f_n\}$ on open Ω such that $(\lambda - T)f_n(\lambda) \to 0$ uniformly on compacts in Ω, a compact K in Ω, $\alpha > 0$ and subsequence $\{f_{n_k}\}$ with

(3.13) $$\max\{\|f_{n_k}(\lambda)\| : \lambda \in K\} \ge \alpha.$$

Let K_k be an increasing sequence of compact sets exhausting Ω such that $K_0 = K$. By hypothesis we may assume that $\|(\lambda - T)f_{n_k}(\lambda)\| \le 1/k^2$ for $\lambda \in K_k$. Write $g_k = kf_k$, then $\|(\lambda - T)g_k(\lambda)\| \le 1/k$ on K_k, hence $(\lambda - T)g_k(\lambda) \to 0$ uniformly on compact sets. But (3.13) implies the contradiction $\max\{\|g_k(\lambda)\| : \lambda \in K\} \ge k\alpha \to \infty$, and this completes the proof.

Let $\pi : L(H) \to L(H)/K(H)$ be the canonical (homomorphic) surjection. Recall that T is Fredholm if $\pi(T)$ is invertible in the quotient algebra $L(H)/K(H)$. In this case T and T^* both have closed ranges of finite co-rank in H. The Fredholm index is $\mathrm{ind}(T) = \dim(\ker T) - \dim(\ker T^*)$. Recall also that the set of Fredholm operators (finite index) is preserved under compact perturbation.

2.3.10 Lemma [38]. Let $T \in L(H)$ have SVEP. Then $\sigma_p(T)$ does not contain a neighborhood of any λ for which $\lambda - T$ is Fredholm.

Proof. Suppose T is Fredholm and $\sigma_p(T)$ contains a disc with center zero. By [27, VI.9.18] the manifold $Y = \cap_1^\infty T^nH$ is closed. If $y \in Y$ write $y = Tx_n$ (with $x_n \in T^nH$). Since $\dim(\ker T) < +\infty$, for $m > k$ large enough $x_m - x_k \in T^kH \cap \ker T = T^mH \cap \ker T \subset T^mH$. Thus x_k

$\in Y$ for some fixed k, hence by Lemma 1.2.5 $\rho(T|Y)$ contains zero ($T|Y$ has SVEP also). For $0 < |\lambda|$ small, $\lambda \in \rho(T|Y)$. But such $\lambda \in \sigma_p(T)$, so clearly also $\lambda \in \sigma_p(T|Y)$, a contradiction.

Proof of Theorem 2.3.1. Since (i) $\Rightarrow$ (ii) $\Rightarrow$ (iii) $\Rightarrow$ (iv) is evident, we prove only (iv) $\Rightarrow$ (v) and (v) $\Rightarrow$ (i).

(iv) $\Rightarrow$ (v). Suppose G (= $i(J)$) contains a disc $D \subset \sigma(T)$. By Weyl's theorem [80, p.7] every $\lambda \in D$ is an eigenvalue of T, but $\lambda - T$ is Fredholm for such λ, hence T does not have SVEP by Lemma 2.3.10. This contradiction proves (iv) $\Rightarrow$ (v).

(v) $\Rightarrow$ (i). We first prove that T has property (β). Let $f_n : \Omega \to H$ be a sequence of analytic functions such that $(\lambda - T)f_n(\lambda) \to 0$ uniformly on compact sets in Ω. Put $h_n(\lambda) = (\lambda - T)f_n(\lambda)$, let $F \subset \Omega$ be a fixed compact set and let V be open with $F \subset V \subset V^- \subset \Omega$. Since $\max\{\|h_n(\lambda)\|: \lambda \in V^-\} \to 0$ (as $n \to \infty$), Th. 2.3.6 applies to h_n for sufficiently large n. Suppose, with no loss of generality, that the boundary of V is a C^2 Jordan curve. We may thus suppose that (3.4) is valid for f_n and that $\Gamma = V^- \cap J$ consists of finitely many subarcs of J. Suppose Γ is a single arc. Then ∂V lies on two sides of J and we denote these parts by J_1, J_2. Let a, b be the endpoints of Γ, and let g_a, h_b, L_a and M_b be the functions and arcs given by Lemma 2.3.7. We may further suppose L_a and M_b to be contained in $J_1 \cup J_2$. If we put $g(\lambda) = (\lambda - a)(\lambda - b)g_a(\lambda)h_b(\lambda)$ for $\lambda \in V$, then by Lemma 2.3.7 g is analytic on V and continuous on V^-. For $n = 1, 2, \ldots$, let $u_n(\lambda) = g(\lambda)f_n(\lambda)$ ($\lambda \in \partial V$). Then u_n is uniformly bounded on $J_1 \cup J_2$ and hence throughout V^- by the maximum modulus principle. Thus some subsequence $\{u_{n_k}\}$ converges to an analytic function u on V. Since $g \neq 0$ anywhere on V, $\{f_{n_k}\}$ converges uniformly on each compact subset of V and so to an analytic function on V. In particular, this sequence is uniformly bounded on F. Since the proof for Γ having more than one arc is similar, we infer from Lemma 2.3.9 that T has property (β).

The hypotheses of Th. 2.3.1 and condition (v) also clearly apply to T^* as well as T, hence T^* also has property (β). By Th 1.2.1 T is decomposable. Now $G \cap \sigma(T)$ is nowhere dense, for otherwise T would not have SVEP by Lemma 2.3.10. Thus T is strongly decomposable by [16].

2.3.11. Remark. In [24, p. 169] it was shown that T

satisfying Th. 2.3.1 is A-scalar if $\sigma(T)$ is contained in the unit circle. The authors of [78] ask whether this conclusion can be extended to operators which have eigenvalues in the open disc. Although we do not answer this question fully, we can give a slight improvement of Th. 2.3.1 if $\sigma(T) \cap i(J)$ is finite.

2.3.12 Corollary. Let $T = A + B$ where A is normal with $\sigma(A)$ contained in a C^2 Jordan curve J and $B \in K_p(H)$. If $\sigma(T) \cap i(J)$ is finite, then T is decomposable relative to the identity. In particular, if $\sigma(T) = \sigma(A)$, then T is decomposable relative to the identity.

Proof. Since $\sigma(T) \cap i(J)$ is finite we can write $T = T_1 \oplus K$ where $\sigma(T_1) \subset J$ and K is defined on a finite-dimensional space [74]. By Lemma 2.3.7 applied to T_1, we see that the hypotheses of [32, Th. 17.20] are satisfied. If $\lambda \in i(J)$, then putting $h_a = f_a^+$ and $g_a = f_a^-$ as in Lemma 2.3.7 and applying (3.12) give

$$\|f_a^{\pm}(\lambda)(\lambda - T)^{-1}x\| \leq M(t)^{-1}M(t) = 1.$$

By [32, Th 17.20] T_1 is decomposable relative to the identity, hence T is also.

Now recall that T is <u>essentially unitary</u> if $\pi(T)$ is unitary: $\pi(TT^*) = \pi(T^*T) = \pi(I)$. Finally the <u>essential spectrum</u> of T is $\sigma_e(T) = \sigma(\pi(T))$. The following is an "elementary" result of Brown, Douglas and Fillmore [21, Th. 3.1].

2.3.13 Theorem. If T is essentially unitary ($\sigma_e(T)$ contained in the unit circle), then $T = V + L$ where $L \in K(H)$ and V is unitary, a shift of multiplicity n or adjoint of such a shift according as $n = \operatorname{ind}(T)$ is zero, negative or positive.

Let D be the open unit disc and suppose $I - T^*T \in K(H)$. Then $\pi(T)$ is an isometry, but if $\sigma(T)$ does not contain D then $\operatorname{ind}(T) = 0$. Suppose that $I - T^*T \in K_p(H)$ and $\sigma(T)$ does not contain D. In the polar decomposition $T = WP$, we may assume W is unitary, and the calculation

$$I - T^*T = I - P^2 = (I - P)(I + P)$$

shows that $I - P \in K_p(H)$ since $I + P$ is invertible. Hence $T = WP = W - W(I - P)$ or $T = W + L$ where W is unitary and $L \in K_p(H)$. We have thus proved

2.3.14 Theorem. Let $T \in L(H)$ such that $I - T^*T \in K_p(H)$ and $\sigma(T)$ does not contain D. Then $T = V + L$ where V is unitary and $L \in K_p(H)$.

We now apply some of the results of this section to the theory of weak contractions. Recall that T is a weak contraction on Hilbert space H if T is a contraction ($\|T\| \leq 1$), $I - T^*T \in K_1(H)$ and $\sigma(T)$ does not contain the unit disc.

2.3.15 Corollary. Let T be a contraction. Then T is a weak contraction iff T is decomposable and $T = W + L$ with W unitary, $L \in K_1(H)$.

Proof. Necessity was proved in Ths. 2.3.1 and 2.3.14. For sufficiency we need to prove $I - T^*T \in K_1(H)$ and $\sigma(T)$ does not contain D. The first is immediate from the decomposition $T = W + L$. If $\sigma(T) = D^-$, then $\operatorname{ind}(T) = 0$ and Lemma 2.3.10 imply that T does not have SVEP. But T does have SVEP, hence $\sigma(T) \neq D^-$. Since $\sigma(T)$ is closed, T is a weak contraction by definition.

2.3.16. Corollary. If T and S are weak contractions, then the following are equivalent.

(i) TS is a weak contraction;
(ii) TS is decomposable;
(iii) TS has SVEP.

Proof. Since (i) $\Rightarrow$ (ii) is Th 2.3.15 and (ii) $\Rightarrow$ (iii) always holds, we suppose (iii). By Cor. 2.3.14 $TS = W + L$ where W is unitary, $L \in K_1(H)$. Hence $\operatorname{ind}(TS) = 0$ and thus by Lemma 2.3.10 $\sigma(TS) \neq D^-$. Since TS is a contraction it is a weak contraction.

2.3.17 Corollary. The product of commuting weak contractions is a weak contraction.

Proof. Let T and S be commuting weak contractions. By Cor 2.3.16 it suffices to prove that TS has SVEP. Clearly $\sigma(T) \cup \sigma(S)$ is nowhere dense in D^-, so we may choose a sequence $\{\alpha_n\}$ in D converging to zero and with no α_n in $\sigma(T) \cup \sigma(S)$. The Cayley transforms

$$T_n = (T - \alpha_n)(I - \bar{\alpha}_n T)^{-1}, \quad S_n = (S - \alpha_n)(I - \bar{\alpha}_n S)^{-1}$$

are invertible, hence T_nS_n is also and $T_nS_n \to TS$. Since the map $\lambda \to (\lambda - \alpha_n)(1 - \overline{\alpha}_n\lambda)^{-1}$ is homeomorphic on D, each $\sigma(T_nS_n)$ is nowhere dense and thus T_nS_n has SVEP. But clearly TS commutes with each T_nS_n, so TS has SVEP by Lemma 2.2.12(i).

2.3.18 Example. The product of noncommuting weak contractions may not be a weak contraction. Let V be the simple bilateral shift on ℓ^2, and let K be a rank-one operator such that $V + K = S \oplus S^*$ where S is a unilateral shift (see [45, p. 295]). Then V and $I + V^*K$ are weak contractions but $V(I + V^*K) = V + K$ clearly is not decomposable, hence not a weak contraction.

The last example shows that compact perturbations of the bilateral shift need not have a spectral decomposition. But the question whether such perturbations have nontrivial hyperinvariant subspaces is important because of Th. 2.3.20 below. Our proof of this theorem depends on the following remarkable generalization of Th. 2.3.3. Recall that T is <u>essentially normal</u> if $\pi(T)$ is normal: $\pi(TT^*) = \pi(T^*T)$.

2.3.19 Theorem [21, p. 118]. If T, S are essentially normal such that $\sigma_e(T) = \sigma_e(S)$ and $\mathrm{ind}(\lambda - T) = \mathrm{ind}((\lambda - S)$ for all $\lambda \notin \sigma_e(T)$, then $T = VSV^* + K$ where V is unitary and $K \in K(H)$.

2.3.20 Theorem. Let B be the simple bilateral shift on separable Hilbert space H. Suppose $B + L$ has a nontrivial hyperinvariant subspace for all $L \in K(H)$. Then every essentially unitary operator which is not a scalar multiple of the identity has a nontrivial hyperinvariant subspace.

Proof. Let T be an essentially unitary operator. We may clearly suppose $\sigma_p(T) = \sigma_p(T^*) = \emptyset$ and that $\sigma(T) = \sigma_e(T)$ is contained in the unit circle. We suppose also that $\sigma(T)$ is connected, for otherwise we may construct the desired subspaces with functional calculus.

We consider two cases: (1) $\sigma(T)$ is a singleton $\{\lambda\}$; in this case $T = \lambda + Q$ with Q quasinilpotent and compact (Th. 2.3.13). By assumption $Q \neq 0$, hence the result follows from Lomonosov's lemma [80, p. 156]. (2) $\sigma(T)$ is an arc. Then $\sigma_e(T^n) = \sigma_e(B)$ is the unit circle for some $n \geq 1$. Since $\mathrm{ind}\,(\lambda - T^n) = \mathrm{ind}(\lambda - B)$ for all $|\lambda| \neq 1$. We have $VT^nV^* = B + K$ for some unitary V and compact K by Th. 2.3.19. By hypothesis $B + K$ has a proper hyperinvariant subspace, hence T^n does also, say M. If $ST = TS$, then S also commutes with T^n so M is S-invariant. This completes the proof.

The Apostol-Foias-Voiculescu theorem [15] states that T is biquasitriangular (BQT) iff $\mathrm{ind}(\lambda - T) = 0$ for all $\lambda \in \sigma_e(T)$. We use this as our definition of BQT.

2.3.21 Theorem. Let $T \in L(H)$. If T and T^* both have SVEP, then T is BQT.

Proof. For $\lambda \in \sigma_e(T)$ we prove $\mathrm{ind}(\lambda - T) = 0$. Suppose $\mathrm{ind}(\lambda - T) > 0$. By the continuity of Fredholm index, $\mathrm{ind}(\mu - T) > 0$ for all μ sufficiently near λ. Hence the point spectrum $\sigma_p(T)$ contains a neighborhood of λ; therefore T does not have SVEP by Lemma 2.3.10. We thus have proved $\mathrm{ind}(\lambda - T) \leq 0$. In a similar way T^* having SVEP implies $\mathrm{ind}(\lambda - T) \geq 0$, and the proof is complete.

2.3.22 Corollary. Every decomposable operator on Hilbert space is BQT.

Proof. Immediate from Th. 2.3.21 and Cor. 1.2.10.

2.3.23 Corollary. If T is essentially normal and decomposable, then $T = N + K$ where N is normal and $K \in K(H)$.

Proof. This follows from Cor. 2.3.22 and a consequence of Th. 2.3.19 [21, p. 125] that every BQT essentially normal operator has the form $N + K$.

The operator $V + K$ in Ex. 2.3.18 is an essentially normal operator which is not decomposable. The unilateral shift is essentially normal but neither decomposable nor of the form $N + K$. Moreover, Cor. 3.4.4 will improve Cor. 2.3.22 significantly.

4 Quasisimilarity

We continue to suppose that our underlying Banach spaces are Hilbert spaces. In this section we study the relationship between spectral decom- position and quasisimilarity of operators; many of these results are consequences of earlier sections of this chapter.

Let H_1 and H_2 be Hilbert spaces. A quasiaffinity $A : H_1 \to H_2$ is a bounded injective operator with dense range. For $T \in L(H_1)$ and $S \in L(H_2)$ we say that T is a <u>quasiaffine transform</u> of S ($T \precsim S$) if $AT = SA$ for some quasiaffinity A; we say T and S are <u>quasisimilar</u> ($T \sim S$) if each is a quasiaffine transform of the other.

2.4.1 Theorem. Let $T \precsim S$. If S has property (β), then $\sigma(S) \subset \sigma(T)$.

Proof. Let $AT = SA$ for some quasiaffinity A. It is easy to see that T has SVEP. Let $x \in H_1$, $T \in L(H_1)$. For $\lambda \in \rho(T)$ we have $x = (\lambda - T)f_x(\lambda)$ where f_x is the local resolvent of T at x, hence $Ax = (\lambda - S)Af_x(\lambda)$. We thus obtain $\rho(T) \subset \rho(Ax, S)$ or $\sigma(y,S) \subset \sigma(T)$ for all y in some dense manifold in H_2. Since S has property (β) it follows that $\sigma(y, S) \subset \sigma(T)$ for all $y \in H_2$. But $\sigma(S)$ is the union of $\sigma(y, S)$ over all $y \in H_2$ by Lemma 1.3.10, so the inclusion $\sigma(S) \subset \sigma(T)$ follows.

2.4.2 Corollary. If $T \sim S$ and T and S both have property (β), then $\sigma(T) = \sigma(S)$.

2.4.3 Corollary. If T and S are decomposable and $T \lesssim S$, then $\sigma(T) = \sigma(S)$.

Proof. By Th. 2.4.1 $\sigma(S) \subset \sigma(T)$. But T^* and S^* are also decomposable by Cor. 1.2.8, and clearly $S^* \lesssim T^*$. Hence $\sigma(T^*) \subset \sigma(S^*)$ by Th. 2.4.1 again, and the result follows.

The next theorem is a somewhat deeper result.

2.4.4 Theorem. If T and S are quasisimilar decomposable operators, then $\sigma_e(T) = \sigma_e(S)$.

Proof. By Cor. 2.4.2 $\sigma(T) = \sigma(S)$. Suppose $\lambda \in \sigma_e(T)$ but $\lambda \notin \sigma_e(S)$, and suppose that λ is isolated in $\sigma(T)$. Hence λ is isolated in $\sigma(S)$, but since $\lambda \notin \sigma_e(S)$ it follows that λ is an eigenvalue of S of finite multiplicity. Thus $H_2(S, \{\lambda\})$ is a finite-dimensional subspace of H_2. Let $A : H_1 \to H_2$ be a quasiaffinity such that $AT = SA$. We then have $AH_1(T,\{\lambda\}) \subset H_2(S,\{\lambda\})$, so $H_1(T,\{\lambda\})$ is also finite-dimensional. By the functional calculus,

$$(\lambda - T)H_1 = (\lambda - T)H_1(T, \mathbf{C}\setminus\{\lambda\}) = H_1(T, \mathbf{C}\setminus\{\lambda\}),$$

and since the last subspace is closed, $\lambda - T$ is Fredholm. From this follows the contradiction $\lambda \in \sigma_e(T)$. Hence λ is not isolated in $\sigma(T)$.

Thus λ lies in a bounded component Ω of $\rho_e(S)$ such that $\Omega \cap \sigma(T) = \Omega$. In this case Ω is contained in the point spectrum of S, i.e. each $\mu \in \Omega$ is an eigenvalue of S. But $\mu - S$ is Fredholm for all μ sufficiently near λ and so it follows from Lemma 2.3.10 that S does not have SVEP, a contradiction. We thus have $\sigma_e(T) \subset \sigma_e(S)$, and by symmetry $\sigma_e(T) = \sigma_e(S)$.

2.4.5 Corollary. Let T and S be quasisimilar decomposable operators. If T and S are both essentially normal, then $UTU^* = S + K$ where U is unitary and K is compact.

Proof. By Th. 2.4.4 $\sigma_e(S) = \sigma_e(T)$, and since T and S are BQT by Cor. 2.3.22, $\text{ind}(\lambda - T) = \text{ind}(\lambda - S) = 0$ for all $\lambda \in \sigma_e(T)$. So the conclusion follows by Th. 2.3.19.

2.4.6 Remarks. (1) By Cor 2.3.22, T and S in the last corollary are both of the form N + K. Hence the normal parts of T and S are unitarily equivalent modulo K(H) by Cor. 2.4.5. (2) Ex. 2.4.11 (below) shows that the decomposability of both operators is necessary in Cor. 2.4.5. (3) Quasisimilarity does not preserve decomposability (Example 2.4.11). How much does it preserve? Some partial answers follow.

2.4.7 Proposition. Let $T \sim S$. If T is decomposable, then S is BQT.

Proof. Since T and T^* both have SVEP and quasisimilarity preserves this property, S and S^* both have SVEP. Hence S is BQT by Th. 2.3.21.

Example 2.4.11 will also show that quasisimilarity may not preserve property (β), but if it does then we obtain a much stronger conclusion than Prop. 2.4.7.

2.4.8 Proposition. Let T be decomposable, and suppose that S has property (β). If $T \sim S$ then S is quasidecomposable.

Proof. By Def. 1.3.1(e) it suffices to prove S has ASD. Suppose that $AT = SA$ for some quasiaffinity A. It is routine to show that $AH_1(T,F) \subset H_2(S,F)$ for each closed F. Let $\{G_1,\dots,G_n\}$ be an open cover of C. Then $H_1 = H_1(T,G_1^-) + \dots + H_1(T,G_n^-)$, and the density of AH_1 and the previous inclusion imply that $H_2(S,G_j^-)$ span H_2. By property (β), $\sigma(S|H_2(S,G_j^-)) \subset G_j^-$ for each j, so the proposition is proved.

2.4.9 Example. Let $H = \ell^2$ and let D(H) be the set of decomposable operators in L(H). Then D(H) is not uniformly closed in L(H). Let V + K be the Ex. 2.3.18. For $0 < \alpha < 1$ the operator $T_\alpha = V + \alpha K$ has spectrum contained in the unit circle. Hence $T_n = V + (1 -$

$n^{-1})K$ is decomposable for each $n = 1, 2, \ldots$, by Th. 2.3.1, but $T_n \to V + K$, which is not decomposable. Hence $D(H)$ is not uniformly closed.

The status of $D(H)$ in BQT of H can better be elucidated using the following result given by Herrero [47]. We use $N(H)$ to denote the set of normal operators on H.

2.4.10 Theorem. For $T \in L(H)$ the following are equivalent:
(i) $T \in$ BQT,
(ii) $T \in \{R \in L(H) : \sigma(R) \text{ totally disconnected}\}^-$,
(iii) $T \in \{R \in L(H) : R \text{ similar to } S \in N(H) + K(H)\}^-$,
where the closures are uniform.

Every operator with totally disconnected spectrum is A-scalar [24, p. 67]. Hence $D(H)^- =$ BQT by Th. 2.4.10. On the other hand, since compact perturbations preserve Fredholm index, $D(H) + K(H) \subset$ BQT so $D(H) + K(H)$ is also dense in BQT. Question: Is this inclusion proper? By [21, Cor. 11.11] $N(H) + K(H)$ is uniformly closed, hence there are decomposable operators not in $N(H) + K(H)$, i.e. some decomposable operators are not essentially normal (cf. Cor. 2.3.22). Thus in Cor. 2.4.5 if the operator S is not essentially normal, then the equality $UTU^* = S + K$ fails for all U and all K.

2.4.11 Example. The following example is given by Stampfli [87, p. 111]. Let $\{e_n\}$ be a bilateral basis for ℓ^2 and for $m = 1, 2, \ldots$, define a weighted bilateral shift A_m with weight sequence $\{a_n^{(m)}\}$ given by

$$a_n^{(m)} = \begin{cases} 1 & \text{for } n \leq 0 \\ 1 - \frac{n}{m} & \text{for } 1 \leq n \leq m \\ 2 & \text{for } m+1 \leq n \leq 2m \\ 1 + \frac{3m-n}{m} & \text{for } 2m+1 \leq n \leq 3m \\ 1 & \text{for } 3m+1 \leq n \end{cases}$$

Then $A_m^*A_m - A_mA_m^*$ is finite-rank with norm $\lesssim 4/m$. Hence $A = \bigoplus_{0 \leq m < \infty} A_m$ is essentially normal. But A_m is similar to the unweighted bilateral shift of multiplicity 1. Hence A is quasisimilar to the (unitary) bilateral shift W of countable multiplicity. Also $\sigma(W) = \sigma_e(W)$ is the unit circle but $\sigma(A) = \sigma_e(A)$ is the annulus with radii 1 and 2. By Cor. 2.4.2 A cannot have property (β). This example shows that the conclusion of Cor. 2.4.5 fails if one of the quasisimilar operators is not decomposable (or at least do not satisfy condition (β)).

Chapter III - WEAKLY DECOMPOSABLE OPERATORS AND AUTOMATIC CONTINUITY

1. Introduction

In this chapter we introduce a new type of spectral decomposition that we call "weakly decomposable relative to the identity," and we prove some of its properties. For example, these operators are quasidecomposable, but they neither contain, nor are contained in, the decomposable operators. Thus, unlike Chapter II, we find two noncomparable classes.

These results are obtained in §2. Section 3 deals with multiplication operators associated with a given operator weakly decomposable relative to the identity and generalizes [67, Th. 3.2]. This extension is proper because of an example from §2. In §4 we study some properties of maximal hyperinvariant chains of subspaces for weakly decomposable operators with "boundedness condition". These can be considered analogs for the case of A-spectral operators in [24, p. 81].

Our principal result of the chapter (Th. 3.5.20) in §5 is a generalization of both [91, Th. 3.5] and [68, Th. 4.3] giving necessary and sufficient conditions that any linear mapping intertwining a decomposable with a weakly decomposable one (in our present sense) must automatically be continuous (bounded). In §§6 and 7 of the chapter we give several applications of Th. 3.5.20 to operators discussed elsewhere in this monograph.

2. Weakly Decomposable Operators.

3.2.1 Definition. Let $T \in L(X)$. We say that T is <u>weakly decomposable relative to the identity</u> if

(i) for every finite open cover $\{G_i: 1 \leq i \leq n\}$ of C there exists a system of T-invariant subspaces $\{X_i\}$ such that $\sigma(T|X_i) \subset G_i$ $(i=1, 2,\dots,n)$.

(ii) for each $i = 1,\dots, n$ there exists a sequence $\{P_{ij}\} \subset \{T\}'$, the commutant of T, such that

$$P_{ij}X \subset X_i \text{ and } \text{wot-lim}_j\,(\Sigma_i P_{ij}) = I.$$

It is easy to see that every operator decomposable relative to the identity satisfies Def. 3.2.1. Below we show that the converse fails.

3.2.2 Theorem. Every operator weakly decomposable relative to the identity has SVEP.

Proof. Let T be weakly decomposable relative to the identity and let $f: D \to X$ be analytic satisfying

$$(2.1) \qquad (\lambda - T)f(\lambda) = 0 \qquad (\lambda \in D)$$

We may also suppose that D is connected. Let G_1, G_2 be open discs in D with $G_1^- \cap G_2^- = \emptyset$. Choose another open set H_1 such that $\{G_1, H_1\}$ covers $\sigma(T)$ and $G_1 \backslash H_1 \neq \emptyset$. By Def. 3.2.1 there are T-invariant subspaces X_1, Y_1 satisfying

$$(2.2) \qquad \sigma(T|X_1) \subset G_1, \qquad \sigma(T|Y_1) \subset H_1.$$

and there exist P_{1j}, P_{2j} (j=1,2,...) such that

$$(2.3) \qquad |\langle x - P_{1j}x - P_{2j}x, u\rangle| < j^{-1}, \quad P_{1j}X \subset X_1, \quad P_{2j}X \subset Y_1$$

for $x \in X$, $u \in X^*$. Now (2.1) implies $(\lambda - T)P_{2j}f(\lambda) = 0$ for $\lambda \in D$, hence by (2.2) $P_{2j}f(\lambda) = 0$ for all j and $\lambda \in G_1 \backslash H_1^-$. By (2.3)

$$|\langle f(\lambda), u\rangle - \langle P_{1j}f(\lambda), u\rangle| < j^{-1}$$

and $P_{1j}f(\lambda) \in X_1$; and since X_1 is closed and thus weakly closed, $f(\lambda) \in X_1$ for $\lambda \in G_1 \backslash H_1^-$. Because D is connected, it follows that

$$(2.4) \qquad f(\lambda) \in X_1 \text{ for } \lambda \in D.$$

By a similar argument we can find X_2 with $\sigma(T|X_2) \subset G_2$ and

$$(2.5) \qquad f(\lambda) \in X_2 \text{ for } \lambda \in D.$$

Then by (2.4) and (2.5) $f(\lambda) \in X_1 \cap X_2$ for all $\lambda \in D$. Since G_1 and G_2 are discs we have $\sigma(T|X_1 \cap X_2) \subset G_1 \cap G_2 = \emptyset$, i.e. $X_1 \cap X_2 = (0)$. Hence f = 0 on D, so the proof is complete.

We now prove that it is closed whenever F is.

3.2.3 Theorem. If T is weakly decomposable relative to the identity, then X(T,F) is closed if F is closed in **C**.

Proof. For $\lambda \notin F$ define

$$G_\lambda = \{\mu : |\mu - \lambda| < (1/2)\text{dist}(\lambda, F)\}$$
$$H_\lambda = \{\mu : |\mu - \lambda| > (1/3)\text{dist}(\lambda, F)\}.$$

Clearly $\{G_\lambda, H_\lambda\}$ covers **C**, so choose T-invariant subspaces X_1, X_2 such that $\sigma(T|X_1) \subset G_\lambda$ and $\sigma(T|X_2) \subset H_\lambda$. By Def. 3.2.1(ii) there are sequences $\{P_{1n}\}$ and $\{P_{2n}\}$ in $\{T\}'$ such that for $x \in X(T,F)$, $u \in X^*$

$$(2.6) \qquad |\langle x - P_{1n}x - P_{2n}x, u\rangle| < n^{-1} \quad \text{and} \quad P_{in}X \subset X_i \ (i=1, 2).$$

But $\sigma(P_{1n}x,T) \subset \sigma(x,T) \cap G_\lambda \subset F \cap G_\lambda = \emptyset$, hence $P_{1n}x = 0$ and (2.6) implies

$$|\langle x, u\rangle - \langle P_{2n}x, u\rangle| < n^{-1} \qquad \text{(all } n\text{)}$$

As in the proof of Th. 3.2.2 one has $x \in X_2$, hence $X(T,F) \subset X_2$. Recalling that X_2 depends on λ, we write $X_2 = X_\lambda$ and obtain

$$(2.7) \qquad X(T,F) \subset \cap\{X_\lambda : \lambda \notin F\}.$$

If x lies in the intersection of (2.7), then we see easily that $\sigma(x,T) \subset \cap\{H_\lambda : \lambda \notin F\} = F$. This proves the reverse inclusion of (2.7), and hence the theorem.

3.2.4 Theorem. Every operator weakly decomposable relative to the identity is quasidecomposable.

Proof. In view of Def. 1.4.1(e) and Ths. 3.2.2 and 3.2.3 it suffices to prove tha

$$(2.8) \qquad X = \vee_{i=1}^{n} X(T,G_i^-)$$

for each finite open cover $\{G_i\}$ of **C**. By Def. 3.2.1 X is the weak closure of the manifold $\Sigma X(T,G_i^-)$, hence (2.8) follows from the fact that weak and strong closures of convex sets coincide.

The last theorem leads naturally to the question whether an

operator weakly decomposable relative to the identity is decomposable. The example we now construct shows the answer to be no.

3.2.5 Spaces $B^j(D^-)$. Example 3.2.11 below is based on the following construction due to Albrecht [4]. Let D be the open unit disc, and let $C_0^\infty(D) \subset C^\infty$ be those C^∞-functions with compact support in D, where C^∞ denotes $C^\infty(\mathbf{C})$. Use $\bar{\partial} = (1/2)(\partial/\partial x + i\partial/\partial y)$.

We also use the following spaces defined in §1.4. Let $B^0(D^-)$ be the set of all continuous $\mathbf{C}$-valued functions on D^-, and let $B^1(D^-)$ be those $f \in B^0(D^-)$ for which there exists $g \in B^0(D^-)$ satisfying

$$\iint (\bar{\partial}\phi) f \, d\eta d\xi = -\iint \phi g \, d\eta d\xi \tag{2.9}$$

for all $\phi \in C_0^\infty(D)$ where the integrals are taken over D. Then g is the $\bar{\partial}$-derivative of in the sense of distributions, and we write $g = \bar{\partial}f$. Inductively define $B^j(D^-)$ to be the set of $f \in B^0(D^-)$ with $\bar{\partial}f \in B^{j-1}(D^-)$. Endowed with the norm

$$\|f\|_j = \max_{0 \le k \le j} \sup_{z \in D^-} |\partial^k f(z)|, \tag{2.10}$$

$B^j(D^-)$ is a Banach space.

3.2.6 Lemma. If $f \in B^j(D^-)$ and $\psi \in C^\infty$, then $\psi f \in B^j(D^-)$ and $\bar{\partial}(\psi f) = (\bar{\partial}\psi)f + \psi(\bar{\partial}f)$. Moreover, the map $(\psi, f) \to \psi f$ from $C^\infty \times B^j(D^-)$ to $B^j(D^-)$ is continuous for $j = 0, 1, 2, \ldots$.

Proof. For $j = 0$ the result is obvious. Let $j = 1$. Since $\psi\phi \in C_0^\infty(D)$ for all $\phi \in C_0^\infty(D)$, we have (for $\eta + i\xi \in D$)

$$\begin{aligned}
\iint f\psi(\bar{\partial}\phi) d\eta d\xi &= \iint f(\bar{\partial}(\psi\phi) - \phi\bar{\partial}\psi)) d\eta d\xi \\
&= -\iint \phi f(\bar{\partial}\psi) d\eta d\xi - \iint \phi\psi(\bar{\partial}f) d\eta d\xi \\
&= -\iint \phi(f\bar{\partial}\psi + \psi\bar{\partial}f) d\eta d\xi
\end{aligned}$$

by (2.9). But as $f(\bar{\partial}\psi)$ and $\psi(\bar{\partial}f) \in B^0(D^-)$ we have $f\psi \in B^1(D^-)$. The continuity of the map $(\psi,f) \to \psi f$ follows from the inequality $\|\psi f\| \leq \|\psi\|\,\|f\|$. The proof for $j > 1$ proceeds by induction.

The following lemma is immediate from the fact that $\bar\partial\phi = 0$ if ϕ is analytic.

3.2.7 Lemma. If $f \in B^j(D^-)$ and g is analytic in a neighborhood of supp f, then

$$\|fg\|_j \leq \|f\|_j \max\{|g(z)| : z \in \text{supp } f\}.$$

Let $D_0 = \{z: |z| < 1/2\}$ and denote by $B_0{}^j(D_0{}^-)$ the set of those $f \in B^j(D^-)$ with supp $f \subset D_0{}^-$. Then $B_0{}^j(D^-)$ is closed in $B_0{}^j(D_0{}^-)$ with norm (2.10) and we have

3.2.8 Lemma. The embedding $B_0{}^j(D_0{}^-) \to B^{j-1}(D^-)$ is compact for $j=1, 2,\dots$.

Proof. Let $j = 1$. In Chap. IV.4 we shall prove the Cauchy-Pompeiu formula, but in fact this also holds for $f \in B^1(D^-)$, hence each $f \in B_0{}^1(D_0{}^-)$ can be expressed as

$$f(z) = \frac{1}{\pi}\int\int(\sigma - z)^{-1}\,\bar\partial f(\sigma)d\eta d\xi \qquad (\sigma = \eta + i\xi),$$

and hence if $z, w \in D^-$, then

$$|f(z) - f(w)| \leq \frac{1}{\pi}\|\bar\partial f\|_0\int\int_D|(\sigma-z)^{-1} - (\sigma-w)^{-1}|d\eta d\xi.$$

An easy estimate of the last integral shows that the map $z \to g_z$ where $g_z(\xi) = (\xi - z)^{-1}$ is continuous from D^- to $L^1(D^-)$. So the unit ball in $B_0{}^1(D_0{}^-)$ is equicontinuous on D^-. This ball is also bounded in $B^0(D^-)$, hence the Arzela-Ascoli theorem implies that the ball is relatively compact in $B^0(D^-)$, i.e. the map $B_0{}^1(D_0{}^-) \to B^0(D^-)$ is compact as a linear transformation. The cases $j = 2, 3,\dots$ follow by induction.

3.2.9 Lemma. Let $z \in D$ such that $|z| < 1/4$ and let $K_n(z) = \{w \in D^-: |w-z| < 1/n\}$ $(n = 2, 3,\dots)$. Then there exists n_0 independent of z such that for all $n \geq n_0$ and all $f \in B^1(D^-)$ with $|f(z)| = \|f\|_0 = 1$

and supp $f \subset K_n(z)$ we have $\|\bar{\partial} f\|_0 \geq 2$.

Proof. Suppose the lemma fails for $z = 0$. For $n = 2, 3, \ldots$, there exist $f_n \in B^1(D^-)$ with $|f_n(0)| = \|f_n\|_0 = 1$, $\text{supp}(f_n) \subset K_n(0)$ and $\|\partial^- f_n\| < 2$. Then $\{f_n\}$ is a bounded sequence in $B_0{}^1(D_0{}^-)$, hence by Lemma 3.2.8 there is a subsequence $\{f_{n_k}\}$ converging in $B^0(D^-)$ to a continuous function g. But clearly $g(0) = 1$ and $g(w) = 0$ for $w \neq 0$, a contradiction to continuity. Hence there exists n_0 such that for $n \geq n_0$ every $f \in B^1(D^-)$ with $|f(0)| = \|f\|_0 = 1$ and $\text{supp}(f) \subset K_n(0)$ satisfies $\|\partial^- f\|_0 \geq 2$.

Now for z arbitrary with $|z| < 1/4$, let $f \in B^1(D^-)$ such that $|f(z)| = \|f\|_1 = 1$ and $\text{supp}(f) \subset K_n(z)$ for $n \geq \max\{4, n_0\}$. Define the function $g \in B^1(D^-)$ by

$$g(w) = \begin{cases} f(z + w) & \text{for } |w| < 1/2 \\ 0 & \text{for } |w| > 1/2. \end{cases}$$

Then $|g(0)| = \|g\|_0 = 1$ and $\text{supp}(g) \subset K_n(0) \subset K_{n_0}(0)$. Hence, by the last paragraph $\|\bar{\partial} f\| = \|\bar{\partial} g\| \geq 2$, and the proof is complete.

3.2.10 Corollary. If $f \in B^j(D^-)$ $(j = 0,1,2,\ldots)$ with $\text{supp}(f) \subset K_{n_0}(0)$, then

$$\|f\|_j \geq 2^j \|f\|_0. \tag{2.11}$$

Where n_0 is as in Lemma 3.2.9.

Proof. We may clearly assume $j > 0$ and $f \neq 0$. Assume (2.11) holds for $j \leq k$ and let $f \in B^{k+1}(D^-)$ such that supp $f \subset K_{n_0}(0)$. Then $|f(z)| = \|f\|_0$ $(\neq 0)$ for some $z \in K_{n_0}(0)$. By Lemma 3.2.9

$$\|\bar{\partial} f\|_0 \geq 2\|f\|_0,$$

and applying the induction hypothesis to $\partial^- f$, one has $\|\bar{\partial} f\|_k \geq 2^k \|\bar{\partial} f\|_0$, hence

$$\|f\|_{k+1} \geq \|\bar{\partial} f\|_k \geq 2^k\|\bar{\partial} f\|_0 \geq 2^{k+1}\|f\|_0,$$

and the proof is complete.

3.2.11. Example. Let X be the ℓ^1-direct sum of the spaces $B^j(D^-)$, i.e

$$X = \{(f_j)_0^\infty : f_j \in B^j(D^-) \text{ and } \sum_{j=0}^{\infty} \|f_j\|_j < \infty\}$$

Then X is a Banach space with norm $\|(f_j)\| = \Sigma \|f_j\|_j$. Define $T: X \to X$ by $T(f_j) = (zf_j)$, multiplication by z in each component. By Lemma 3.2.7

$$\|T(f_j)\| = \Sigma_j \|zf\|_j \leq \sup_D |z|(\Sigma \|(f_j)\|_j) = \|(f_j)\|;$$

T is bounded. In a similar way one shows that $\sigma(T) = D^-$.

3.2.12 Lemma. T has SVEP.

Proof. Let $f: U \to X$ be analytic such that

$$(w - T)f(w) = 0 \quad (w \in U) \tag{2.12}$$

Then $f(w) = (f_j(\cdot,w))$ where each $f_j(\cdot,w) \in B^j(D^-)$ depends analytically on w. By (2.12), for every j, $(w - z)f_j(z, w) = 0$ $(w \in U)$. Thus $f_j(z, w) = 0$ for all $w \neq z$, hence identically by continuity. It follows that $f(w) = 0$ for all $w \in U$, and the lemma is proved.

3.2.13 Lemma. For $(f_j) \in X$ the local spectrum

$$\sigma((f_j),T) = (\cup_{j=0}^{\infty} \operatorname{supp}(f_j))^-.$$

Proof. Let $F = (\cup_{j=0}^{\infty} \operatorname{supp}(f_j))^-$. Then it is easy to see that $F \subset \sigma((f_j),T)$. For the reverse inclusion, let $w_0 \in \mathbb{C}\setminus F$ and let U be an open neighborhood of w_0 with $U^- \cap F = \emptyset$. For $w \in U$ let

$$f_j(z, w) = \begin{cases} 0, & \text{if } z \notin \operatorname{supp}(f_j) \\ f_j(z)/(w - z), & \text{if } z \in \operatorname{supp}(f_j). \end{cases}$$

By Lemma 3.2.7 $(f_j(\cdot,w)) \in X$ for all $w \in U$ and the map $w \to (f_j(\cdot,w))$ is analytic on U. But clearly $(w - T)(f_j(\cdot,w)) = (f_j)$ on U, hence $w_0 \in \rho((f_j),T)$, and thus $\sigma((f_j),T) \subset F$.

3.2.14 Lemma. If F is closed in **C**, then

$$X(T,F) = \{(f_j) \in X : \cup_j \operatorname{supp}(f_j) \subset F\}$$

and is closed.

Proof. Let $\{(f_j^{(k)})\}$ be a sequence in X(T,F) converging to (f_j). Then $f_j^{(k)} \to f_j$ uniformly on D^- for j = 0, 1, 2,... . Since $f_j^{(k)} = 0$ on $D^-\backslash F$ for all j, it follows that each $f_j = 0$ on $D^-\backslash F$ also. Hence $\operatorname{supp}(f_j) \subset F$ for each j, and therefore $(f_j) \in X(T,F)$. Thus X(T,F) is closed in X.

3.2.15 Theorem. T is weakly decomposable relative to the identity, but T is not decomposable.

Proof. Let $\{G_1,..., G_m\}$ be an open cover of $\sigma(T) = D^-$. Then we may choose a corresponding system of C^∞-functions $\{\phi_1,..., \phi_m\}$ such that $\operatorname{supp}(\phi_k) \subset G_k^-$, k = 1, 2,...,m, and $\Sigma\, \phi_k = 1$ on a neighborhood of D^-. For each $(f_j) \in X$ define

$$f_j^{(n)} = \begin{cases} f_j & \text{for } n \geq j \\ 0 & \text{for } j > n \end{cases}$$

The sequence $(f_j^{(n)}) \subset X$ converges to (f_j) in the norm of X. For $1 \leq k \leq m$ and n = 1, 2,..., define $P_{kn} \in L(X)$ by $P_{kn}(f_j) = (\phi_k f_j^{(n)})$. Then P_{kn} commutes with T and

$$\Big(\sum_{k=1}^{m} P_{kn}\Big)(f_j) = (f_j^{(n)}) \to (f_j) \quad (n \to \infty),$$

i.e. $\Sigma P_{kn} \to I$ in the strong operator topology, hence $\Sigma P_{kn} \to I$ (wot). Moreover, $\mathrm{supp}(\phi_k) \subset G_k^-$ implies by Lemma 3.2.14 that $P_{kn}X \subset X(T, G_k^-)$. Since T thus satisfies Def. 3.2.1, it is weakly decomposable relative to the identity.

We now prove that T is not decomposable. Suppose it is and choose $m = n_0$ in Lemma 3.2.9. As $\{K_m(0), C\backslash K_{m+1}(0)^-\}$ is an open cover of C, there are T-invariant subspaces X_1, X_2 such that

$$
\begin{aligned}
X &= X_1 + X_2, \\
\sigma(T|X_1) &\subset K_m(0), \qquad\qquad (2.13)\\
\sigma(T|X_2) \cap K_{m+1}(0)^- &= \emptyset.
\end{aligned}
$$

Define $(f_j) \in X$ by $f_0 = 0$, $f_j(z) = j^{-2}$ $(j > 0)$. By (2.13) we may find $(g_j) \in X_1$ and $(h_j) \in X_2$ such that $(f_j) = (g_j + h_j)$. Now $\sigma((g_j), T) \subset K_m(0)$ implies $\mathrm{supp}(g_j) \subset K_m(0)$ (all j), and $\sigma((h_j),T) \cap K_{m+1}(0)^- = \emptyset$ implies $\mathrm{supp}(h_j) \cap K_{m+1}(0) = \emptyset$ (all j). But $g_j(0) = g_j(0) + h_j(0) = f_j(0) = j^{-2}$, so it follows that $\|g_j\|_0 \geq j^{-2}$. From Cor. 3.2.10 we obtain

$$
\|g_j\|_j \geq 2^j j^{-2} \to \infty \qquad (\text{as } j \to \infty).
$$

This contradicts the convergence of $\Sigma \|g_j\|_j$, hence T is not decomposable.

The reader will recall from Chapter I.4 that the various classes of spectral decomposition considered there are all comparable: they are totally ordered by containment. The next example shows that the two classes of operators, decomposable and weakly decomposable relative to the identity, are not comparable in that sense.

3.2.16 Example. The operator $V = T + A(1)$, shown in Th. 1.4.14 to be strongly decomposable, is not weakly decomposable relative to the identity. Let $\{G_1, G_2\}$ be an open cover of C, and suppose we can find sequences $\{P_{1n}\}$ and $(P_{2n}\}$ in $\{T\}'$ such that $P_{1n} + P_{2n} \to I$ (wot) and $P_{jn}X \subset X(V, G_j^-)$ (all n, j=1, 2). By Th. 1.4.14 each P_{jn} has form $Q(a) + A(b)$. By (4.2) of Chap. I, $P_{in}P_{jk} = P_{jk}P_{in} = 0$ for all n, k and i,j = 1,2. Put $S_n = P_{1n} + P_{2n}$, so $S_nS_k = 0$ for all n,k. Now fix k, $x \in X$, $u \in X^*$ so that $\langle S_kx, u\rangle = 1$. Then, as $n \to \infty$,

$$0 = \langle S_n S_k x, u \rangle \to \langle S_k x, u \rangle = 1.$$

This contradiction proves that V is not weakly decomposable relative to the identity.

3. Multiplication Operators

Let $A \subset L(X)$ be an algebra closed in the weak operator topology (wot), and let A' be its commutant. For $T \in A \cap A'$ we define $T^\bullet$ on A by the formula

$$T^\bullet(U) = TU \quad (U \in A) \tag{3.1}$$

Clearly $T^\bullet \in L(A)$.

3.3.1 Definition. If for each finite open cover $\{G_i\}_1^m$ of $\sigma(T^\bullet)$ there exist corresponding spectral maximal spaces $A_i \subset A$ for which

$$\sigma(T^\bullet|A_i) \subset G_i \qquad (i=1, \ldots, m)$$
$$A = (A_1 + \ldots + A_m)^{wot},$$

then we say $T^\bullet$ is a <u>weakly decomposable multiplication operator</u>.

3.3.2 Theorem. Suppose $T \in L(X)$ is weakly decomposable relative to the identity. Let $A = \{T\}'$. Then $T^\bullet$ is a weakly decomposable multiplication operator on A having property (κ).

Proof. The proof proceeds in three steps. (A). For F closed in C define

$$A_F = \{U \in A : UX \subset X(T,F)\} \tag{3.2}$$

Then A_F is closed and $T^\bullet$-invariant since $X(T,F)$ is closed and each $U \in A$ commutes with T. We prove $\sigma(T^\bullet|A_F) \subset F$. Let $\lambda \notin F$ and $V \in A_F$. Then $U = (\lambda - T|X(T, F))^{-1}V \in A_F$ and

$$(\lambda - T^\bullet)U = (\lambda - T)U = (\lambda - T)(\lambda - T|X(T,F))^{-1}V = V,$$

so $\lambda - T^\bullet$ is surjective on A_F. For injectivity, let $U \in A_F$ with $(\lambda - T^\bullet)U = 0$. Then $(\lambda - T)Ux = 0$ for all $x \in X$, and since $\lambda \notin \sigma(T|X(T,F))$ it follows that $Ux = 0$. Hence $U = 0$, so $\lambda \in \rho(T^\bullet|A_F)$. This proves $\sigma(T^\bullet|A_F) \subset F$.

(B). We now prove that $T^\bullet$ has SVEP such that $A_F = A(T^\bullet,F)$, so T has property (κ) too. Let $U: \Omega \to A$ be analytic with $(\lambda - T^\bullet)U(\lambda) = 0$ for $\lambda \in \Omega$. Then $(\lambda - T)U(\lambda)x = 0$ for $x \in X$, and since T has SVEP $U(\lambda)x = 0$. Thus $U(\lambda) = 0$ on W, and so $T^\bullet$ has SVEP. It is clear from (A) that $A_F \subset A(T^\bullet,F)$. To prove the opposite inclusion, let $V \in A(T^\bullet,F)$ so that $(\lambda - T^\bullet)V(\lambda) = V$ $(\lambda \notin F)$ where $V(\cdot)$ is the local resolvent of V. Then $(\lambda - T)V(\lambda)x = Vx$ for $x \in X$, hence $\sigma(Vx,T) \subset F$ or $VX \subset X(T,F)$. By (3.2) $V \in A_F$, thus $A_F = A(T^\bullet,F)$.

(C). Let $\{G_i\}_1^m$ be an open cover of C. By hypothesis there are m sequences $\{P_{jn}\}$ $(j=1, \ldots, m)$ in $\{T\}'$ such that

$$(3.3) \qquad \Sigma_j P_{jn} \to I \text{ (wot)} \quad \text{and} \quad P_{jn}X \subset X(T,G_j^-) \quad (1 \leq j \leq m)$$

Now (3.3) implies $P_{jk} \in A(T^\bullet,G_j^-)$ so $A = (\Sigma_j A(T^\bullet,G_j^-))^{wot}$, and the theorem is proved.

We next prove a converse of Th. 3.3.2. Suppose $A \subset L(X)$ is a wot-closed subalgebra containing the identity, and let $T \in A \cap A'$ and let $T^\bullet$ be given by (3.1). We say $T^\bullet$ has wot-ASD if for each open cover $\{G_i\}_1^m$ of C there are $T^\bullet$-invariant subspaces A_i such that

$$(3.4) \qquad \sigma(T^\bullet|A_i) \subset G_i \text{ and } A = (A_1 + \ldots + A_m)^{wot}$$

If, in addition, for each cover $\{G_1,G_2\}$ of $\sigma(T^\bullet)$ there are uniformly bounded sequences $\{P_n\} \subset A_1$ and $\{Q_n\} \subset A_2$ with $P_n + Q_n \to I$ (wot), then we say $T^\bullet$ (or T) has the <u>boundedness condition</u>.

3.3.3 Theorem. Let $A \subset L(X)$ be a wot-closed subalgebra with identity. Suppose $T \in A \cap A'$ such that $T^\bullet$ has wot ASD with boundedness condition and property (κ). Then T is weakly decomposable relative to the identity.

Proof. We reach the conclusion in three steps. (A) For F closed in $\mathbf{C}$ let

$$X_F = \{x \in X : Vx = 0 \text{ for } V \in A \text{ with } \sigma(V,T^\bullet) \cap F = \emptyset\}.$$

Evidently X_F is closed; to see that it is T-invariant, let $V \in A$ with $\sigma(V,T^\bullet) \cap F = \emptyset$ and let $x \in X_F$. Then $VTx = TVx = 0$, and hence X_F is T-invariant.

(B.) Now we prove $\sigma(T|X_F) \subset F$. Let $\lambda \notin F$ and let $\{G_1,G_2\}$ be an open cover of $\mathbf{C}$ with $F \subset G_1$, $F \cap G_2^- = \emptyset$ and $\lambda \notin G_1$. By the boundedness condition there are bounded sequences $\{P_n\}$ in $A(T^\bullet,G_1^-)$ and $\{Q_n\}$ in $A(T^\bullet,G_2^-)$ such that $P_n + Q_n \to I$ (wot). Since $\sigma(Q_n,T^\bullet) \cap F \subset G_2^- \cap F = \emptyset$, we have $Q_n|X_F = 0$ (all n), hence

$$P_n|X_F \to I|X_F \qquad \text{(wot)}. \tag{3.5}$$

But $\{P_n\}$ is bounded, so for $x \in X_F$ and $n = 1, 2, \ldots$.

$$\begin{aligned} \|P_n x\| &= \|(\lambda - T^{\bullet\wedge})^{-1}(\lambda - T^\bullet)P_n x\| \\ &= \|(\lambda - T^{\bullet\wedge})^{-1}P_n(\lambda - T)x\| \\ &\leq M\|(\lambda - T^{\bullet\wedge})^{-1}\| \, \|(\lambda - T)x\| \end{aligned}$$

where $T^{\bullet\wedge} = T^\bullet|A(T^\bullet,G_1^-)$ and $M = \sup_n\|P_n\|$. It follows by the inequality $\|x\| \leq \underline{\lim}\,\|P_n x\|$ (cf. (3.5)) that

$$\|x\| \leq M_1\|(\lambda - T)x\| \tag{3.6}$$

where $M_1 = M\|(\lambda - T^{\bullet\wedge})^{-1}\|$. Let Y be the linear span of $\{P_n x : x \in X_F, n=1, 2,\ldots\}$. By (3.5) Y is dense in X_F. Moreover, for each n and $x \in X_F$ we have

$$P_n x = (\lambda - T)[(\lambda - T^{\bullet\wedge})^{-1}P_n]x,$$

hence for each $z \in Y$ there exists $y \in Y$ with

$$z = (\lambda - T)y. \tag{3.7}$$

Since $\lambda - T|X_F$ is surjective by (3.7) and bounded below on a dense linear manifold in X_F by (3.6), we see that $\lambda \notin \sigma(T|X_F)$ and this proves $\sigma(T|X_F) \subset F$.

(C) Finally we prove T is weakly decomposable relative to the identity. Let $\{G_j\}_1^m$ be an open cover of $\sigma(T)$. Since $T^\bullet$ has wot-ASD and property (κ), we can find m sequences $\{P_{jn}\} \subset A(T^\bullet, G_j^-)$ ($1 \le j \le m$) such that $\Sigma_j P_{jn} \to I$ (wot). Fix j and put $F = G_j^-$. For $V \in A$ with $\sigma(V,T^\bullet) \cap F = \emptyset$ we have

$$(\lambda - T^\bullet)V(\lambda)P_{jn} = VP_{jn} \quad (\lambda \notin \sigma(V,T^\bullet)), \text{ and}$$
$$(\lambda - T^\bullet)V(\lambda - T^\bullet|A(T^\bullet,F))^{-1}P_{jn} = VP_{jn} \quad (\lambda \notin F)$$

Thus $V(\lambda)P_{jn}$ is continuable to a function analytic on all **C** and since $V(\lambda)P_{jk} \to 0$ as $|\lambda| \to \infty$, $VP_{jn} = 0$ by Liouville's theorem. Hence $P_{jn}X \subset X_F$; and since $\sigma(T|X_F) \subset F = G_j^-$ by (B), T satisfies Def. 3.3.1 and the proof is complete.

3.3.4 Corollary. In Theorem 3.3.3 $X_F = X(T,F)$ for each closed F.

Proof. By (B) of the previous proof $X_F \subset X(T,F)$. Let $x \in X(T,F)$ and let f_x be its local resolvent. Let $V \in A$ with $\sigma(V,T^\bullet) \cap F = \emptyset$. Then

$$(\lambda - T)V(\lambda)x = Vx \qquad \text{for } \lambda \notin \sigma(V,T^\bullet) \text{ and}$$
$$(\lambda - T)Vf_x(\lambda) = Vx \qquad \text{for } \lambda \notin F.$$

As in (C) above, $Vx = 0$ since $V(\lambda)x$ is continuable to an entire function and $V(\lambda)x \to 0$ as $\lambda \to \infty$. Hence $x \in X_F$, or equivalently, $X_F = X(T,F)$, as required.

4. Hyperinvariant Chains.

In this section we study some properties of a maximal hyperinvariant subspace chain of an operator weakly decomposable relative to the identity. We begin with some preliminary results for such operators.

3.4.1 Theorem. Let $T \in L(X)$ be weakly decomposable relative to the identity such that T has the boundedness condition. Let $X_1 \subset X_2$ be hyperinvariant under T. Then $T|X_1$ and $(T|X_2)/X_1$ are both weakly decomposable relative to the identity and have the boundedness condition.

Proof. Let $A = \{T\}'$. For $V \in A$ let V' denote the operator coinduced by $V|X_2$ on X_2/X_1, and for F closed in C define

$$X_F' = \{x' \in X_2/X_1 : V'x' = 0 \text{ if } V \in A \text{ such that } \sigma(V,T^\bullet) \cap F = \emptyset\}$$

where $T^\bullet$ is given by (3.1). As in the proof of Th. 3.3.3 we prove that X_F' is closed, W'-invariant for $W \in A$ and $\sigma(T'|X_F') \subset F$. Let $\{G_i : 1 \le i \le k\}$ be an open cover of $\sigma(T)$. There are then bounded sequences $\{P_{in}\} \subset A$ $(i=1,\ldots,k)$ with $\Sigma_i P_{in} \to I$ (wot) and $P_{in}X \subset X(T,G_i^-)$ $(i=1,\ldots,k)$. As in the proof of Th. 3.3.2 one shows $P_{in} \in A_{G_i^-}$ and $P_{in}'X' \subset (X_{G_i^-})'$. Since it is easy to see that $\Sigma P_{in}' \to I'$ (wot), T' is weakly decomposable relative to the identity and has the boundedness condition. This completes the proof.

We say that T is <u>strongly quasidecomposable</u> [51] if $T|X(T,F)$ is quasidecomposable for each closed F.

3.4.2 Corollary. Let $T \in L(X)$ be weakly decomposable relative to the identity with the boundedness condition. Then

(i) if Y is hyperinvariant to T, then $T|Y$ and T/Y are weakly decomposable relative to the identity; in particular, T is strongly quasidecomposable;

(ii) every Y hyperinvariant to T is T-analytic;

(iii) if $X_1 \subset X_2$ are hyperinvariant, then $\sigma(T|X_1) \subset \sigma(T|X_2)$.

Proof (i). For the case $T|Y$ apply Th. 3.4.1 to $X_2 = Y$, $X_1 = 0$, and for T/Y take $X_2 = X$, $X_1 = Y$. Then $T|Y$ and T/Y are weakly decomposable relative to the identity and hence strongly quasidecomposable.
(ii). Since T/Y is weakly decomposable relative to the identity, T/Y has SVEP by Th. 3.2.2 and Y is analytically invariant. (iii). If $X_1 \subset X_2$, then X_1 is analytically invariant for $T|X_2$ by part (ii), hence $\sigma(T|X_1) \subset \sigma(T|X_2)$ [30, Prop. 1.15].

The most general spectral decomposition we know of for which the <u>adjoint</u> is known to have SVEP is given in the following proposition. The next corollary is immediate from this.

3.4.3 Proposition. If T has ASD, then T^* has SVEP. (ASD is defined in Chap I.4).

Proof. Let $D \subset \mathbf{C}$ be open, and let $f: D \to X^*$ be analytic satisfying $(\lambda - T^*)f(\lambda) = 0$ on D; we may also assume that D is connected. Let $\{G_1, G_2\}$ be an open cover of $\mathbf{C}$ such that $G_1^- \subset D$ and $D \backslash G_2^- \neq \emptyset$. The ASD implies that there exists T-invariant subspaces X_1, X_2 with

$$X = X_1 \vee X_2 \quad \text{and} \quad \sigma(T|X_j) \subset G_j \quad (j=1,2).$$

For $x_1 \in X_1$ and $\lambda \in D \backslash G_1^-$,

$$\begin{aligned} \langle x_1, f(\lambda)\rangle &= \langle(\lambda - T)(\lambda - T|X_1)^{-1}x_1, f(\lambda)\rangle \\ &= \langle(\lambda - T|X_1)^{-1}x_1, (\lambda - T^*)f(\lambda)\rangle = 0. \end{aligned}$$

Hence $\langle x_1, f(\lambda)\rangle = 0$ on all of D by analytic continuation. Similarly, one has $\langle x_2, f(\lambda)\rangle = 0$ on D for each $x_2 \in X_2$ since $D \backslash G_2^- \neq \emptyset$. It follows that $\langle x, f(\lambda)\rangle = 0$ on D for all x in the norm-dense manifold $X_1 + X_2$. Thus $f(\lambda) = 0$ on D, so T^* has SVEP.

3.4.4 Corollary. Let X be a Hilbert space. If T is analytically decomposable, then T is BQT.

Proof. Use Cor. 3.4.3 and Prop. 1.4.30.

3.4.5 Remark. So far we do not know if every Hilbert space operator with ASD is biquasitriangular.

3.4.6 Definition. For the complex Banach space X, let $\mathcal{N} = \{X_\alpha : \alpha \in B\}$ be a set of closed subspaces such that

(i) B is totally ordered,

(ii) if $\alpha, \beta \in B$ with $\alpha < \beta$, then $X_\alpha \subset X_\beta$ properly.

We call $\mathcal{N}$ a <u>chain</u>. If each $X_\alpha \in \mathcal{N}$ is T-hyperinvariant, then $\mathcal{N}$ is a hyperinvariant subspace chain. By Zorn's lemma every such chain can be imbedded in a maximal one, so henceforth we assume $\mathcal{N}$ is <u>maximal</u>. By [24, Prop. 3.4.2] for each $\alpha \in B$ there exists $\beta \in B$ with

$$X_\beta = \vee_{\gamma < \alpha} X_\gamma,$$

and we write $\beta = \alpha^-$. If $\alpha = \alpha^-$, $\mathcal{N}$ is said to be continuous at α. If $\mathcal{N}$ is not continuous at α, write $X_\alpha' = X_\alpha / X_{\alpha^-}$ and T_α' for the corresponding operator on X_α'.

The next lemma is a consequence of maximality.

3.4.7 Lemma. (i) If $\mathcal{N}$ is not continuous at α, then T_α' has no proper hyperinvariant subspace. (ii) $\sigma(T_\alpha') \subset \sigma(T)$ and is connected.

Now follows our principal result of the section.

3.4.8 Theorem. Suppose $T \in L(X)$ is weakly decomposable relative to the identity with boundedness condition. If $\mathcal{N} = \{X_\alpha : \alpha \in B\}$ is a maximal hyperinvariant subspace chain, then the following hold.

(i) For $\beta < \alpha$ in B

$$\sigma(T|X_\beta) \subset \sigma(T|X_{\alpha^-}) \subset \sigma(T|X_\alpha) \text{ and } \sigma(T|X_{\alpha^-}) = (\cup_{\beta<} \sigma(T|X_\beta))^-$$

(ii) If $\mathcal{N}$ is discontinuous at α, then $\sigma(T_\alpha')$ contains exactly one point $\eta_\alpha \in \sigma(T)$ and T_α' is the scalar η_α or $T_\alpha' - \eta_\alpha$ is quasinilpotent but not nilpotent.

(iii) (a) If $\mathcal{N}$ is discontinuous at α with $\sigma(T|X_{\alpha^-}) \neq \sigma(T|X_\alpha)$, then

$$X_\alpha = X_{\alpha^-} \oplus Y_\alpha \tag{4.1}$$

where Y_α is hyperinvariant for T. (b) If X_α' is finite-dimensional, then Y_α is finite-dimensional and $P_\alpha T P_\alpha = \eta_\alpha I$ on X_α where P_α is the projection of X_α onto Y_α along X_{α^-}.

Proof. (i) The first two inclusions follow from Cor. 3.4.2. As for the third relation, let F denote its right side. Then $F \subset \sigma(T|X_{\alpha^-})$. If this inclusion is proper let $\eta \in \sigma(T|X_{\alpha^-}) \setminus F$ and let $\{G_1, G_2\}$ be an open cover of C with

$$F \subset G_2, \quad G_1^- \cap F = \emptyset \quad \text{and} \quad \eta \notin G_2^-.$$

By Cor. 3.4.2 we can find spectral maximal subspaces $Y_i \subset X_{\alpha^-}$ such that

$$X_{\alpha^-} = Y_1 \vee Y_2 \text{ and } \sigma(T|Y_i) \subset G_i$$

By [30, Th. 10.3] $\sigma(T|X_{\alpha^-}) = \sigma(T|Y_1) \cup \sigma(T|Y_2)$, so

$$F \subset \sigma(T|X_{\alpha^-})\backslash G_2^- \subset \sigma(T|X_{\alpha^-})\backslash\sigma(T|Y_2) \subset \sigma(T|Y_1),$$

and $X_\beta \subset Y_1$ for all $\beta<\alpha$. But $Y_1 \subset X_{\alpha^-}$ properly, a contradiction. Thus

$$\sigma(T|X_{\alpha^-}) = (\cup_{\beta<\alpha} \sigma(T|X_\beta))^-.$$

(ii) By Th. 3.4.1 T_α' is weakly decomposable relative to the identity. If $\sigma(T_\alpha')$ has more than one point, then T_α' has a proper hyperinvariant subspace and Lemma 3.4.7 (i) is contradicted. By part (ii) of that lemma $\sigma(T_\alpha') = \{\eta_\alpha\}$ for some $\eta_\alpha \in \sigma(T)$. If $T_\alpha' - \eta_\alpha$ is nilpotent, then $(T_\alpha' - \eta_\alpha)^m = 0$ but $(T_\alpha' - \eta_\alpha)^{m-1} \neq 0$ for some $m > 1$. But in this case $((T_\alpha' - \eta_\alpha)^{m-1}X_\alpha')^-$ is a proper hyperinvariant subspace of T_α', another contradiction.

(iii) If $\sigma(T|X_{\alpha^-}) \neq \sigma(T|X_\alpha)$, then $\sigma(T|X_\alpha) = \sigma(T|X_{\alpha^-}) \cup \sigma(T_\alpha') = \sigma(T|X_{\alpha^-}) \cup \{\eta_\alpha\}$. Thus $\{\eta_\alpha\}$ is a (Dunford) spectral set in $\sigma(T|X_\alpha)$. Let P_α be the Riesz projection for $\{\eta_\alpha\}$, and put $Y_\alpha = P_\alpha X_\alpha$, $M_\alpha = (I - P_\alpha)X_\alpha$. Since Y_α, M_α are spectral maximal, it follows that $M_\alpha = X_{\alpha^-}$, so (4.1) is proved. Moreover, $T|Y_\alpha$ and T_α' are similar. If X_α' is finite dimensional with basis $\{x_1',\dots, x_p'\}$ $(x_i \in X_\alpha)$, we may choose $y_i \in x_i' \cap Y_\alpha$ $(i=1,\dots, p)$ which span Y_α. Then $P_\alpha TP_\alpha|Y_\alpha$ is also similar to T_α', so $P_\alpha TP_\alpha = \eta_\alpha I$ or $P_\alpha TP_\alpha - \eta_\alpha I$ is quasinilpotent on Y_α, but the latter is not possible. Thus (b) follows.

5. Automatic Continuity

In this section we shall generalize some results of [67] on automatic continuity of intertwining of decomposable operators (Th. 3.5.20). Specifically we prove that (under suitable restrictions) any linear map intertwining a decomposable operator and one weakly decomposable relative to the identity must necessarily be continuous. We begin with the notion of separating space.

3.5.1. Definition. Let $\theta: X \to Y$ be a linear (not necessarily continuous) map between Banach spaces. The _separating space_ of θ is the subset of Y defined by

$$s(\theta) = \{y \in Y : y = \lim \theta x_n \text{ for } x_n \to 0 \text{ in } X\}.$$

It follows from the closed graph theorem that θ is continuous iff $s(\theta) = (0)$. Other properties of $s(\theta)$ are given by

3.5.2 Lemma. For the linear map $\theta: X \to Y$, $s(\theta)$ has the following properties.

(i) $s(\theta)$ is a closed subspace of Y.

(ii) Let $T \in L(X)$, $S \in L(Y)$ such that $\theta T = S\theta$. Then $Ss(\theta) \subset s(\theta)$.

(iii) If $R: Y \to Z$ is continuous and linear, then $R\theta$ is continuous iff $Rs(\theta) = (0)$; moreover, in this case $\|R\theta\| \leq M\|R\|$ for some $M > 0$ independent of R, Z.

(iv) With R as in (iii) $[Rs(\theta)]^- = s(R\theta)$.

Proof. (i) Since $s(\theta)$ is clearly a linear manifold, let $\{y_n\} \subset s(\theta)$ with $y_n \to y$. Choose $x_n \in X$ such that $\|x_n\| < n^{-1}$ and $\|\theta x_n - y_n\| < n^{-1}$ for each n. Then $x_n \to 0$ and $\theta x_n \to y$, hence $y \in s(\theta)$ and (i) is proved.

(ii) If $y \in s(\theta)$ then let $x_n \in X$ with $x_n \to 0$ and $\theta x_n \to y$. Hence $Tx_n \to 0$ while $\theta Tx_n = S\theta x_n \to Sy$, so $Sy \in s(\theta)$.

(iii) Suppose $R\theta$ is continuous. For $y \in s(\theta)$ there exists $\{x_n\} \subset X$ with $x_n \to 0$ and $\theta x_n \to y$. Hence $R\theta x_n \to 0$ and $R\theta x_n \to Ry$, so $Ry = 0$. Conversely, suppose $Rs(\theta) = (0)$. We show $R\theta$ continuous by commutativity of the diagram

$$\begin{array}{ccccc} & \theta & & R & \\ X & \to & Y & \to & Z \\ & & Q\downarrow & \nearrow R_0 & \\ & & Y/s(\theta) & & \end{array}$$

where $Q: Y \to Y/s(\theta)$ is the canonical surjection and $R_0(y^\wedge) = Ry$ for $y^\wedge = y + s(\theta)$. In this case $R\theta$ is continuous if $Q\theta$ is. Let $x_n \to 0$ in X with $Q\theta x_n \to y^\wedge \in Y/s(\theta)$. We can then find $y_n \in s(\theta)$ with $\|\theta x_n - y - y_n\| \to 0$. Next choose $w_n \in X$ so that $\|w_n\| < n^{-1}$ and $\|\theta w_n - y_n\| < n^{-1}$. Then $x_n - w_n \to 0$ and $\theta(x_n - w_n) \to y$, i.e. $y \in s(\theta)$. Hence $Qqx_n \to 0$, so $Q\theta$ is continuous. For the last assertion note that $\|R\| = \|R_0\|$, so $\|R\theta\| = \|R_0Q\theta\| \leq \|R_0\| \, \|Q\theta\| = \|R\| \, \|Q\theta\|$; so put $M = \|Q\theta\|$.

(iv) Since $x_n \to 0$ and $\theta x_n \to y$ imply $R\theta x_n \to Ry$, we have $Rs(\theta) \subset s(R\theta)$ and thus $[Rs(\theta)]^- \subset s(R\theta)$ by (i). To prove the reverse inclusion let Q_0: $Z \to Z/[Rs(\theta)]^-$ be the canonical map. Clearly $Q_0Rs(\theta) = (0)$, hence $Q_0s(R\theta) = (0)$, so $Q_0R\theta$ is continuous, and it follows that $s(R\theta) \subset [Rs(\theta)]^-$.

3.5.3 Definition. Let $T \in L(X)$ and $S \in L(Y)$. We say θ: $X \to Y$ <u>intertwines</u> T with S (or θ is an intertwining of T with S) if $\theta T = S\theta$.

3.5.4 Definition. Let $T \in L(X)$. For F closed in $\mathbf{C}$, define $E_T(F)$ to be the linear span of all manifolds Z in X satisfying

$$(5.1) \qquad (\lambda - T)Z = Z \qquad \text{for all } \lambda \notin F.$$

Moreover, the manifold Z X is T-<u>divisible</u> if

$$(5.2) \qquad (\lambda - T)Z = Z \qquad \text{for all } \lambda \in \mathbf{C},$$

We observe that $E_T(F)$ is maximal with respect to (5.1), and $E_T(\emptyset)$ is the maximal T-divisible manifold.

3.5.5 Lemma. If $T \in L(X)$, then a T-divisible manifold Z is zero iff Z is closed.

Proof. Only sufficiency needs proof. Let Z be closed, and suppose $\lambda_0 \in \sigma(T|Z)$ is an eigenvalue. Put $S = T|Z$ and $V = \ker(\lambda_0 - S)$. Then $(\lambda_0 - S)/V$ is bijective. Since V is S-hyperinvariant, by [30, p. 12-13] $\sigma(S) = \{\lambda_0\} \cup \sigma(S/V)$, hence $\lambda_0 \notin \sigma(S/V)$ means that λ_0 is isolated in $\sigma(S)$. Let W be the direct summand of Z associated with $\{\lambda_0\}$ by the functional calculus. Then W is T-divisible, but since $S|W$ clearly has SVEP we have $\lambda_0 \in \sigma(T|Z)$ by Lemma 1.2.5. This contradiction proves that $\sigma_p(S) = \emptyset$. Since Z is closed, it follows that $Z = (0)$.

3.5.6 Theorem. Let $T \in L(X)$ be weakly decomposable relative to the identity. Then the following are equivalent:

(i) $E_T(\emptyset) = (0)$;

(ii) $E_T(F) = X(T,F)$ for each closed F;

(iii) $E_T(F)$ is closed for each closed F.

Proof. (i) $\Rightarrow$ (ii). Let U, V be open neighborhoods of F such that $F \subset U \subset U^- \subset V$. Then $\{V, \mathbf{C}\setminus U^-\}$ is an open cover of $\mathbf{C}$, hence there are sequences $\{P_n\}$, $\{Q_n\}$ in $\{T\}'$ such that

$$P_n + Q_n \to I \text{ (wot)}$$
$$P_nX \subset X(T,V^-) \text{ and } Q_nX \subset X(T,\mathbf{C}\setminus U).$$

Let Z be maximal submanifold of $M = X(T, \mathbf{C}\setminus U)$ satisfying (5.1). We claim that Z is T-divisible in X. For let $\mu \in F$ and $\lambda \notin F$. Then

$$(\lambda - T)(\mu - T|M)^{-1}Z = (\mu - T|M)^{-1}Z,$$

so by maximality of Z, $(\mu - T|M)^{-1}Z \subset Z$ or $Z \subset (\mu - T)Z$. On the other hand, if we write $\mu - T = (\mu - \lambda) + (\lambda - T)$ it follows that $(\mu - T)Z \subset Z$, so (5.1) holds for all $\lambda \in \mathbf{C}$. Then $Z = (0)$ by hypothesis.

By Def. 3.5.4 for each Q_n (n=1, 2,....) we have

$$(\lambda - T)Q_nE_T(F) = Q_nE_T(F) \qquad (\lambda \notin F),$$

hence $Q_nX \subset M$ implies $Q_nE_T(F) \subset M$, and thus $Q_nE_T(F) = (0)$. It follows that for $x \in E_T(F)$

$$x = \lim_n (P_n + Q_n)x = \lim_n P_nx \in X(T,V^-).$$

As $V \supset F$ is arbitrary, $E_T(F) \subset X(T,F)$. But the reverse inclusion follows from Def. 3.5.4, so the proof of (i) $\Rightarrow$ (ii) is complete.

(ii) $\Rightarrow$ (iii) is clear because T is weakly decomposable relative to the identity, and (iii) $\Rightarrow$ (i) follows by Lemma 3.5.5.

3.5.7 Example. Not every operator weakly decomposable relative to the identity has trivial divisible manifold. Let $X = C[0,1]$ and define $T \in L(X)$ by

$$(Tf)(s) = \int_0^s f(t)dt \qquad (f \in X,\ 0 \le s \le 1).$$

Then it is straightforward to verify that each $f \in X$ for which $f^{(k)}(0) = 0$ for $k = 0, 1, 2,...$, is an element of a T-divisible subspace. But it is clear that T is (weakly) decomposable relative to the identity because it is compact and hence A-scalar ([24, p. 67] and Th. 1.4.8).

On the other hand, a generalized scalar operator T always has $E_T(\emptyset) = (0)$. Vrbova [92, p. 495] proved that for a generalized scalar T there exists an integer $p > 0$ such that

$$X(T,F) = \cap\{(\lambda - T)^p X: \lambda \notin F\} \tag{5.3}$$

3.5.8 Corollary. If T is generalized scalar, then $E_T(\emptyset) = (0)$ and $X(T,F) = E_T(F)$ for all F closed.

Proof. The second equality follows from the first and Lemma 3.5.5. Let Z be T-divisible, so that $Z = (\lambda - T)Z$ for all λ. Thus $Z = (\lambda - T)^2 Z = (\lambda - T)^p Z \subset (\lambda - T)^p X$ for all λ, so by (5.3) $Z \subset X(T,\emptyset) = (0)$. This yields $E_T(\emptyset) = (0)$.

3.5.9 Theorem. Suppose $T \in L(X)$ has SVEP, and suppose $S \in L(Y)$ is weakly decomposable relative to the identity with $E_S(\emptyset) = (0)$. Then every $\theta: X \to Y$ intertwining T with S satisfies

$$\theta X(T,F) \subset Y(S,F) \tag{5.4}$$

for each closed F.

Proof. By (5.1) $\lambda \notin F$ implies

$$\theta E_T(F) = \theta(\lambda - T)E_T(F) = (\lambda - S)\theta E_T(F),$$

so $\theta E_T(F) \subset E_S(F)$. By Def. 3.5.4 and Th. 3.5.6,

$$\theta X(T,F) \subset \theta E_T(F) \subset E_S(F) = Y(S,F),$$

hence (5.4) is proved.

We digress briefly to give an interesting consequence of Th. 3.5.9. Let X be a Hilbert space. Then there exists a linear transformation θ on X which has no closed nontrivial invariant subspace [52]. But any weakly decomposable operator T relative to the identity with $\sigma(T)$ having at least two points must have closed proper subspace such as $M = X(T,F)$. If $E_T(\emptyset) = (0)$, then M is θ-invariant for any linear θ such that $\theta T = T\theta$.

See Ex. 3.5.18 below for a case where (5.4) fails if $E_S(\emptyset) \neq (0)$.

We now reach the pivotal result on which our main theorem depends. We follow the general idea in [84, pp. 12-18], but, confining our argument to Banach spaces, we are able to simplify the proofs. The next lemma is a special case of [9, p. 257].

3.5.10 Lemma. Let X, Y be Banach spaces, and let $\{X_n\}$ be a decreasing sequence of closed subspaces of X. Let $\{Y_n\}$ be some other sequence of Banach spaces and $\Pi_n : Y \to Y_n$ continuous linear maps. Suppose $\theta: X \to Y$ is linear. If $\Pi_n\theta: X_n \to Y_n$ is continuous for all n, then $\Pi_n\theta: X_k \to Y_n$ is continuous for some k and all n.

3.5.11 Definition. Let $T \in L(X)$ and $S \in L(Y)$ both have property (κ), and let $\theta: X \to Y$ be a linear intertwining of T with S. Let $\lambda \in C$. Then λ is called a _discontinuity point_ of θ of _type I_ if $\theta|X(T,F)$ is not continuous for each closed neighborhood F of λ, and λ is a _discontinuity point_ of θ of _type II_ if the composition $p_V\theta : X \to Y/Y(S,C\backslash V)$ is not continuous for all open V containing λ where p_V is the surjection $p_V: Y \to Y/Y(S,C\backslash V)$. Denote the discontinuity points of type I [II] by $\Lambda_1(\theta)$ [$\Lambda_2(\theta)$].

3.5.12 Theorem. Let T be decomposable, and let S be weakly decomposable relative to the identity. If θ is a linear map intertwining T with S, then $\Lambda_1(\theta) = \Lambda_2(\theta)$ and is finite.

Proof. We first prove $\Lambda_1(\theta) \subset \Lambda_2(\theta)$. Suppose $\lambda \in C\backslash\Lambda_2(\theta)$. Then by Def. 3.5.11 there is some open V containing λ such that $p_V\theta: X \to Y/Y(S, C\backslash V)$ is continuous. Let U be open such that $\lambda \in U \subset U^- \subset V$. Since $Y(S, U^- \cup (C\backslash V))$ is closed and $U^- \cap (C\backslash V) = \emptyset$, by [12, p.1484]

$$Y(S, U^- \cup (C\backslash V)) = Y(S, U^-) \oplus Y(S,C\backslash V),$$

hence the restriction ρ of p_V to $Y(S, U^-)$ is an isomorphism onto the space $Y[S,U^-\cup(C\backslash V)]/Y(S,C\backslash V)$. by (5.4) $\theta X(T, U^-) \subset Y(S, U^-)$, so our assumption implies that $\theta|X(T,U^-) = \rho^{-1}p_V\theta|X(T, U^-)$ is continuous. Hence $\lambda \notin \Lambda_1(\lambda)$, so $\Lambda_1(\theta) \subset \Lambda_2(\theta)$.

Now suppose $\Lambda_2(\theta)$ is not finite. Since C is a regular topological space, we select a sequence $\{V_n\}$ of open sets with pairwise disjoint closures such that $V_n \cap \Lambda_2(\theta) \neq \emptyset$ for each n. For each n choose $\mu_n \in V_n \cap \Lambda_2(\theta)$ and define spaces and surjections as follows:

$$X_n = X(T, C\setminus\cup_{k=1}^{\infty} V_k)$$

$$Y_n = Y/Y(S,C\setminus V_n)$$

$$\Pi_n = P_{V_n} : Y \to Y_n.$$

Now $\Pi_n\theta|X_n = 0$ for all n. By Lemma 3.510 there exists m with $\Pi_n\theta: X_m \to Y_n$ continuous for all n. Put n = m+1. Since V_n are pairwise disjoint and T is decomposable

$$X = X(T,C\setminus V_{m+1}) + X_m \tag{5.5}$$

Moreover, $\Pi_{m+1}\theta$ is continuous on X_m and zero on $X(T,C\setminus V_{m+1})$ so $\Pi_{m+1}\theta$ is continuous by (5.5) and an easy application of the closed graph theorem. This proves $\mu_{m+1} \notin \Lambda_2(\theta)$ by Def. 3.5.11, a contradiction which proves $\Lambda_2(\theta)$ is finite.

Finally we prove $\Lambda_2(\theta) \subset \Lambda_1(\theta)$. Let $\lambda \notin \Lambda_1(\theta)$. Then $\theta|X(T,U^-)$ is continuous for some neighborhood U of λ. Let V be another open set with $\lambda \in V \subset V^- \subset U$. Then $X = X(T, U^-) + X(T,C\setminus V)$ because T is decomposable; but since $p_V\theta$ is continuous on $X(T, U^-)$ and zero on $X(T,C\setminus V)$, $p_V\theta$ is continuous on X. Hence $\lambda \notin \Lambda_2(\theta)$, and thus we have $\Lambda_1(\theta) = \Lambda_2(\theta)$.

3.5.13 Corollary. Let T, S, θ, X, Y be as in Th. 3.5.12. Then $s(\theta) \subset Y(S,\Lambda_1(\theta))$.

Proof. For $\lambda \notin \Lambda_1(\theta)$, by Th. 3.5.11 there is an open set V with $\lambda \in V$ such that $p_V\theta$ is continuous on X. We claim $s(\theta) \subset Y(S,C\setminus V)$. For if $y \in s(\theta)$ then $\theta x_n \to y$ for some $x_n \to 0$ in X. Since $p_V\theta x_n \to 0$ and $p_V\theta x_n \to p_V y$ we must have $p_V y = 0$, hence $s(\theta) \subset Y(S,C\setminus V)$. Let $\{U_n\}$ be the decreasing sequence of closed n^{-1}-neighborhoods of $\Lambda_1(\theta)$. Fix n. There are sequences $\{\lambda_j\}$ and $\{V_j\}$ with V_j open and $\lambda_j \in V_j$ such that $C = U_n \cup \{V_1 \cup V_2 \cup \ldots\}$. Hence for each n

$$s(\theta) \subset \cap_j Y(S,C\backslash V_j) = Y(S, C\backslash \cup V_j) \subset Y(S,U_n)$$

hence $s(\theta) \subset Y(S,\Lambda_1(\theta))$, so the proof is finished.

Denote by $C[\chi]$ the ring of polynomials over C.

3.5.14 Theorem. Let $T \in L(X)$ be decomposable, and let $S \in L(Y)$ be weakly decomposable relative to the identity. Let $\theta: X \to Y$ intertwine T with S. If $E_S(\emptyset) = (0)$, then $p(S)\theta$ is continuous for some nonzero polynomial $p \in C[\chi]$ with zeros in $\Lambda_1(\theta)$.

Proof. By Lemma 3.5.2 it suffices to prove $p(S)s(\theta) = (0)$ for some polynomial $p(\lambda)$. By Th. 3.5.12 we may write $\Lambda_1(\theta) = \{\lambda_1, \lambda_2, \ldots, \lambda_k\}$, so by Cor. 3.5.13 we can use the functional calculus to write $s(\theta) = W_1 \oplus \ldots \oplus W_k$ where $\sigma(S|W_j) = \{\lambda_j\}$ (all j). For j fixed write $\lambda = \lambda_j$, $V = W_j$. We shall prove that $(\lambda - S)^m V = (0)$ for some $m > 0$. To do so define

$$W_0 = \cap\{(\lambda - S)^n V : n = 1, 2, \ldots\},$$
$$W_k = ((\lambda - S)^k V)^- \qquad (k = 1, 2, \ldots).$$

Since $\sigma(S|V) = \{\lambda\}$ it is easy to see that $(\mu - S)W_0 = W_0$ for all $\mu \in C$, hence $W_0 \subset E_S(\emptyset) = (0)$. Suppose also that we can prove there is m fixed such that

$$W_m = W_n \qquad \text{for all } n \geq m. \tag{5.6}$$

By the Mittag-Leffler theorem $W_0^- = \cap W_k$ [53, p. 535]. Hence (5.6) yields

$$(\lambda - S)^m V \subset W_m = \cap_{k>0} W_k = W_0^- = (0).$$

For each j we thus have m_j with $(\lambda_j - S)^{m_j} W_j = (0)$. Clearly $p(\lambda) = \Pi(\lambda - \lambda_j)^{m_j}$ is the desired polynomial.

It remains to prove (5.6). Suppose $W_k \subset W_{k-1}$ is proper for all k. Let Q_k be the surjection from W_{k-1} onto W_{k-1}/W_k. Then $Q_n(\lambda - S)^n\theta$ is continuous but $Q_n(\lambda - S)^{n-1}\theta$ is not (all n). Define $\{x_n\} \subset X$ with

$$\|x_n\| \leq (2\|\lambda - T\|)^{-n} \qquad \text{(all } n)$$
$$\|Q_n(\lambda - S)^{n-1}\theta x_n\| \geq n + \|Q_n(\lambda - S)^n\theta\| + \|Q_n\theta(\Sigma_j(\lambda - T)^{j-1}x_j)\|$$

Then $z = \Sigma_{j=1}(\lambda-T)^{j-1}x_j \in X$ but a simple calculation [84, p.12] shows $\|\theta z\| \geq n$ for all n. Hence $W_k = W_{k-1}$ for some k, so the proof is complete.

We now recall that λ is a critical eigenvalue for the pair (T,S) if λ is an eigenvalue of S such that $(\lambda - T)X$ has infinite codimension in X.

3.5.15 Lemma. If (T,S) has a critical eigenvalue, then there is a discontinuous $\theta: X \to Y$ with $\theta T = S\theta$.

Proof. Let λ be a critical eigenvalue of (T,S). Since $X/(\lambda - T)X$ is (algebraically) infinite-dimensional, by Zorn's lemma we can find a discontinuous linear functional f on X vanishing on $(\lambda - T)X$. Let $y \neq 0$ with $Sy = \lambda y$ and define $\theta: X \to Y$ by $\theta x = f(x)y$ for $x \in X$. Then $\theta(\lambda - T)x = \theta(\lambda x - Tx) = f(\lambda x - Tx)y = 0$. But $(\lambda - S)\theta x = f(x)(\lambda - S)y = 0$, so the result follows.

3.5.16 Lemma. Let $U: X \to Y$ be continuous and linear. If UX has finite codimension in Y, then UX is closed.

Proof. Let $Y = UX \oplus W$ where W is finite dimensional and let $\psi: X \oplus W \to Y$ by $\psi(x,w) = Ux + w$ and give $X \oplus W$ norm $\|(x,w)\| = \|x\| + \|w\|$. Then $X \oplus W$ is a Banach space and ψ is a bounded surjection, hence open. Since $A = \{(x,w) : w \neq 0\}$ is open in $X \oplus W$, its image $\psi(A) = \{Ux + w : w \neq 0\}$ is open in Y. Thus $Y \backslash \psi(A) = UX$ is closed.

We now recall some algebraic notions. If $T \in L(X)$, $S \in L(Y)$, we may regard X, Y as $\mathbf{C}[\chi]$-modules by defining for $p \in \mathbf{C}[\chi]$ the ring multiples $p \cdot x = p(T)x$ [resp. $p \cdot y = p(S)y$]. Moreover, θ intertwines T with S iff θ is a $\mathbf{C}[\chi]$-homomorphism from X to Y.

Recall also that T is algebraic if $p(T) = 0$ for some nonzero polynomial p(z). A submodule $Z \subset X$ is an infinitely generated free submodule if Z can be decomposed into a direct sum with components of the form $\mathbf{C}[\chi] \cdot u_\beta$ where $\{u_\beta\}$ is an indexed set in X and each such component is ring-isomorphic to $\mathbf{C}[\chi]$ under the map $p \cdot u_\beta \to p$ [84, p. 20].

The following purely algebraic result is proved in [84, p. 21].

3.5.17 Lemma. If $T \in L(X)$ is not algebraic, then X contains an infinitely generated free $\mathbf{C}[\chi]$-module.

3.5.18 Theorem. Let $T \in L(X)$, $S \in L(Y)$. If T is not algebraic and $E_S(\varnothing) \neq (0)$, then there exists a discontinuous $\theta: X \to Y$ intertwining T with S.

Proof. By Lemma 3.5.17 X contains an infinitely generated free $\mathbf{C}[\chi]$-module W with components the form $\mathbf{C}[\chi]\cdot u_k$ where $\{u_1, u_2, \ldots\} \subset X$. Clearly we may suppose $\|u_k\| = 1$ for all k. Let $y \in E_S(\varnothing)$ be nonzero, and define $\theta: W \to Y$ by $\theta[p(T)u_k] = kp(S)y$ on the k-th component of W. Then θ maps W into $E_S(\varnothing)$ and is obviously unbounded.

To extend θ to X we proceed as follows. Let (Z,ψ) be a pair with Z a submodule of X such that $W \subset Z \subset X$ and $\psi: Z \to E_S(\varnothing)$ with $\psi|W = \theta$. If we partially order $\{(Z, \psi)\}$ by extension where $Z \subset Z'$ implies $\psi'|Z = \psi$, then there is a maximal pair (Z_0, ψ_0) by Zorn's lemma. We prove $Z_0 = X$.

Suppose not. Let $u \in X \backslash Z_0$ and let $J = \{p \in \mathbf{C}[\chi] : p\cdot u \in Z_0\}$. Then J is an ideal in $\mathbf{C}[\chi]$. Define $\eta: J \to E_S(\varnothing)$ by $\eta(p) = \psi_0(p\cdot u)$. Choose $q \in \mathbf{C}[\chi]$ such that $1 = \alpha p_0 + \beta q$ where p_0 generates J, q is relatively prime to p_0 and $\alpha, \beta \in \mathbf{C}[\chi]$. Extend η to $\eta': \mathbf{C}[\chi] \to E_S(\varnothing)$ by $\eta'(s + rq) = \eta(s) + ry$ $(s \in J, r \in \mathbf{C}[\chi])$. Put $Z_1 = Z_0 + \mathbf{C}[\chi]\cdot u$ and define $\psi_1: Z_1 \to E_S(\varnothing)$ by $\psi_1(z_1 + r\cdot u) = \psi_0(z_0) + r\,\eta'(1)$. If ψ_1 is well-defined then it is a nontrivial extension of ψ_0, and the maximality of ψ_0 is contradicted. To see this let $z_0 + r\cdot u = z_0' + r'\cdot u \in Z_1$. Then $(r - r')\cdot u = z_0' - z_0 \in Z$, so

$$(5.7) \qquad \psi_0(z_0' - z_0) = \psi_0[(r - r')\cdot u] = \eta(r - r') = (r - r')\eta'(1).$$

Rewriting (5.7) shows ψ_1 to be well-defined. The resulting contradiction implies ψ_0 extends θ to X.

Finally, since ψ_0 is a $\mathbf{C}[\chi]$-homomorphism, for $x \in X$ we have $\psi_0 Tx = \psi_0(p(T)x) = p(S)(\psi_0 x) = S\psi_0 x$, where $p(\lambda) = \lambda$. Hence ψ_0 intertwines T with S. This completes the proof.

3.5.19 Example. For a continuous θ such that $\theta T = S\theta$, (5.4) above always holds, but (5.4) may fail for some discontinuous θ if $E_S(\emptyset) \neq (0)$. Let T_1 be the Volterra operator constructed in Ex. 3.5.6, let $X = X_1 \oplus X_2$ where each $X_i = C[0,1]$ and put $T = T_1 \oplus (I - T_1)$ on X. Let Z be the maximal divisible manifold of T_1. Since $Z \neq (0)$ and $p(T_1)Z = Z$ for all polynomials p(z), T_1 is not algebraic and clearly $I-T_1$ has divisible space Z. By Th. 3.5.17 there is a discontinuous linear map $\theta: X_1 \to X_2$ with $(I - T_1)\theta = \theta T_1$. Let $\psi: X \to X$ be given by $\psi(x_1,x_2) = (0,\theta x_1)$. Then ψ is discontinuous on X and ψ commutes with T since

$$\begin{aligned}\psi T(x_1,x_2) &= \psi(T_1x_1, x_2 - T_1x_2) = (0,\theta T_1x_1)\\ &= (0,(I - T_1)\theta x_1) = T(0,\theta x_1)\\ &= T\psi(x_1,x_2).\end{aligned}$$

Now X(T,F) is defined for all F, and $X(T,\{0\}) = X_1$ but y does not map X_1 into itself. If so, then $\psi(x_1,0) = (0,\theta x_1) = (0,0)$ or θ is zero, contradicting the discontinuity of θ. We observe also that $E_T(\{1\}) = Z \oplus X_2$ whereas $X(T,\{1\}) = X_2$ in contrast to Th. 3.5.6.

We now state the main theorem of the section.

3.5.20 Theorem. Let $T \in L(X)$ and let $S \in L(Y)$. Let T be decomposable, and let S be weakly decomposable relative to the identity. Then the following assertions are equivalent.

(i) Every linear map $\theta: X \to Y$ which intertwines T with S is continuous.

(ii) (T,S) has no critical eigenvalues, and either T algebraic or $E_S(\emptyset) = (0)$.

Proof. (i) $\Rightarrow$ (ii). This follows from Th. 3.5.18 and Lemma 3.5.15. (ii) $\Rightarrow$ (i). Let θ: $X \to Y$ be C-linear. By Lemma 3.5.2(iii) $p(S)\theta$ is continuous if we can find a polynomial $p \neq 0$ for which $p(S)s(\theta) = (0)$. From this we shall prove continuity of θ. Moreover, we may assume that every zero of p is an eigenvalue of S. For if $p(\mu) = 0$ and μ has multiplicity k but is not an eigenvalue of S, then since $p(S) = (\mu - S)^k p_1(S)$, the injectivity of $\mu - S$ yields $p_1(S)s(\theta) = (0)$.

To construct p we consider two cases. First suppose that T is algebraic with $p(T) = 0$. Then

$$p(S)s(\theta) \subset p(S)(\theta X)^{-} \subset [p(S)\theta X]^{-} = [\theta p(T)X]^{-} = (0)$$

where the first inclusion follows from Lemma 3.5.2 and the second from the continuity of S. Second, suppose $E_S(\emptyset) = (0)$. By Th. 3.5.14 $p(S)\theta$ is continuous for some $p(\lambda)$ in this case also.

We finish by showing θ continuous. Let $p(z) = (\lambda_1 - z)p_1(z)$ where λ_1 is an eigenvalue of S. The mapping

$$x + (\lambda_1 - T)X \rightarrow p_1(T)x + (\lambda_1 - T)p_1(T)X$$

is a surjection from $X/(\lambda_1 - T)X$ onto $p_1(T)X/p(T)X$, hence $p_1(T)X/p(T)X$ is finite dimensional. By induction it follows that $p(T)X$ has finite codimension in X, thus

(5.8) $$X = p(T)X \oplus Z$$

where Z is finite-dimensional. Thus $p(T)X$ is closed by Lemma 3.5.16. Let $x\hat{}$ be the coset of $x \in X$ in $X/(\ker p(T))$. By the open mapping theorem there exists $M > 0$ such that

(5.9) $$\|x\hat{}\| \leq M\|p(T)x\| \quad (x \in X).$$

From this we prove $\theta|p(T)X$ is continuous. Let $p(T)x_n \rightarrow 0$ for $\{x_n\}$ in X. By (5.9) $x_n\hat{} \rightarrow 0$, hence for each n choose $u_n \in \ker p(T)$ with $\|x_n - u_n\| \rightarrow 0$. Then $\theta p(T)x_n = p(S)\theta(x_n - u_n) \rightarrow 0$, i.e. θ is continuous on $p(T)X$. Since continuity of $\theta|Z$ is obvious, θ is continuous on all X by (5.8).

6. Applications

Here we give several applications of Th. 3.5.20. We start with some results on "subscalar" operators. In Chapter IV we define $T \in L(X)$ to be <u>subscalar</u> if there is a Banach space Y containing X and a generalized scalar operator $S \in L(Y)$ such that $T = S|X$. We can extend Cor. 3.5.8 to the case of subscalar operators as follows.

3.6.1 Lemma. If T is subscalar, then $E_T(F) = X(T,F)$ for each closed F.

Proof. Let $Z = E_T(F)^-$, and let $S \in L(Y)$ be a generalized scalar extension of T. By Cor. 3.5.8

$$(6.1) \qquad Z \subset E_S(F) \cap X = Y(S,F) \cap X \subset Y(S,F)$$

Then $\sigma(T|Z) \subset F$. For if $\lambda \notin F$ then from (6.1) it is easy to see that $\lambda - T$ is bijective on Z. Hence $\lambda \in \rho(T|Z)$ so $\sigma(T|Z) \subset F$. Now T inherits property (β) from the decomposable S, hence X(T,F) is closed and

$$E_T(F) \subset Z \subset X(T,F) \subset E_T(F).$$

This completes the proof.

By a careful review of the proofs of Ths. 3.5.14 and 3.5.20, the reader will note that the crucial facts used in the situation $\theta T = S\theta$ were that the spectral subspaces of T and S obey (5.4) and Y(S,F) be closed for all closed F. But Lemma 3.6.1 clearly implies both these conditions. Hence we can state the following analog of Th. 3.5.20.

3.6.2 Theorem. Let $T \in L(X)$ be decomposable, and let $S \in L(Y)$ be subscalar. Then the following are equivalent:

(i) every linear map $\theta: X \to Y$ such that $\theta T = S\theta$ is continuous;

(ii) (T,S) has no critical eigenvalue.

3.6.3 Corollary. Let $T \in L(X)$ be decomposable, let $S \in L(Y)$ be subscalar and let $\theta: X \to Y$ intertwine T with S. If θ has range dense in Y and (T,S) has no critical eigenvalue, then $\sigma(S) \subset \sigma(T)$.

Proof. By Th. 3.6.2 θ is continuous. The proof now follows that of Th. 2.4.1.

3.6.4 Corollary. Let T be hyponormal (see Chap. 4.4) on Hilbert space H. If $T \neq \lambda I$ for all λ, then one of the following holds:

(i) T has a nontrivial hyperinvariant subspace;

(ii) if X is a Banach space, then every linear $\theta: X \to H$ intertwining any decomposable operator on X with T is zero.

Proof. Suppose (i) fails. We may suppose $\sigma_p(T) = \emptyset$ and $E_T(F) = (0)$ for all F proper in $\sigma(T)$ by Lemma 3.6.1. Let V be decomposable on X and let $\theta: X \to H$ be linear with $\theta V = T\theta$. Since T is subscalar by Th. 4.4.7, θ is continuous by Th. 3.6.2. Let $\{G_1, G_2\}$ be an open cover of $\mathbf{C}$ with $\sigma(T)\backslash G_i^- \neq \emptyset$ for i =1, 2. Then

$$\theta X = \theta X(V,G_1^-) + \theta X(V,G_2^-) \subset H(T,G_1^-) + H(T,G_2^-) = (0)$$

3.6.5 Remarks. (1) The question whether hyponormal operators have proper hyperinvariant subspaces can be rephrased as follows using Cor. 3.6.4: given hyponormal T can we find decomposable $V \in L(X)$ and nonzero θ intertwining V with T? (Note that θ need only be linear.)

(2) The unilateral shift S on ℓ^2 has the property that $E_S(F) = (0)$ for every F proper in $\sigma(S)$. Thus S satisfies (ii) of Cor. 3.6.4. Thus the answer to the previous question may be no. Of course, S shows that (i) and (ii) are not exclusive of each other.

We now give a corollary in a slightly different direction. In [75] Radjabalipour constructed subnormal operators which are decomposable but not normal. Such operators have "decomposable operational calculi" but are not generalized scalar [77, p. 43]. However, these operators, which we call "R-subnormal", are subscalar. With these remarks the next result follows immediately from Th. 3.6.2.

3.6.6 Corollary. Let V, T be R-subnormal. If θ is a linear map intertwining V with T, then θ is continuous iff (V,T) has no critical eigenvalue.

3.6.7 Remark. We observe that when the roles of T and S are reversed in Th. 3.6.2 the result is quite different. Let S be the unilateral shift, let T be the Volterra operator of Ex. 3.5.7. Of course, T satisfies $E_T(\emptyset) \neq (0)$ and S is clearly nonalgebraic. Hence Th. 3.5.18 guarantees existence of discontinuous θ with $\theta S = T\theta$.

In the applications considered so far, the full force of Th. 3.5.20 has not been used. As our final result we show that Ex. 3.2.11 is subject to Th. 3.5.20, which is thus a proper generalization of [67, Th. 4.3].

3.6.8 Proposition. Let $T \in L(X)$ be Ex. 3.2.11. Let $V \in L(Y)$ be decomposable, and suppose $\theta: Y \to X$ is linear with $\theta V = T\theta$. Then θ is continuous.

Proof. It suffices to show (V,T) has no critical eigenvalues and $E_T(\emptyset) = (0)$. Clearly T has no eigenvalues. Let Z be T-divisible in X. In particular, $(\lambda - T)Z = Z$ for all $|\lambda| < 1$. Hence for each $f = (f_j) \in Z$ there is $(g_j) \in Z$ with $(\lambda - z)g_j(z) = f_j(z)$ for all $|z| \leq 1$ and $j=1, 2,\ldots$. But if we suppose $f_k(\lambda) = 1$ for some k and some $|\lambda| < 1$, then g_k is discontinuous at $z = \lambda$. This contradiction proves $Z = (0)$, so the proof is complete.

Finally, we consider intertwinings with restriction of weakly decomposable operators.

3.6.9 Theorem. Let $T \in L(X)$ be weakly decomposable relative to the identity, with $E_T(\emptyset) = (0)$. Let M be a T-invariant subspace and let $S = T|M$. Then every linear map $\theta: Y \to M$ intertwining a decomposable operator $V \in L(Y)$ with S is continuous if and only if (V, S) has no critical eigenvalues.

Proof. Since T is quasidecomposable (Th. 3.2.4), the conclusion of Lemma 3.6.1 holds for S. Hence the result follows as before (Th. 3.6.2).

7. Other Considerations

In this section we study some further properties related to automatic continuity. In particular, we give some general necessary conditions for automatic continuity. We use some facts below on the manifolds $E_T(F)$.

3.7.1 Lemma. If $T \in L(X)$ and $E_T(\emptyset) = (0)$, then T has SVEP.

Proof. Let $(\lambda - T)f(\lambda) = 0$ for some analytic $f: D \to X$. Fix $\lambda \in D$ and suppose D is connected. Let M be the linear span of $\{f(\lambda), f'(\lambda), \ldots\}$. An easy inductive argument shows.

$$
\begin{aligned}
Tf(\lambda) &= \lambda f(\lambda) \\
Tf'(\lambda) &= \lambda f'(\lambda) + f(\lambda) \\
&\cdots\cdots\cdots\cdots \\
Tf^{(k)}(\lambda) &= \lambda f^{(k)}(\lambda) + kf^{(k-1)}(\lambda) \ .
\end{aligned}
\tag{7.1}
$$

From (7.1) it follows that $M = (\mu - T)M$ for all $\mu \in \mathbf{C}$. Hence $M \subset E_T(\emptyset) = (0)$. Thus $f^{(n)}(\lambda) = 0$ for $n = 0, 1, 2, \ldots$, hence $f(\mu) = 0$ on some disc in D. Since D is connected, $f = 0$ on D.

3.7.2 Definition. Let $T \in L(X)$, $S \in L(Y)$. For a linear map $\theta: X \to Y$ write $C(T,S)(\theta) = S\theta - \theta T$ and inductively $C^n(T,S)(\theta) = C(T,S)(C^{n-1}(T, S)(\theta))$. We say θ n-<u>intertwines</u> T with S if $C^n(T,S)(\theta) =$

0 for some $n = 1, 2, \ldots$ In case $T = S$, we write $C^n(T)(\theta) = C^n(T,S)(\theta)$.

3.7.3 Lemma. If θ n-intertwines T with S, then $\theta E_T(F) \subset E_S(F)$ for all F.

Proof. Since $C^n(T, S)\theta = C(T,S)C^{n-1}(T,S)\theta = 0$ it suffices to prove that if $SC^k(T,S)\theta - (C^k(T,S)\theta)T$ maps $E_T(F)$ into $E_S(F)$ then $C^k(T,S)\theta$ does also. Let $A = C^k(T,S)\theta$, and let $x \in E_T(F)$, $y \in E_S(F)$, $\lambda \notin F$. Since $SA-AT = (S - \lambda)A - A(T - \lambda)$,

$$(S - \lambda)(Ax + y) = A(T - \lambda)x + (S - \lambda)y + y_1 \in AE_T(F) + E_S(F)$$

where $y_1 = (SA - AT)x \in E_S(F)$. Thus $(S - \lambda)(AE_T(F) + E_S(F)) \subset AE_T(F) + E_S(F)$. A similar argument proves the reverse inclusion for all $\lambda \notin F$. By maximality $AE_T(F) + E_S(F) \subset E_S(F)$, so $AE_T(F) \subset E_S(F)$. Inductively we arrive at $\theta E_T(F) \subset E_S(F)$.

3.7.4 Theorem. If $T \in L(X)$ does not have the SVEP, then there exists a discontinuous $\theta: X \to X$ such that $\theta T = T\theta$.

Proof. Under the hypothesis T is not algebraic and $E_T(\emptyset) \neq (0)$ by Lemma 3.7.1. Hence there is a desired θ by Th. 3.5.17.

We now use the results above to show that operators with the automatic continuity property are close to being decomposable. We begin with

3.7.5 Definition. Let $T \in L(X)$, $S \in L(Y)$. We say that the pair (T, S) has the <u>automatic continuity property</u> (ACP) if $\theta T = S\theta$ for $\theta: X \to Y$ implies θ is continuous. We say T has ACP iff (T, T) does and T has <u>dual</u> ACP if T, T^* both have ACP.

3.7.6 Theorem. If (T,S) has ACP, then either T or S has SVEP.

Proof. By Th. 3.5.17 either T is algebraic or $E_S(\emptyset) = (0)$. Then either T has SVEP since $\sigma(T)$ is finite or S has SVEP by Lemma 3.7.1.

3.7.7 Corollary. If T has ACP, then T has SVEP.

On Hilbert space we have

3.7.8 Theorem. Let T be bounded on Hilbert space H. If T has dual ACP, then T is BQT.

Proof. This is clear from Cor. 3.7.7 and Th. 2.3.21.

3.7.9 Remark. On Hilbert space H the last result shows by Th. 2.3.21 that if T, T^* both have ACP, then T is arbitrarily near a decomposable operator in $L(H)$.

The following results are due to Laursen [67].

3.7.10 Lemma. Let $T \in L(X)$. Then

(i) if $\lambda \in F$ and $(\lambda - T)x \in E_T(F)$, then $x \in E_T(F)$;

(ii) $E_T(F) \cap E_T(H) = E_T(H \cap F)$.

Proof. (i) Let $\lambda \in F$, $(\lambda - T)x \in E_T(F)$. Then for $\mu \notin F$, $(\lambda - T)x = (\lambda - \mu)x + (\mu - T)x = (\mu - T)y$ for $y \in E_T(F)$. Hence $x = (\mu - T)(\alpha y - \alpha x))$ where $\alpha = (\lambda - \mu)^{-1}$. It follows that $(\mu - T)$ is surjective on the manifold $E_T(F) + \mathbf{C}x$, hence $E_T(F) + \mathbf{C}x \subset E_T(F)$ by maximality. Thus $x \in E_T(F)$.

(ii) It is easily seen that $E_T(F) = E_T(F \cap \sigma(T))$. Suppose $F \cup H = \sigma(T)$, let $\mu \notin F \cap H$. If $\lambda \notin F$ we may suppose $\lambda \in H$. Let $x \in M = E_T(F) \cap E_T(H)$. Then $x = (\lambda - T)y$ for $y \in E_T(H)$, but $y \in E_T(F)$ by part (i). Hence $\lambda - T$ is surjective on M. By maximality

$$M \subset E_T(F \cap H) \tag{7.2}$$

By symmetry (7.2) holds if $\lambda \notin H$, and the reverse inclusion of (7.2) holds trivially, so (ii) holds. In case $F \cup H \neq \sigma(T)$, let $G = H \cup (\sigma(T) \backslash F)$. Clearly $F \cup G = \sigma(T)$ and $F \cap G = F \cap H$. Hence

$$\begin{aligned} E_T(F) \cap E_T(H) &\subset E_T(F) \cap E_T(G) \\ &= E_T(F \cap G) = E_T(F \cap H), \end{aligned}$$

and since $E_T(F \cap H) \subset E_T(F) \cap E_T(H)$ is trivial, (ii) holds in this case also.

3.7.11 Proposition. Let $T \in L(X)$ have SVEP, let $S \in L(Y)$ such that $E_S(F)$ is closed for all closed F. If $\theta: X \to Y$ n-intertwines T with S, then $\theta X(T,F) \subset Y(S,F)$.

Proof. By Lemma 3.7.3, $\theta X(T,F) \subset \theta E_T(F) \subset E_S(F)$. It remains to prove

$E_S(F) \subset Y(S,F)$, which makes sense because of Lemmas 3.7.1 and 3.3.5. It suffices to prove $\sigma(S|E_S(F)) \subset F$. If $\lambda \notin F$, then $\lambda - S$ is surjective on $E_S(F)$. Hence $(\lambda-S)|E_S(F)$ is bijective by Lemmas 3.7.1 and 3.5.5. It follows that $\sigma(S|E_S(F)) \subset F$ and the proof is complete.

Prop. 3.7.11 is one element needed in the proof of the next theorem [69, Th. 4.1].

3.7.12 Theorem. Let T be decomposable on X, and let $S \in L(Y)$ with $E_S(F)$ closed for all F. Then every $\theta: X \to Y$ which n-intertwines T with S is continuous iff (T,S) has no critical eigenvalue.

3.7.13 Corollary. Let $T \in L(X)$ and $S \in L(Y)$ be generalized scalar such that (T,S) has no critical eigenvalues. If $\theta: X \to Y$ is linear, then the following are equivalent.

(i) θ n-intertwines T with S;

(ii) θ is continuous with $\theta X(T,F) \subset Y(S,F)$ for each closed F;

(iii) θ is continuous with $\|C^n(T,S)(\theta)\|^{1/n} \to 0$ $(n\to\infty)$.

Proof. The implication (i) $\Rightarrow$ (ii) follows from Cor. 3.5.8, Th. 3.7.3 and Th. 3.7.12. Implications (ii) $\Rightarrow$ (iii) $\Rightarrow$ (i) are in [24, Th. 4.4.5]

For completely regular generalized scalar operators. We having the following striking result. Recall that a generalized spectral operator $T \in L(X)$ is <u>completely regular</u> if $A \in L(X)$ such that $AX(T,F) \subset X(T,F)$ implies that A commutes with the spectral distribution of T.

3.7.14 Corollary. Let $T \in L(X)$ be a completely regular generalized scalar operator without critical eigenvalue. Then every $\theta: X \to X$ commutes with T whenever $C^n(T)(\theta) = 0$ for some n.

Proof. If $C^n(T)(\theta) = 0$ for some n, then θ is continuous and $\theta X(T,F) \subset X(T,F)$. Hence $\theta T = T\theta$ by complete regularity.

3.7.15 Lemma. Every boundedly decomposable operator is completely regular.

Proof. Let T be boundedly decomposable, let $\theta \in L(X)$ and suppose $\theta X(T,F) \subset X(T,F)$ for F closed. For $x \in X(T,\sigma(x,T))$ we have $\sigma(\theta x,T) \subset \sigma(x,T)$. By Th. 1.4.4 (proof) $\theta f(S) = f(S)\theta$ where $T = S + Q$, $f \in C$. Hence T is completely regular.

3.7.16 Remark. According to the hierarchy of Chapter I.4, Lemma 3.7.15 is best possible: There are generalized scalar operators that are not even regular [2]. On the other hand, this lemma is a nontrivial extension of [24, Ex. 4.3.12], which states that spectral operators are completely regular.

3.7.17 Theorem. Let $T \in L(X)$ be boundedly decomposable with decomposition $T = S + Q$. Then the following are equivalent for $\theta \in L(X)$:

(i) $\|C^n(T)(\theta)\|^{1/n} \to 0$;

(ii) $\theta X(S,F) \subset X(S,F)$ (all F);

(iii) $\theta S = S\theta$;

(iv) $C^m(S)(\theta) = 0$ (for some $m \geq 1$);

(v) $\|C^n(S)(\theta)\|^{1/n} \to 0$;

(vi) $\theta T = T\theta$.

Proof. The equivalence of (i) and (ii) is well known [24] as well as that of (ii), (iv), (v) [24, Th. 4.4.5] since S is generalized scalar (Prop. 1.4.7). But (ii) $\Rightarrow$ (iii) follows from Lemma 3.7.15, while (iii) $\Rightarrow$ (iv) is trivial. Equivalence of (iii) and (vi) follows from the complete regularity of both S and T.

Chapter IV - INVARIANT SUBSPACES FOR SUBDECOMPOSABLE OPERATORS

1. Introduction

An operator on a Banach space is called subdecomposable if it is the restriction of some decomposable operator. (This is a natural generalization of subnormal operators.) We shall be principally concerned in this chapter with the existence of invariant subspaces for subdecomposable operators. This question is dealt with in §§3-5. In preparation for this we first discuss in §2 some properties of uniform algebras. In §§4 and 5 we discuss the special class of hyponormal operators.

2. Dirichlet Algebras

Let X be a compact Hausdorff space and let C(X) be the space of all continuous complex-valued functions defined on X; C(X) is endowed with the supremum norm

$$\|f\|_X = \sup_{x \in X} |f(x)|. \tag{2.1}$$

4.2.1 Definition. A norm closed subalgebra A of C(X) which contains the constants and separates the points of X is called a uniform algebra on X.

Let K be a compact subset of the complex plane C. Associated with K there are three uniform algebras in which we shall be especially interested.

(1) The algebra P(K) consists of the functions in C(K) which can be approximated uniformly by polynomials in z.

(2) The algebra R(K) consists of the functions in C(K) which can be approximated uniformly by rational functions with poles off K.

(3) The algebra A(K) consists of the functions in C(K) which are analytic on the interior of K.

Evidently $P(K) \subset R(K) \subset A(K)$.

Consider A in Definition 4.2.1. Endowed with the norm (2.1), A becomes a Banach algebra. Let $C_R(X)$ be the set of all real-valued elements in C(X), i.e. $C_R(X) = \text{Re } C(X)$. Furthermore, Re(A) will denote the set of real parts of elements in A.

4.2.2 Definition. The uniform algebra A is said to be Dirichlet on X if Re(A) is dense in $C_R(X)$ in the topology of C(X); A is said to be Dirichlet if $Re(A)|_{\partial K}$ is dense in $C_R(\partial K)$ in the norm topology of $C(\partial K)$.

Let D be a region in the plane **C** with the boundary ∂D consisting of a finite number of smooth closed curves. Let $u \in C_R(\partial D)$ be continuously differentiable on ∂D. Then the following Dirichlet problem,

$$\frac{\partial^2 v}{\partial x^2} + \frac{\partial^2 v}{\partial y^2} = 0, \tag{2.2}$$

$$u = v|\partial D, \tag{2.3}$$

has a unique solution v_0 which is called the harmonic extension of u and minimizes the Dirichlet integral

$$\iint_D \left[\left(\frac{\partial v}{\partial x}\right)^2 + \left(\frac{\partial v}{\partial y}\right)^2\right] dxdy$$

among all continuously differentiable functions v on D^- which coincide with u on ∂D. The existence of v_0 for the problem (2.2), (2.3) is called the Dirichlet principle.

4.2.3 Lemma. Let K be a compact subset of the complex plane. If $f \in C(K)$ and f extends to an analytic function in a neighborhood of K, then $f \in R(K)$.

Proof. Since

$$f(z) = \frac{1}{2\pi i} \int_\Gamma \frac{f(\zeta)}{\zeta - z} d\zeta, \tag{2.4}$$

where Γ is an appropriate contour surrounding K, the Riemann sums which approximate integral (2.4) are rational functions which approximate f uniformly on K.

4.2.4 Runge's Theorem. Let K be a compact subset of the plane with connected complement. Then every function analytic in a neighborhood of K can be approximated uniformly on K by polynomials in z.

Proof. Let U be the set of all $\lambda \in \mathbf{C}\backslash K$ such that $(z-\lambda)^{-1} \in P(K)$. If $|\lambda|$ is large enough, the series

$$(z-\lambda)^{-1} = \sum_{n=0}^{\infty} \frac{(-1)^{n+1}}{\lambda^{n+1}} z^n$$

converges uniformly on K, therefore $\lambda \in U$ for such λ, and U is thus nonempty.

Now suppose $\lambda_n \in U$ and $\lambda_n \to \lambda \in \mathbf{C}\backslash K$. Then $(z-\lambda_n)^{-1}$ tends to $(z-\lambda)^{-1}$ uniformly on K. Hence $\lambda \in U$, and U is a closed subset of $\mathbf{C}\backslash K$. Now suppose $\lambda_0 \in U$, so that $(z-\lambda_0)^{-n} \in P(K)$ for all $n \geq 1$. If $|\lambda-\lambda_0|$ is sufficiently small, the series

$$(z-\lambda)^{-1} = \sum_{n=0}^{\infty} \frac{(\lambda-\lambda_0)^n}{(z-\lambda_0)^{n+1}}$$

converges uniformly on K. Consequently, $\lambda \in U$ if $|\lambda-\lambda_0|$ is sufficiently small, so U is open. Since U is open, closed and nonempty in $\mathbf{C}\backslash K$, we have $U = \mathbf{C}\backslash K$.

4.2.5 Theorem (Walsh-Lebesgue). Let K be a compact subset of $\mathbf{C}$ with connected complement. Then every continuous real-valued function on ∂K can be approximated uniformly on ∂K by harmonic polynomials in x and y.

Proof. Let F be compact such that $K \subset \operatorname{Int} F$ and let $F = \cap_{n=1}^{\infty} K_n$, where K_n is compact, $K_{n+1} \subset \operatorname{Int} K_n$, ∂K_n consists of a finite number of smooth disjoint closed curves and $\mathbf{C}\backslash K_n$ is connected.

Let u be a real-valued continuously differentiable function on $\mathbf{C}$. By the Dirichlet principle, we may choose the harmonic extension u_n of $u|\partial K_n$ to K_n. Let v_n be the harmonic conjugate of u_n. Then $u_n + iv_n$ is analytic in a neighborhood of F. By the existence of a conformal mapping from a closed simply connected domain with smooth boundary to the closed unit disc and the uniform boundedness of $\{u_n\}$ [42, p. 33], we may assume that $\{v_n\}$ is uniformly bounded. Specifically, $\{u_n + iv_n\}$ is uniformly bounded on F, and hence there exists a subsequence, namely, $\{u_n + iv_n\}$ itself, converging uniformly on every compact subset of Int F and hence converging uniformly on K. By Runge's theorem, $u_n + iv_n$ can be approximated uniformly on K by polynomials in z. The real parts of these polynomials are harmonic polynomials in x and y and converge uniformly to u_n on ∂K. By the definition of $\{u_n\}$, $u_n \to u$ on ∂K. It follows that u, and hence any function in $C_R(\partial K)$, is uniformly approximable by harmonic polynomials on K.

4.2.6 Corollary. Let K be a compact subset of the plane. Then P(K) is a Dirichlet algebra on the outer boundary of K, where "outer" means the mutual boundary of K and the unbounded component of $\mathbf{C}\setminus K$.

Proof. If the complement $\mathbf{C}\setminus K$ is connected, then the conclusion is an easy consequence of Th. 4.2.5. Now suppose $\mathbf{C}\setminus K$ is not connected, and let $\hat{K}$ be the union of K and all the bounded components of $\mathbf{C}\setminus K$, Then $\hat{K}$ is compact, $\mathbf{C}\setminus\hat{K}$ is connected and $\partial\hat{K}$ is the outer boundary of K.

If f is a polynomial in z, then $\|f\|_K = \|f\|_{\hat{K}}$ by the maximum modulus principle. Hence every sequence of polynomials converging uniformly on K to $g \in P(K)$ will also converge uniformly to an extension $\hat{g} \in P(\hat{K})$ of g such that $\|\hat{g}\|_{\hat{K}} = \|g\|_K$. The isometric isomorphism $g \to \hat{g}$ allows us to identify the algebras P(K) and $P(\hat{K})$. Since $P(\hat{K})$ is Dirichlet by Th. 4.2.5, so is P(K).

4.2.7 Lemma. Let $\{K_n\}$ be a decreasing sequence of compact sets with $K = \cap_1^\infty K_n$. If $R(K_n)$ is Dirichlet for each n, then R(K) is also Dirichlet.

Proof. Let u be a real-valued continuously differentiable function in a neighborhood of K_1. for each n let u_n be the solution of the Dirichlet problem (2.2), (2.3), with D replaced by K_n, where without loss of generality we assume that ∂K_n consists of a system of smooth closed curves. Because $R(K_n)$ is Dirichlet, there is a function $u_n \in R(K_n)$ with

$$|u_n - \operatorname{Re} h_n| \leq 1/n \tag{2.5}$$

on ∂K_n. By the maximum modulus principle for harmonic functions, (2.5) holds throughout K_n. Hence $\{\operatorname{Re} h_n\}$ converges to u uniformly on ∂K, since we can prove that $u_n \to u$ uniformly on ∂K. Therefore, R(K) is a Dirichlet algebra.

The proof of the following lemma is omitted [43, p. 395].

4.2.8 Lemma. Let K be a compact subset of K such that $R(K)$ is a Dirichlet algebra. Let L be a compact subset of K such that the boundary of each component of $(\text{Int } K)\backslash L$ meets ∂K. Then $R(L)$ is also a Dirichlet algebra.

4.2.9 Definition. Let $G \subset \mathbf{C}$ be an open set. A subset $K \subset G$ is called dominating in G if

$$\|f\|_K = \|f\|_G$$

for all f bounded and analytic on G.

4.2.10 Theorem. Let K be a compact subset of the plane with the property that, for all nonempty open $G \subset \mathbf{C}$, the set $K \cap G$ is not dominating in G. Then $R(K) = C(K)$.

Proof. Let B be the collection of all subsets δ of $\mathbf{C}$ satisfying

(i) $\delta \in B$ is the union of K with some bounded components of $\mathbf{C}\backslash K$;

(ii) $R(\delta)$ is a Dirichlet algebra.

By Th. 4.2.5, $B \neq \emptyset$. Choose a sequence $\{L_j\}_1^\infty \subset B$ such that $L_1 \supset L_2 \supset \ldots$ and there is no $W \in B$ such that

$$m_2(W) < \inf\{ m_2(L_j): j = 1, 2, \ldots\},$$

where m_2 denotes planar Lebesgue measure. Let $L = \cap_1^\infty L_j$. Then $R(L)$ is a Dirichlet algebra by Lemma 4.2.7. Put $G = \text{Int } L$.

If $G \neq \emptyset$, then by hypothesis there is a bounded analytic function f on G with $\|f\|_{G\cap K} < \|f\|_G$. Let Q be a component of the open set $\{z \in G: |f(z)| > \|f\|_{G\cap K}\}$. By the maximum modulus principle, $\partial Q \cap \partial L \neq \emptyset$. Since $Q \cap K = \emptyset$, there is a unique component V of $\mathbf{C}\backslash K$ for which $V \supset Q$. Then $\partial V \cap \partial L \neq \emptyset$, hence $R(L\backslash V)$ is Dirichlet by Lemma 4.2.8. Since $m_2(L\backslash V) < m_2(L)$, we obtain a contradiction to the minimality of L. Hence $G = \emptyset$, so $L = K$. Because $R(K)$ is a Dirichlet algebra, $R(K) = C(K)$ by the Stone-Weierstrass theorem.

We need one more lemma whose proof is omitted [42, p. 47].

4.2.11 Lemma. Let K be a nowhere dense compact subset of C. If R(K) is a Dirichlet algebra on K, then R(K) = C(K).

3. Invariant Subspaces for Subdecomposable Operators

In this section we discuss the existence of invariant subspaces for subdecomposable operators. We begin with some results on weak* topologies.

Let X_1, X_2 be two Banach spaces. We say that a linear mapping S from X_1^* to X_2^* is continuous in the weak* topology if for every net $\{x_\alpha^*\}$ converging weak* to zero in X_1^*, $\{Sx_\alpha^*\}$ converges weak* to zero in X_2^*.

The following theorem offers an easy criterion for S to be weak* continuous.

4.3.1 Theorem. The linear mapping S from X_1^* to X_2^*, with X_1 separable, is continuous in the weak* topology if and only if for every sequence $\{x_n^*\}$ in X_1^* converging weak* to zero, the sequence $\{Sx_n^*\}$ converges weak* to zero in X_2^*.

Proof. Only the "if" part is nontrivial. Now suppose that $\{Sx_n^*\}$ converges weak* to zero whenever $\{x_n^*\}$ does so in X_1^*. For any $y \in X_2$, the mapping

$$\phi_y: x^* \to \langle y, Sx^* \rangle \quad (x^* \in X_1^*)$$

defines a linear functional on X_1^*, To prove that ϕ_y is weak* continuous, it suffices to show that ϕ_y has weak* closed kernel [27, Cor. V.3.11]. Let B^* be the norm closed unit ball of X_1^*. Since X_1 is separable, by [27, Th. V.5.1] B^* is metrizable and hence the weak* closure of $E_y = B^* \cap \ker(\phi_y)$ coincides with its weak* sequential closure. Now assume that $\{x_n^*\} \subset E_y$ converges weak* to x^*. Then $x^* \in B^*$. By the definition of E_y, one has $\langle y, Sx_n^* \rangle = 0$ for all n. Since $\langle y, Sx^* \rangle = \lim_{n\to\infty} \langle y, Sx_n^* \rangle = 0$, we see that $x^* \in E_y$ and so E_y, and in turn $\ker(\phi_y)$, is weak* closed by the Krein-Smulyan theorem. Thus ϕ_y is weak* continuous, so ϕ_y can be identified with a unique element $x \in X_1$. Therefore,

(3.1) $$\langle x,x^*\rangle = \langle y,Sx^*\rangle.$$

Let $\{x_\alpha^*\} \subset X_1^*$ be a net weak* convergent to zero. Then (3.1) implies

$$\lim_\alpha \langle y,Sx_\alpha^*\rangle = \lim_\alpha \langle x,x_\alpha^*\rangle = 0,$$

and $\{Sx_\alpha^*\}$ is thus weak* convergent to zero. This concludes the proof.

Recall that a set G in a vector space X is said to be balanced if $\lambda G \subset G$ for all $|\lambda| \le 1$. The absolutely convex hull of a set G is the smallest convex and balanced set containing G. Alternatively, it is the collection of all linear combinations $a_1x_1 + \dots + a_nx_n$ of elements $x_1,\dots,x_n$ in G with $\Sigma_i |a_i| \le 1$.

4.3.2 Lemma. Let X be a complex Banach space and let G be a subset of the closed unit ball B of X such that for all $x^* \in X^*$

$$\|x^*\| = \sup_{x \in G} |\langle x, x^*\rangle|.$$

Then the norm closure of the absolutely convex hull of G is the unit ball B.

Proof. Let E be the norm closure of the absolutely convex hull of G. Clearly $E \subset B$. Suppose $E \neq B$ and let $x_1 \in B\backslash E$. By the Hahn-Banach theorem there exists an element $x^* \in X^*$ and a real number c such that

$$\text{Re}\,\langle x,x^*\rangle \le c \text{ for all } x \in E,$$
$$\text{Re}\,\langle x_1,x^*\rangle > c.$$

Since E contains zero, the number c in nonnegative. For any $x \in E$, write $|\langle x,x^*\rangle| = \lambda\langle x, x^*\rangle$ with $|\lambda| = 1$. Then

$$|\langle x,x^*\rangle| = \langle \lambda x,x^*\rangle = \text{Re}\,\langle \lambda x,x^*\rangle \le c.$$

By the hypothesis on G, we have $\|x^*\| \le c$, contradicting the fact that

$$\|x^*\| \ge |\langle x_1,x^*\rangle| \ge \text{Re}\,\langle x_1,x^*\rangle > c.$$

Hence E = B.

Now we introduce some special spaces which are needed for the later discussion. Let Q be a nonempty bounded open connected subset of **C** and let $H^\infty(Q)$ be the algebra of all bounded analytic functions defined on Q with norm

$$(3.2) \qquad \|f\| = \sup_{\lambda \in Q} |f(\lambda)| \qquad (f \in H^\infty(Q)).$$

Then $H^\infty(Q)$ is a subspace of $L^\infty(Q)$. Let ${}^\perp[H^\infty(Q)]$ be the preannihilator of $H^\infty(Q)$ in $L^1(Q)$, i.e.

$${}^\perp[H^\infty(Q)] = \{f \in L^1(Q): \langle f,g\rangle = 0 \text{ for all } g \in H^\infty(Q)\}.$$

Then $\{{}^\perp[H^\infty(Q)]\}^\perp$ is weak* closed and hence it is the dual space of $M(Q) = L^1(Q)/{}^\perp[H^\infty(Q)]$, as we now show.

4.3.3 Theorem. $H^\infty(Q) = \{{}^\perp[H^\infty(Q)]\}^\perp$, that is, $H^\infty(Q)$ is the dual space of $M(Q)$.

Proof. It suffices to show that $H^\infty(Q)$ is weak* closed in $L^\infty(Q)$. In view of the Krein-Smulyan theorem, we only have to show that the norm-closed unit ball B of $H^\infty(Q)$ is weak* closed. Since $L^1(Q)$ is separable, $M(Q)$ is also, thus as we did in Th. 4.3.1 it suffices to show that B is sequentially weak* closed.

Let $\{f_n\}$ be a sequence in B and suppose that $f \in L^\infty(Q)$ such that $f_n \to f$ (weak*). Then by the definition of weak* convergence,

$$(3.3) \qquad \int_Q f_n g \, d\mu \to \int_Q fg \, d\mu$$

for every $g \in L^1(Q)$, where μ is planar Lebesgue measure. Fix $z \in Q$, let D be a disc $\{\zeta: |\zeta - z| \le r\}$ such that $D \subset Q$, and define

$$g(\zeta) = \begin{cases} (\zeta - z)^{-1} \exp[i \arg(\zeta - z)], & \text{f } \zeta \in D \\ 0, & \text{f } \zeta \notin D. \end{cases}$$

Then $g \in L^1(Q)$ and

(3.4) $$\int_D f_n g \, d\mu = \int_0^r \Big[\int_{|\zeta - z| = \rho} \frac{f_n(\zeta)}{\zeta - z} d\zeta \Big] d\rho = 2\pi i r f_n(z).$$

Hence (3.3) and (3.4) imply that $\{f_n(z)\}$ converges. Since $z \in Q$ is arbitrary and $\{f_n\}$ is bounded in $H^\infty(Q)$, it follows that $\{f_n\}$ converges uniformly on compact subsets of Q to a function $h \in H^\infty(Q)$. Thus

$$\int_Q hg \, d\mu = \int_Q fh \, d\mu$$

for every $g \in L^1(Q)$, hence $h = f$ almost everywhere $[\mu]$. This completes the proof.

In the sequel we assume that X is a not necessarily separable Banach space.

4.3.4 Definition. $T \in L(X)$ is said to be <u>unconditionally</u> decomposable if it is decomposable and there exists $A_T > 0$ such that for every finite system $F_1, F_2, \ldots, F_n$ of pairwise disjoint closed sets one has

(3.5) $$\sup\{\|\Sigma \beta_j x_j\| : |\beta_j| = 1 \text{ for all } j\} \le A_T \|\Sigma x_j\|$$

for all $x_j \in X(T, F_j)$ $(j = 1, 2, \ldots, n)$.

4.3.5 Examples. (a) Let $T \in L(X)$ be a spectral operator with spectral measure E. Then T is decomposable and for each E-measurable function f which is E-essentially bounded

$$f(T) = \int_{\sigma(T)} f(\lambda) \, dE(\lambda),$$

hence by [27, Th. IV.10.8]

(3.6) $$\|f(T)\| \le \{\|E\|(\sigma(T))\}(\text{ess sup } |f|),$$

where $\|E\|(\sigma(T))$ is the semi-variation of E [27, Def. IV.10.3].

Let $\{F_j\}$ be a finite system of disjoint closed sets where, without loss of generality,we take $F_j \subset \sigma(T)$ (all j) and suppose $y_j \in X(T, F_j)$. Let χ_j be the characteristic function of F_j and put

$$f(\lambda) = \Sigma_j \beta_j \chi_j(\lambda), \qquad |\beta_j| = 1, \text{ each } j.$$

Then for $x = \Sigma y_j$, one has from (3.6) that

$$\|f(T)x\| \le \{\|E\|(\sigma(T))\}\|x\|.$$

Since $f(T)x = \Sigma \beta_j \chi_j(T)y_j = \Sigma \beta_j y_j$, T is unconditionally decomposable.

(b) Let $G \subset C$ be open, connected and let X denote the Banach space of bounded m-times continuously differentiable functions defined on G with norm

$$(3.7) \qquad \|x\| = \max \sup_{\lambda \in G} \left| \frac{\partial^{m+k}}{\partial^m \eta \partial^k \zeta} x(\lambda) \right|,$$

where the max runs over all m, k with $0 \le m + k \le n$ and $\lambda = \eta + i\zeta$. Define T on X by the multiplier

$$(Tx)(\lambda) = \lambda x(\lambda), \qquad x \in X.$$

Then T is easily seen to be a generalized scalar operator, hence decomposable, with spectral manifolds $X(T, F) = \{x \in X: \text{ supp } x \subset F\}$. If $\{F_j\}$ is a finite set of disjoint closed sets in G, then for $x_j \in X(T, F_j)$ and $|\beta_j| = 1$, one has from (3.7) that

$$\|\Sigma \beta_j x_j\| = \|\Sigma x_j\|.$$

Therefore T is unconditionally decomposable with $A_T = 1$.

(c) Every boundedly decomposable operator T is unconditionally decomposable. By Th. 1.4.4, $T = S + Q$ where S is C-scalar ($C = C(\sigma(T))$) and Q is quasinilpotent, $QS = SQ$. Let U be the C-spectral function of S and let $\{F_j\}$ be a finite system of disjoint closed sets in $\sigma(T)$. Let $f \in C(\sigma(T))$ satisfy

$$\|f\| = 1 \text{ and } f(\lambda) = \beta_j \quad (|\beta_j| = 1)$$

for λ in some neighborhood of F_j. Then for $x_j \in X(T,F_j) = X(S,F_j)$,

$$U_f[\Sigma_j x_j] = \Sigma_j \beta_j x_j$$

and hence

$$\|\Sigma_j \beta_j x_j\| \le \|U_f\|\|\Sigma_j x_j\| \le \|U\| \|\Sigma_j x_j\|.$$

Thus T is unconditionally decomposable.

(d) This last example we might call "conditionally" decomposable. Let $X = C[0,1]$, and let $y_n(t) = (-1)^n t^n$ for $n = 0,1,\ldots,$ and $t \in [0,1]$. Define $Ty_n = (n+1)^{-1} y_n$ and extend T to all X by Weierstrass' theorem.. Then T is compact, hence decomposable, but it not unconditionally decomposable. In fact, $y_k \in X(T,\{1/(k+1)\})$ with $\{1/(k+1)\}$ disjoint $(k = 0,1,\ldots)$; but for $\beta_k = (-1)^k$ one has

$$\left\|\sum_{k=0}^{n} \beta_k y_k\right\| = \left\|\sum_{k=0}^{n} t^k\right\| = n + 1 \quad (n = 0, 1, \ldots.),$$

whereas for all n, $\|\Sigma_{0\le k\le n} y_k\| = \|\Sigma_{0\le k\le n}(-1)^k t^k\| \le 1$. Hence there is no A_T satisfying (3.5).

4.3.6 Lemma. Suppose $T \in L(X)$ is unconditionally decomposable. Let F be closed, G open such that $F \cap G = \emptyset$. Let $Z \supset X(T,F)$ be a closed subspace of X in which $X(T,F)$ has finite codimension. Then for each $\varepsilon > 0$, there exists a closed $\delta \subset G$ such that if $z \in Z$ and $\sigma(x,T) \subset G\backslash\delta$ then

$$\|x\| \le (2A_T + \varepsilon)\|x + z\|.$$

Proof. Let Y be a finite dimensional subspace such that $Z = Y \oplus X(T,F)$. If the lemma fails, then we can find sequences $\{x_n\} \subset X$, $\{y_n\} \subset Y$ and $\{z_n\} \subset Z$ such that

$$(3.8)\qquad \left\{\begin{array}{l} \|x_n\| = 1 \text{ for each } n \\ G \text{ contains } \sigma(x_n,T) \text{ for each } n \text{ and } \sigma(x_n, T) \text{ are pairwise disjoint} \\ y_n \to y_0 \text{ and} \\ (2A_T + \varepsilon)\|x_n + y_n + z_n\| < 1 \text{ for some } \varepsilon > 0. \end{array}\right\}$$

Set $u_{kn} = x_k - z_n + z_k$. Then

$$2 \le \|x_n - u_{kn}\| + \|x_n + u_{kn}\|$$

$$\le (1 + A_T)\|x_n - u_{kn}\|$$

$$= (1 + A_T)\|x_n + z_n - x_k - z_k\|$$

$$\leq (1 + A_T)(\|x_n + y_n + z_n\| + \|x_k + y_k + z_k\| + \|y_n - y_k\|)$$

$$\leq (1 + A_T)\left\{\frac{2}{2A_T + \varepsilon} + \|y_n - y_k\|\right\}.$$

Therefore

$$\|y_n - y_k\| \geq \frac{2}{1 + A_T} - \frac{2}{2A_T + \varepsilon},$$

but this contradicts the convergence $y_n \to y_0$ of (3.8).

The following lemma will be useful; we sketch the proof.

4.3.7 Lemma. Let $T \in L(X)$ be decomposable. If G is open and $\lambda \in \sigma(T) \cap G$, then $\lambda \notin \sigma(T/X(T,G^-))$.

Proof. Let H be a second open set with $H \cup G = \mathbf{C}$, $\lambda \notin H$, and let V be a third open set such that $\lambda \in V^- \subset G$ and $\{V,H\}$ covers $\mathbf{C}$. Let $x = x_1 + x_2$ with $\sigma(x_1,T) \subset V$, $\sigma(x_2,T) \subset H$. It follows that $(\lambda - T)/X(T,G^-)$ is surjective. By the proof of Th. 1.4.2 [(9) $\Rightarrow$ (10)] $T/X(T,G^-)$ has SVEP, hence the conclusion follows by Lemma 1.2.5.

4.3.8 Corollary. Suppose $T \in L(X)$ is decomposable, $F \subset \mathbf{C}$ is closed and $\mu \in \operatorname{Int} F \cap \sigma(T)$. Let $\{x_n\}$ be a sequence of unit vectors in X with $\|(\mu - T)x_n\| \to 0$. Then $\hat{x}_n \to 0$ where $\hat{x}_n$ is the coset $x_n + X(T,F)$.

Proof. By Lemma 4.3.7 we have $\mu \in \rho(T/X(T,F))$, hence $\|(\mu - T)\hat{x}_n\| \to 0$ implies $\|\hat{x}_n\| \to 0$.

Throughout the remainder of this section, $T \in L(X)$ will denote a fixed unconditionally decomposable operator and Y is a T-invariant subspace. Let $Q \subset \mathbf{C}$ be open and let F_Q be the collection of closed sets $\delta \subset \mathbf{C}$ satisfying $\operatorname{Int} \delta \cup Q = \mathbf{C}$. For given $\delta \in F_Q$, the annihilator $X(T,\delta)^\perp$ is isometrically isomorphic to the dual of $X^\delta = X/X(T,\delta)$. To simplify the notation we will view them as the same. Thus each $x^* \in X(T,\delta)^\perp$ defines a bounded linear functional $x^* = x_\delta^*$ on the space X^δ. Let T^δ be the operator induced by T on X^δ. For every pair $x \in X$, $x^* \in X(T,\delta)^\perp$, define the bounded linear functional $x \otimes_\delta x^*$

on $H^\infty(Q)$ by the equality

$$(3.9) \qquad (x \otimes_\delta x^*)(f) = x_\delta^*(f(T^\delta)x^\delta) \qquad (f \in H^\infty(Q)),$$

where $f(T^\delta)$ is defined by the Riesz-Dunford functional calculus, whose validity is assured by the inclusion $\sigma(T^\delta) \subset C\setminus \text{Int } \delta \subset Q$ (Lemma 4.3.7).

It is easy to see that if $x^* \in X(T,\delta)^\perp$ then for any $\sigma \in F_Q$ such that $\sigma \subset \delta$, we have $x^* \in X(T,\sigma)^\perp$ and

$$x \otimes_\sigma x^* = x \otimes_\delta x^*.$$

We denote point evaluation by ε_μ, i.e. if $f \in H^\infty(Q)$ then $\varepsilon_\mu(f) = f(\mu)$ for $\mu \in Q$.

4.3.9 Lemma. For $\delta \in F_Q$, $x \in X$ and $x^* \in X(T,\delta)^\perp$, we have $x \otimes_\delta x^* \in M(Q)$. Moreover, if $\{x_n\}$ is a bounded sequence in X and $\lim_n \|(T-\mu)x_n\| = 0$ for some $\mu \notin \sigma(T)\setminus \text{Int } \delta$, then

$$(3.10) \qquad \lim_{n\to\infty} \|x^*(x_n)\varepsilon_\mu - x_n \otimes_\delta x^*\| = 0$$

uniformly on bounded sets in $X(T,\delta)^\perp$.

Proof. Let $\{f_n\}$ be a sequence in $H^\infty(Q)$ converging weak* to zero. By the proof of Th. 4.3.3, $f_n \to 0$ uniformly on compact sets in Q. Since $\sigma(T^\delta) \subset Q$ by the remark above, $f_n(T^\delta) \to 0$ in the uniform operator topology of $L(X^\delta)$. Hence $(x \otimes_\delta x^*)(f_n) \to 0$ by (3.9). This proves that $x \otimes_\delta x^* \in M(Q)$ by Th. 4.3.1.

To prove (3.10), let $f \in H^\infty(Q)$ and for $\mu \notin \sigma(T)\setminus \text{Int } \delta$, define $g_\mu \in H^\infty(Q)$ by

$$f(\lambda) - f(\mu) = g_\mu(\lambda)(\lambda - \mu) \quad (\lambda \in Q).$$

Since $\sigma(T^\delta) \subset \sigma(T)\setminus \text{Int } \delta$ and $\mu \notin \sigma(T)\setminus \text{Int } \delta$, we have $\|g_\mu(T^\delta)\| \le M\|f(T^\delta)\|$ for some $M > 0$. Hence

$$\begin{aligned}
|x^*(x_n)\varepsilon_\mu(f) - x_n\otimes_\delta x^*)(f)| &= |x^*(x_n)f(\mu) - x^*(f(T^\delta)x_n{}^\delta)| \\
&= |x^*(g_\mu(T^\delta)(\mu - T^\delta)x_n{}^\delta)| \le \|x^*\|\,\|g_\mu(T^\delta)\|\,\|(\mu - T)x_n\| \\
&\le M\|x^*\|\,\|f(T^\delta)\|\,\|(\mu - T)x_n\| \le M'\|x^*\|\,\|f\|\,\|(\mu - T)x_n\|,
\end{aligned}$$

where $M' > 0$ is another constant arising from the functional calculus. Since the last expression tends to zero an $n \to \infty$, (3.10) is proved.

4.3.10 Lemma. Let $\delta \in F_Q$, $y \in Y$ and $x^* \in X(T,\delta)^\perp$ be given. Then for any $\varepsilon > 0$, there exist two subspaces $Y' \subset Y$ and $X_0{}^* \subset X(T,\delta)^\perp$ such that $\dim(Y/Y') < \infty$, $\dim(X(T,\delta)^\perp/X_0{}^*) < \infty$, $X_0{}^*$ is weak* closed and

$$\|y' \otimes_\delta x^*\| < \varepsilon\|y'\| \quad (\text{all } y' \in Y'), \tag{3.11}$$

$$\|y \otimes_\delta x_0{}^*\| < \varepsilon\|x_0{}^*\| \quad (\text{all } x_0{}^* \in X_0{}^*). \tag{3.12}$$

Proof. Set

$$M = \{f(T^\delta)y^\delta\colon f \in H^\infty(Q),\ \|f\| = 1\} \quad (\, y^\delta = y + X(T,\delta)),$$
$$M' = \{f(T^\delta)^*x_\delta{}^*\colon f \in H^\infty(Q),\ \|f\| = 1\}.$$

Since the unit ball in $H^\infty(Q)$ is weak* compact, it follows that M, M' are compact in $X/X(T,\delta)$, $X(T,\delta)^\perp$, resp. Thus we can find two subspaces $Z \subset X$, $Z' \subset X^*$ such that

$$Z \supset X(T,\delta), \qquad \dim(Z/X(T,\delta)) < \infty,$$
$$Z' \subset X(T,\delta)^\perp, \qquad \dim Z' < \infty$$

and such that

$$\begin{aligned}
&\operatorname{dist}(x, Z^\delta) < \varepsilon \text{ for all } x \in M, \\
&\operatorname{dist}(z_\delta{}^*, Z') < \varepsilon \text{ for all } z_\delta{}^* \in M',
\end{aligned} \tag{3.13}$$

where $Z^\delta = Z/X(T,\delta)$. Taking

$$Y' = \{y \in Y: \ z^*(y) = 0 \text{ for all } z^* \in Z'\}$$
$$X_0^* = Z^\perp,$$

we obtain the required subspaces. In fact, for $f \in H^\infty(Q)$, $\|f\| = 1$ and $y' \in Y'$, one has

$$|(y' \otimes_\delta x^*)(f)| = |x_\delta^* f(T^\delta) y'^\delta| = |(f(T^\delta)^* x_\delta^*)(y'^\delta)|$$
$$= |(f(T^\delta)^* x_\delta^* - z^*)(y'^\delta)|$$

for some $z^* \in Z'$. By (3.13) we may choose $z^* \in Z'$ such that

$$\|f(T^\delta)^* x_\delta^* - z^*\| < \varepsilon,$$

and hence $\|y' \otimes_\delta x^*\| < \varepsilon \|y'\|$. This proves (3.11), and in a similar way we can prove (3.12).

4.3.11 Lemma. Let V be an m-dimensional space with a basis $\{e_k\}_1^m$ of unit vectors such that

$$\Big\| \sum_{k=1}^m \beta_k \lambda_k e_k \Big\| \le a \Big\| \sum_{k=1}^m \lambda_k e_k \Big\|$$

for some fixed constant $a \ge 1$ and any $\lambda_k, \beta_k \in C$ with $|\beta_k| = 1$. Let $\{c_k\}_1^m \subset C$ be such that $\Sigma\, |c_k| = 1$. Then there exist $\{\mu_k\}_1^m$ and $y^* \in V^*$ such that

$$\mu_k y^*(e_k) = c_k, \quad \|y^*\| \le a \text{ and } \Big\| \sum_{k=1}^m \mu_k e_k \Big\| \le 1.$$

Proof. Assume first that $a = 1$ and the norm of V is strictly convex and smooth. Set

$$S = \left\{ \sum_{k=1}^m a_k e_k: \ a_k \ge 0, \ \Big\| \sum_{k=1}^m a_k e_k \Big\| = 1 \right\},$$

Then for any $x \in S$ there exists a unique $y_x^* \in V^*$ such that $\|y_x^*\| = \|y_x^*(x)\| = 1$. Because the norm of V is invariant under rotations on the axis determined by the basis $\{e_k\}$, we obtain

$$y_x^*(e_k) \geq 0 \text{ for all } x \in S \quad (1 \leq k \leq m).$$

Let $\{e_k^*\}$ be biorthogonal to $\{e_k\}$, i.e. $\|e_k^*\| = e_k^*(e_k) = 1$ and $e_k^*(e_j) = 0$ for $k \neq j$. Then we have

$$\left\|\sum_{k=1}^{m} \beta_k \lambda_k e_k^*\right\| = \left\|\sum_{k=1}^{m} \lambda_k e_k^*\right\|$$

for any $\lambda_k \in \mathbf{C}$, $|\beta_k| = 1$ and the functions

$$\lambda_j \to \left\|\sum_{k=1}^{m} \lambda_k e_k\right\|, \qquad \lambda_j \to \left\|\sum_{k=1}^{m} \lambda_k e_k^*\right\|$$

increase with $|\lambda_j|$ if $|\lambda_k|$ remain fixed for $k \neq j$. This, together with the strict convexity of the norm in both V and V^*, implies the following equivalence: $y_x^*(e_k) = 0$ if and only if $e_k^*(x) = 0$.

Now suppose that our lemma holds for the dimension of V not exceeding $m-1$. For any $t \in [0,1]$ put

$$e_k(t) = \begin{cases} e_k, & \text{if } k < m-1 \\ te_{m-1} + f(t)e_m, & \text{if } k = m-1 \end{cases}$$

such that $\|e_{m-1}(t)\| = 1$ and $f: [0,1] \to [0,1]$ is continuous with $f(0) = 1$, $f(1) = 0$. By the inductive hypothesis on dimension, the Hahn-Banach theorem and the invariance under rotations on axes for $0 \leq t \leq 1$ let us find $x(t) \in S$ with $x(t) = \Sigma_1^{m-1} a_k(t)e_k(t)$ and $y_t^* = (y_{x(t)})^* \in V^*$ such that

$$\|x(t)\| = \|y_t^*\|$$
$$a_k(t)y_t^*(e_k(t)) = |c_k| \text{ for } k < m-1, \text{ and}$$
$$a_{m-1}(t)y_t^*(e_{m-1}(t)) = |c_{m-1}| + |c_m|.$$

The vector-valued functions $x(t)$ and y_t^* are uniquely determined. For let $\hat{x}(t)$, $\hat{y}_t^*$ be another solution of the previous equations where

$$\hat{x}(t) = \sum_{k=1}^{m-1} \hat{a}_k(t)e_k(t);$$

then we have

$$\hat{y}_t^*(x(t)) + y_t^*(\hat{x}(t))$$

$$= \sum_{k=1}^{m-1} \left[a_k(t)\hat{y}_t^*(e_k(t)) + \hat{a}_k(t)y_t^*(e_k(t)) \right] \leq 2.$$

Put

$$r_k(t) = \frac{\hat{a}_k(t)}{a_k(t)} = \frac{y_t^*(e_k(t))}{\hat{y}_t^*(e_k(t))} \quad (\text{if } a_k(t) \neq 0).$$

Since $|c_k| = a_k(t)y_t^*(e_k(t)) = \hat{a}_k(t)\hat{y}_t^*(e_k(t))$, we have

$$\hat{y}_t^*(\hat{x}(t)) + y_t^*(\hat{x}(t)) =$$

$$\sum_{a_k(t) \neq 0} \left[r_k(t) + \frac{1}{r_k(t)} \right] a_k(t)y_t^*(e_k(t)) \leq 2$$

which is possible only if $r_k(t) = 1$, $x(t) = \hat{x}(t)$ and $y_t^* = \hat{y}_t^*$. This implies also the continuity of the functions $a_k(t)$ and y_t^*. Specifically, the function $y_t^*(ta_{m-1}(t)e_{m-1})$ is continuous. Since $y_0^*(0) = 0$ and $y_2^*(a_{m-1}(1)e_{m-1}) = a_{m-1}(1)y_1^*(e_{m-1})) = |c_{m-1}| + |c_m|$, we can find $s \in [0,1]$ such that $y_s^*(a_{m-1}(s)e_{m-1}) = |c_{m-1}|$. Write $c_k = \beta_k|c_k|$ with $|\beta_k| = 1$ and set

$$\begin{aligned} \mu_k &= \beta_k a_k(s), \quad \text{if } k < m-1, \\ \mu_{m-1} &= s\beta_{m-1}a_{m-1}(s), \\ \mu_m &= \beta_{m-1}a_{m-1}(s)f(s), \\ y^* &= y_s^*. \end{aligned}$$

Then we have a solution satisfying the conclusions in the lemma. Since the lemma is trivial for the case dim $V = 1$, the proof is completed by induction.

Now suppose that $a = 1$ and the norm in V is not strictly convex nor smooth. Then for any $0 < \varepsilon < 1$, we introduce a new norm in V^*:

(3.14)
$$\left\|\sum_{k=1}^{m}\lambda_k e_k^*\right\|_{\varepsilon,1} \le (1-\varepsilon)\left\|\sum_{k=1}^{m}\lambda_k e_k^*\right\| + \varepsilon\left(\sum_{k=1}^{m}|\lambda_k|^2\right)^{1/2}$$

and let $\|\ \|_\varepsilon$ demote the norm in V determined by duality with (3.14). Furthermore, we define one more norm in V by the equality

$$\left\|\sum_{k=1}^{m}\lambda_k e_k\right\|_\varepsilon = \left\|\sum_{k=1}^{m}\lambda_k e_k\right\|_{\varepsilon,1} + \varepsilon\left(\sum_{k=1}^{m}|\lambda_k|^2\right)^{1/2}.$$

Since $\|\ \|_{\varepsilon,1}$ in V^* is strictly convex, the norm $\|\ \|_{\varepsilon,1}$ is smooth. Then the norm $\|\ \|_\varepsilon$ in V is both smooth and strictly convex as well as invariant under rotations of axes. Thus we can find a solutions x_ε, x_ε^* as above for each $\varepsilon > 0$. Let $\varepsilon \to 0$. Then there exist solutions $\{m_k\}_1^m$ and y^* satisfying the conditions in the lemma.

Finally, assume that $a > 1$. Let us introduce the norm

$$\left\|\sum_{k=1}^{m}\lambda_k e_k\right\|_0 = \sup\left\{\left\|\sum_{k=1}^{m}\beta_k\lambda_k e_k\right\| : |\beta_k| = 1\right\}.$$

From the previous argument, we may assume that $x = \Sigma_1^m \mu_k e_k$, $y^* \in V^*$, $\|x\|_0 \le 1$, $\|y^*\|_0 \le 1$, is a solution. Then $\|x\| \le 1$, $\|y^*\| \le 1$, so the assumption $a = 1$ is not essential. The lemma is thus proved.

4.3.12 Lemma. Let T be unconditionally decomposable. Suppose that the restriction $A = T|Y$ satisfies conditions

(a) $\sigma(A)$ is connected but not a singleton;

(b) $\sigma_p(A) = \sigma_p(A^*) = \varnothing$.

Suppose also that $\delta \in F_Q$, $y \in Y$ and $x^* \in X(T,\delta)^\perp$, as in Lemma 4.3.10, and that $\{\mu_k\}_1^m \subset \sigma(A) \cap Q \cap \operatorname{Int}\delta$ and $\{c_k\}_1^m \subset \mathbf{C}$ are given. Then for every $\varepsilon > 0$, there exist $\sigma \in F_Q$ with $\sigma \subset \delta$, $u \in Y$ and $u^* \in X(T,\delta)^\perp$ such that

(i) $\|y - u\| \le \left(\sum_{k=1}^{m}|c_k|\right)^{1/2}$;

$$\text{(ii)} \quad \|y^* - u^*\| \le 4A_T^2\left(\sum_{k=1}^{m}|c_k|\right)^{1/2};$$

$$\text{(iii)} \quad \|u \otimes_\sigma u^* - y \otimes_\delta x^* - \Sigma_1^m c_k\varepsilon_{\mu_k}\| < \varepsilon.$$

Proof. Without loss of generality we may suppose $\Sigma\,|c_k| = 1$. Fix $r > 0$ such that

$$D_k = \{\lambda \in \mathbf{C}:\ 0 < |\lambda - \mu_k| < r\} \subset \operatorname{Int}\delta \cap Q,$$

and put

$$G_0 = \cup_{k=1}^{m} D_k, \qquad \sigma = \delta \backslash G_0.$$

We may further assume that r is sufficiently small so that the D_k^- are pairwise disjoint and

$$\left\|\sum_{k=1}^{m} c_k(\varepsilon_{\mu_k} - \varepsilon_{\zeta_k})\right\| < \frac{\varepsilon}{4} \quad \text{for all } \zeta_k \in D_k^-.$$

Applying Lemma 4.3.10, we can find two subspaces $Y' \subset Y$ and $X_0^* \subset X(T,\sigma)^\perp$ such that $\dim(Y/Y') < \infty$ and $\dim(X(T,\sigma)^\perp/X_0^*) < \infty$ and

$$\|y' \otimes_\delta x^*\| < \frac{\varepsilon}{4}\|y'\| \quad (\text{all } y' \in Y'),$$

$$\|y \otimes_\delta x_0^*\| < \frac{\varepsilon}{4A_T^2}\|x_0^*\| \quad (\text{all } x_0^* \in X_0^*).$$

Set

$$Z = \{z \in X:\ x_0^*(z) = 0 \ \text{ for all } x_0^* \in X_0^*\}.$$

Then $Z \supset X(T,\sigma)$ and $\dim Z^\sigma = \dim(Z/X(T,\sigma)) < \infty$. By Lemma 4.3.6 one can find a closed set $\delta_0 \subset G_0$ such that

(3.15) $\|x\| \leq A_T\|x + z\|, \; z \in Z$ and $\sigma(x,T) \subset G_0\backslash\delta_0$.

The assumptions on A imply that we can pick

$$\zeta_k \in [(\sigma(A)\cap D_k)\backslash\delta_0], \qquad y_k(\eta) \in Y' \quad (\eta > 0),$$

such that $\|y_k(\eta)\| = 1$ and $\|(T - \zeta_k)y_k(\eta)\| < \eta$. Taking η sufficiently small and applying Lemma 4.3.6, we may suppose that dist($y_k(\eta)$, $X(T,D_k^-)$) is so small that for $|\beta_k| = 1$ we have

$$\Big\|\sum_{k=1}^{m} \beta_k\alpha_k y_k(\eta)\Big\| \leq \frac{4}{3}A_T\Big\|\sum_{k=1}^{m} \alpha_k y_k(\eta)\Big\|.$$

Let Y_η denote the subspace spanned by $\{y_k(\eta)\}$. Then Lemma 4.3.11 implies the existence of $y_\eta^* \in Y_\eta^*$ and $\{\alpha_k(\eta)\} \subset C$ such that

$$\Big\|\sum_{k=1}^{m} \alpha_k(\eta)y_k(\eta)\Big\| \leq 1,$$

$$\|y_\eta^*\| \leq \frac{3}{4}A_T, \quad \text{and}$$

$$\alpha_k(\eta)y_k^*(\eta)(y_k(\eta)) = \alpha_k.$$

Let $\delta' \subset G_0\backslash\delta_0$ be a closed set such that $\zeta_k \in \operatorname{Int}\delta'$. Applying Lemma 4.3.6 again, we have

$$0 = \overline{\lim}_{\eta>0}\{\operatorname{dist}(y_\eta,X(T,\delta')\colon\; y_\eta \in Y_\eta, \|y_\eta\| \leq 1\}.$$

By (3.15) we may thus suppose that

(3.16) $\|y_\eta\| \leq 3A_T\|y_\eta + z\|$ for $z \in Z$ and all $y_\eta \in Y_\eta$.

Define $z_\eta^*(y_\eta + z) = y_\eta^*(y_\eta)$ for $y_\eta \in Y_\eta$ and $z \in Z$. Then (3.16) implies that $z_\eta^* \in (Y_\eta + Z)^*$. Extend z_η^* to an element $v_\eta^* \in X^*$ by the Hahn-Banach theorem. Then it is easy to see that

$$\|z_\eta^*\| \leq 4A_T^2 \quad \text{and} \quad z_\eta^*(z) = 0,$$

and hence

(3.17) $$v_\eta^* \in X_0^* \quad \text{and} \quad \|v_\eta^*\| \le 4A_T^2.$$

Let $v(\eta) = \Sigma_1^m \alpha_k(\eta) y_k(\eta)$. Then by Lemma 4.3.9 we may suppose

$$\left\|\sum_{k=1}^m c_k \varepsilon_{\zeta_k} - v(\eta) \otimes_\sigma v_\eta^*\right\| < \frac{\varepsilon}{4}.$$

Putting $u = y + v(\eta)$ and $u^* = x^* + v_\eta^*$, we obtain

$$\left\|u \otimes_\sigma u^* - y \otimes_\delta x^* - \sum_{k=1}^m c_k \varepsilon_{\mu_k}\right\|$$

$$\le \left\|v(\eta) \otimes_\delta x^*\right\| + \left\|y \otimes_\delta v_\eta^*\right\|$$

$$+ \left\|v(\eta) \otimes_\sigma v_\eta^* - \sum_{k=1}^m c_k \varepsilon_{\zeta_k}\right\| + \left\|\sum_{k=1}^m c_k (\varepsilon_{\mu_k} - \varepsilon_{\zeta_k})\right\|$$

$$< \varepsilon.$$

This proves (iii); (i) is clear, and (ii) follows from (3.17).

4.3.13 Lemma. Let A be the restriction to Y of the unconditionally decomposable operator T. Suppose $\sigma(A) \cap Q$ is dominating in Q, $\sigma(A)$ is connected and $\sigma_p(A) = \sigma_p(A^*) = \emptyset$, and let $\mu \in Q$ and $0 < b < 1$ be given. Then there exist sequences $\{y_n\} \subset Y$ and $\{\delta_n\} \subset F_Q$ and $x_n^* \in X(T,\delta_n)^\perp$ (each n) such that

$$\|y_{n+1} - y_n\| < b^{n-1},$$

$$\|x_{n+1}^* - x_n^*\| < 4A_T b^{n-1} \quad \text{and}$$

$$\|\varepsilon_\mu - y_n \otimes x_n^*\| < b^{2(n-1)}.$$

Proof. We assume that y_j, δ_j and x_j^* have been determined up through $j \le n$ with $y_0 = 0$, $\delta_0 = C$ and $x_0^* = 0$. By the maximum modulus principle for analytic functions, $\delta' = \sigma(A) \cap Q \cap \operatorname{Int} \delta_n$ is clearly dominating in Q. It follows from Lemma 4.3.2 that there exist $\{c_k\}_1^m \subset C$, $\{\mu_k\}_1^m \subset \delta'$ such that

$$\sum_{k=1}^{m} |c_k| < b^{2(n-1)}$$

and

$$\|\varepsilon_\mu - y_n \otimes x_\eta^* - \sum_{k=1}^{m} c_k \varepsilon_{\mu_k}\| < b^{2n}.$$

Then by Lemma 4.3.12 we can determine y_{n+1}, $\delta_{n+1} = \sigma$ and x_{n+1}^*. This completes the proof.

4.3.14 Lemma. Suppose A is the restriction of the unconditionally decomposable operator T. If $\sigma(A) \cap Q$ is dominating in Q, then there exists a proper subspace in Y invariant under $(\lambda - A)^{-1}$ for $\lambda \notin \sigma(A) \cup Q^-$, and hence Y is invariant under A

Proof. Assume that A has no proper invariant subspace. Consequently, $\sigma(A)$ is connected and $\sigma_p(A) = \sigma_p(A^*) = \emptyset$. Let $\mu \in Q$ and $0 < b < 1$ be given, and let $\{y_n\} \subset Y$, $\{\delta_n\} \subset F_Q$ and $x_n^* \in X(T,\delta_n)^\perp$ be determined as in Lemma 4.3.13. Putting

$$y' = \lim_{n\to\infty} y_n, \quad y = (A - \mu)y', \text{ and } x^* = \lim_{n\to\infty} x_n^*,$$

we have

$$1 = \varepsilon_\mu(1) = \lim_{n\to\infty} x_n^*(y_n) = x^*(y').$$

Thus $y' \neq 0$ and $x^* \neq 0$. Let Y_0 be the subspace generated by $(\lambda - A)^{-1}y$ for $\lambda \notin \sigma(A) \cup Q^-$. Let f be a rational function with poles off $\sigma(A) \cup Q^-$ and let $g_\mu(\lambda) = (\lambda)(\lambda - \mu)$. Then f, g_μ are in $H^\infty(Q)$, and by (3.9)

$$x^*(f(A)y) = x^*(g_\mu(A)y') = \lim_{n\to\infty} x_n^*(g_\mu(A)y_n) = g_\mu(\mu) = 0.$$

Consequently, $y' \notin Y_0$, and so $y \in Y_0$ by the definition of Y_0. Hence Y_0 is a proper subspace. Evidently, Y_0 is invariant under $(\lambda - A)^{-1}$ for all $\lambda \notin \sigma(A) \cup Q^-$ and hence is invariant under A. This contradicts our assumption, and the lemma is proved.

4.3.15 Theorem. Let $A = T|X$ be the restriction of an unconditionally decomposable operator T. If $\mathrm{Int}\,\sigma(A) \neq \emptyset$, then there exists a proper subspace Y_0 of X such that Y_0 is invariant under $(\lambda - A)^{-1}$ for all $\lambda \notin \sigma(A)$ and hence Y_0 is invariant under A.

Proof. Let Q be a connected component of $\mathrm{Int}\,\sigma(A)$. Then Lemma 4.3.14 is applicable.

4.3.16 Theorem. Let A be the restriction of the unconditionally decomposable operator T. If G is any simply connected set such that $R((\sigma(A) \cup G)^-) \neq C((\sigma(A) \cup G)^-)$, then there exists a proper subspace Y_0 of X such that Y_0 is invariant under $(\lambda - A)^{-1}$ for $\lambda \notin \sigma(A) \cup G^-$ and hence it is invariant under A.

Proof. By Th. 4.3.15, we may assume that $\mathrm{Int}\,\sigma(A)$ is empty. Set $K = (\sigma(A) \cap G)^-$. Then $K \neq \emptyset$ but $\mathrm{Int}\,K = \emptyset$. By Lemma 4.2.11, R(K) is not a Dirichlet algebra. The following is similar to the proof of Th. 4.2.5. Let F denote the collection of all compact sets δ in **C** satisfying

(i) δ is the union of K with some bounded components of $\mathbf{C}\backslash K$;

(ii) $R(\delta)$ is a Dirichlet algebra.

It follows from Th. 4.2.5 that F is nonempty. Let us order F by inclusion. If $\{\delta_\alpha\} \subset F$ is totally ordered, then by Lemma 4.2.7, $\cap\delta_\alpha \in F$. By Zorn's lemma we can

find a minimal $\delta_0 \in F$. Let Ω denote a component of $\mathrm{Int}\,\delta_0$ with $\Omega \neq \emptyset$. We shall prove that $\sigma(A) \cap \Omega$ is dominating in Ω, or equivalently, $K \cap \Omega$ is dominating in Ω. In fact, in the contrary case we can find $f \in H^\infty(\Omega)$ such that

$$\sup_{\lambda \in K \cap \Omega} |f(\lambda)| < \sup_{\lambda \in \Omega\backslash K} |f(\lambda)| = \|f\|,$$

and hence $\Omega\backslash K$ has a component Ω_0 such that $\partial\Omega_0 \cap \partial\Omega \neq \emptyset$. It is easy to check that Ω_0 is a component of $\mathbf{C}\backslash K$, thus $L = \delta_0\backslash\Omega_0$ is a member of F and hence $R(L)$ is a Dirichlet algebra by Lemma 4.2.8, contradicting the minimality of δ_0. Therefore $\sigma(A) \cap \Omega$ is dominating in Ω and hence $\Omega \subset \operatorname{Int} G^-$. Application of Lemma 4.3.14 concludes the proof.

4. Hyponormal Operators

Let H be a complex Hilbert space, and let L(H) be the algebra of all bounded linear operators on H. An operator $T \in L(H)$ is called hyponormal if $TT^* \leq T^*T$, or equivalently, $\|T^*h\| \leq \|Th\|$ for all $h \in H$.

Referring to Chapter 1, Def. 1.4.1(d), we say that $S \in L(H)$ is generalized scalar of order m if there is a homomorphism $U: C^m(\mathbf{C}) \to L(H)$ with $U(1) = I$ and $U(id) = S$, where $C^m(\mathbf{C})$ is the algebra of continuously differentiable functions through order m ($m = 0, 1,\ldots$); U is called a spectral distribution of S of order m. A restriction $A = S|M$ to an invariant subspace M is called subscalar. It is our purpose in this section to prove that hyponormal operators are subscalar of order 2. To do this we need some vector-valued function spaces. Let μ denote planar Lebesgue measure, and let $Q \subset \mathbf{C}$ be a bounded open set. We shall denote by $L^2(Q;H)$ the Hilbert space of H-valued measurable functions satisfying

$$\|f\|_{2,Q} = \left(\int_Q \|f(z)\|^2 \, d\mu\right)^{1/2} < +\infty.$$

Let $A^2(Q;H)$ be the set of all those $f \in L^2(Q;H)$ which are analytic in Q.

Next let $L^\infty(Q;H)$ denote the Banach space of μ-essentially bounded H-valued measurable functions on Q. As Q is bounded, there is a natural continuous imbedding $L^\infty(Q;H) \to L^2(Q;H)$.

Finally, let $C^m(Q^-;H)$ denote the space of continuously differentiable functions on Q^- of order $m \geq 0$. A simple calculation shows that

$$\int_{|\zeta| \leq 1} \frac{d\mu}{|\zeta|} = 2\pi,$$

hence for fixed $z \in \mathbf{C}$ the function $|\zeta - z|^{-1}$ is integrable $[\mu]$ on every bounded domain with z in its interior.

4.4.1 Theorem (Cauchy-Pompeiu formula). If D is an open disc, then every $f \in C^1(D^-;H)$ can be evaluated according to the formula

$$f(z) = \frac{1}{2\pi i}\int_{\partial D}\frac{f(\zeta)}{\zeta - z}d\zeta - \frac{1}{\pi}\int_{D}\frac{\bar{\partial} f(\zeta)}{\zeta - z}d\mu(\zeta), \tag{4.1}$$

where $z \in D$ and $\bar{\partial}$ represents the derivative operator introduced on Chap I.4..

Proof. Fix $z \in D$ and let $\Delta_r = \{w: |w - z| \le r\}$ be small enough so that $\Delta_r \subset D$. Applying Green's formula to the function in ζ, $(\zeta - z)^{-1}f(\zeta)$, on $D\backslash\Delta_r$, one has

$$2\int_{D\backslash\Delta_r}\bar{\partial}\frac{f(\zeta)}{\zeta - z}d\mu(\zeta) = \frac{1}{i}\int_{\partial D}\frac{f(\zeta)}{\zeta - z}d\zeta - \int_0^{2\pi} f(z + re^{i\theta})\, d\theta.$$

Since $\bar{\partial}(\zeta - z)^{-1} = 0$, we obtain $\bar{\partial}[(\zeta - z)^{-1}f(\zeta)] = (\zeta - z)^{-1}\bar{\partial}f(\zeta)$. By the remark above, the function $|\zeta - z|^{-1}$ is integrable $[\mu]$ on D, so (4.1) follows by letting $r \to 0$.

The first term on the right-hand side of (4.1),

$$g(z) = \frac{1}{2\pi i}\int_{\partial D}\frac{f(\zeta)}{\zeta - z}d\zeta,$$

is clearly analytic on D, while the other terms are continuous on D^-. Thus $g \in A^2(D;H)$.

For the reader's convenience, we include a proof of the following well-known result.

4.4.2 Lemma. For the given disc D and compact set $K \subset D$, there exists $C > 0$ for which

$$\|f\|_{\infty,K} < C\|f\|_{2,D} \qquad (f \in A^2(D;H)).$$

Proof. Let $r_0 = \mathrm{dist}(K, \partial D)$. For any $z \in K$, D also contains the disc Δ

$= \{\zeta\colon\ |z - \zeta| < r_0\}$, and if $0 < r < r_0$ one has

$$f(z) = \frac{1}{2\pi i}\int_{|\zeta - z| = r} \frac{f(\zeta)}{\zeta - z}\, d\zeta .$$

But since the integral depends on z only, we have also

$$r_0^2 f(z) = \frac{1}{\pi i}\int_0^{r_0} r\, dr \int_{|z-\zeta|=r} \frac{f(\zeta)d\zeta}{\zeta - z} =$$

$$= \frac{1}{\pi}\int_\Delta f(z + re^{i\theta})\, d\mu.$$

The Cauchy-Schwartz inequality then implies

$$r_0^2|f(z)| \le \frac{r_0}{\sqrt{\pi}}\left\{\int_\Delta |f(z + re^{i\theta})|^2 d\mu\right\}^{1/2},$$

and the result follows with $C = 1/r_0\sqrt{\pi}$.

We shall need a special Sobolev-type space. By $C_0^\infty(Q)$ we denote the set of all infinitely differentiable functions with compact support in Q. Define $W^m(Q;H)$ to be the set of those functions $f \in L^2(Q;H)$ whose derivatives $\bar\partial f, \bar\partial^2 f, \ldots, \bar\partial^m f$ (in the sense of distributions) also lie in $L^2(Q;H)$; i.e. for each $\phi \in C_0^\infty(Q)$

$$\int_Q (\bar\partial^i f)\phi\, d\mu = (-1)^i \int_Q f(\bar\partial^i \phi)\, d\mu$$

and $\bar\partial^i f \in L_2(Q;H)$ for $i = 0, 1, \ldots, m$. If given the norm

$$\|f\|^2_{W^2} = \sum_{i=0}^m \|\bar\partial^i f\|^2_{2,Q},$$

$W^m(Q;H)$ becomes a Hilbert space. Clearly, the natural imbedding $W^m(Q;H) \to L^2(Q;H)$ is continuous.

The elements of $C_0^\infty(Q)$ are usually called "test functions." A classical example of a test function on $\mathbf{C}$ is $\phi(z) = \exp[1/(|z|^2 - 1)]$ for $|z| < 1$ and $\phi = 0$ otherwise. On multiplication by a suitable constant factor, one has

$$\phi \in C_0^\infty(C), \quad \int \phi \, d\mu = 1, \quad \operatorname{supp} \phi = \{z\colon |z| \le 1\}. \tag{4.2}$$

Starting with an arbitrary function satisfying (4.2), we construct an infinitely differentiable H-valued function by convolution of ϕ with an arbitrary integrable H-valued function ψ: for $\eta \neq 0$, let

$$\psi_\eta(z) = \int \psi(z - \eta\xi)\phi(\xi)\, d\mu.$$

If ψ has compact support in Q, then $\psi_\eta \in C_0^\infty(Q;H)$ for $|\eta|$ sufficiently small. Using such ψ_η, one may prove the following theorem; we omit the proof (see [48, §1.2.6]).

4.4.3 Theorem. $C_0^\infty(Q;H)$, and hence $C^\infty(Q^-;H)$, is a dense subspace of $W^m(Q;H)$.

In the following, it is important to note that, if $f, g \in L^2(Q;H)$, then $f = (z - T)g$ is to be interpreted as

$$f(z) = (z - T)g(z) \qquad (z \in Q).$$

4.4.4 Proposition. Let D be an open disc. Then there exists $C_D > 0$ such that for an arbitrary operator $T \in L(H)$ and $f \in W^2(D;H)$ we have

$$\|(I - P)f\|_{2,D} \le C_D\left\{\|(z - T)^*\bar\partial f\|_{2,D} + \|(z - T)^*\bar\partial^2 f\|_{2,D}\right\}, \tag{4.3}$$

where P is the orthogonal projection of $L^2(D;H)$ onto $A^2(D;H)$.

Proof. Let $\{f_n\}$ be a sequence in $C^\infty(D;H)$ converging to f in the norm topology of $W^2(D;H)$. For n fixed

$$\bar\partial(f_n(z) - (z - T)^*\bar\partial f_n(z)) = -(z - T)^*\bar\partial^2 f_n(z).$$

By the Cauchy-Pompieu formula (4.1), we have

$$f_n(z) - (z - T)^*\bar{\partial} f_n(z) - \frac{1}{2\pi i}\int_{\partial D} \frac{f_n(\zeta) - (\zeta - T)^*\bar{\partial} f_n(\zeta)}{\zeta - z} d\zeta$$

$$= \frac{1}{\pi}\int_{D} \frac{(\zeta - T)^*\bar{\partial}^2 f_n(\zeta)}{\zeta - z} d\mu(\zeta).$$

Denoting the first of these integrals as $g_n(z)$, we obtain $g_n \in A(D;H)$, and hence

$$\|f_n - g_n\|_{2,D} \le \|(z - T)^*\bar{\partial} f_n\|_{2,D} + 4R\|(z - T)^*\bar{\partial}^2 f_n\|_{2,D}, \tag{4.4}$$

where the second integral above has been majorized by a convolution with an L^1-function and R is the radius of D. From (4.4) we have

$$\begin{aligned} \|f - Pf\|_{2,D} & \\ &\le \|f - g_n\|_{2,D} \\ &\le \|f - f_n\|_{2,D} + \|f_n - g_n\|_{2,D} \\ &\le \|f - f_n\|_{2,D} + \|(z - T)^*\bar{\partial} f_n\|_{2,D} + 4R\|(z - T)^*\bar{\partial}^2 f_n\|_{2,D}. \end{aligned}$$

In the limit, $\|f - f_n\|_{2,D} \to 0$, so we get (4.3) with $C_D = \max\{1, 4R\}$.

4.4.5 Corollary. If T is hyponormal, then

$$\|f - Pf\|_{2,D} \le C_D\{\|(z - T)\bar{\partial} f\|_{2,D} + \|(z - T)\bar{\partial}^2 f\|_{2,D}\}.$$

Proof. This follows from the inequality $\|(z - T)^*h\| \le \|(z - T)h\|$, since $z - T$ is also hyponormal.

For the hyponormal operator $T \in L(H)$, define the quotient space

$$H_Q = W^2(Q;H)/[(z - T)W^2(Q;H)]^-.$$

Then H_Q is a Hilbert space if it is endowed with the quotient norm

$$\|\tilde{f}\| = \inf \{\|f + g\|: \ g \in [(z - T)W^2(Q;H)]^-\}.$$

For any operator $A \in L(H)$ for which $[(z - T)W^2(Q;H)]^-$ is invariant, we shall use $\tilde{\mathbf{A}}$ to denote the operator on H_Q coinduced by A.

The multiplication operator M on $W^m(Q;H)$ defined by

$$(Mf)(z) = zf(z) \quad (z \in Q \text{ and } f \in W^m(Q;H))$$

is continuous with spectral distribution of order m given by the relation

$$U(\phi)f = \phi f \quad (\phi \in C^m(\mathbf{C}), \ f \in W^m(Q;H)).$$

The spectral maximal spaces of M are given by

$$[W^m(Q;H)](M, F) = \{f \in W^m(Q;H): \ \operatorname{supp} f \subset F\},$$

where F is closed in $\mathbf{C}$. In our present case of interest we take m = 2.

Because the multipliers defining M and U on $W^2(Q;H)$ are scalar-valued, they both commute with T, hence ran(z - T) is invariant under M and U. Therefore, the coinduced operator $\tilde{\mathbf{M}}$ as defined above is also a generalized scalar operator with spectral distribution $\tilde{\mathbf{U}}$ of order 2.

For $h \in H$ and scalar function $f: Q \to \mathbf{C}$, let $f \otimes h$ denote the function $f(\cdot)h$. Then $1 \otimes h$ is a constant in $W^2(Q;H)$. Now define $V: H \to H_Q$ by the formula $Vh = (1 \otimes h)^\sim$. It follows that

$$VT = \tilde{\mathbf{M}}V. \tag{4.5}$$

In fact, since $Th = (T - z)h + zh$ for $h \in H$, we have $(1 \otimes Th)^\sim = (\mathrm{id} \otimes h)^\sim$. Hence

$$\begin{aligned} VTh = (1 \otimes Th)^\sim &= (\mathrm{id} \otimes h)^\sim \\ = (\mathrm{id}\cdot h)^\sim &= \tilde{\mathbf{M}}(1 \otimes h)^\sim \\ &= \tilde{\mathbf{M}}Vh, \end{aligned}$$

so (4.5) is proved.

Before proving Lemma 4.4.6, we extend the Riesz-Dunford

functional calculus. For an open neighborhood Q of $\sigma(T)$, define the map Φ: $A(Q;H) \to H$ by

$$\Phi(f) = \frac{1}{2\pi i}\int_{\Gamma} R(z;T)f(z)\,dz,$$

where Γ is an oriented closed contour in Q with $\sigma(T)$ in its interior. It is easy to see that for G fixed there exists $a > 0$ such that

$$\|\Phi(f)\| \leq a\|f\|_{\infty,D} \tag{4.6}$$

where the last norm is the uniform one on $D = \Gamma \cup i(\Gamma)$. If $h: Q \to H$ is analytic and $f(z) = (z - T)h(z)$, then

$$\Phi(f) = 0 \tag{4.7}$$

from the definition of Φ.

4.4.6 Lemma. Let D be a disc containing $\sigma(T)$. Then the map $V{:}H \to H_D$ (as above) is injective with closed range.

Proof. To prove the lemma it suffices to prove the following assertion: if $h_n \in H$ and $f_n \in W^2(D;H)$ are sequences satisfying

$$\lim_n \|(z - T)f_n + 1 \otimes h_n\| = 0, \tag{4.8}$$

then $h_n \to 0$. For if h_n is bounded below, then (4.8) implies that $(1 \otimes h_n)^{\sim}$ is also bounded below, i.e. V is a topological isomorphism into H_D.

By the definition of the W^2 norm,

$$\lim_{n\to\infty} \{\|z - T)\bar\partial f_n\|_{2,D} + \|(z - T)\bar\partial^2 f_n\|_{2,D}\} = 0.$$

So by Cor. 4.4.5

$$\lim_{n\to\infty} \|f_n - Pf_n\|_{2,D} = 0. \tag{4.9}$$

Thus (4.8) and (4.9) imply

$$\lim_{n\to\infty} \|(z - T)Pf_n + 1 \otimes h_n\|_{2,D} = 0.$$

For every compact domain K with $\sigma(T) \subset K \subset D$, Lemma 4.4.2 implies

$$(4.10) \qquad \lim \|(z - T)Pf_n + 1 \otimes h_n\|_{\infty,K} = 0,$$

and we infer from (4.6) and (4.7) that

$$\begin{aligned} \|h_n\| &= \|\Phi(z - T)Pf_n + 1 \otimes h_n\| \\ &\leq a\|(z - T)Pf_n + 1 \otimes h_n\|_{\infty,K}, \end{aligned}$$

hence the lemma follows from (4.10).

4.4.7 Theorem. Every hyponormal operator is subscalar of order 2.

Proof. By Lemma 4.4.6 V is a topological isomorphism from H onto VH. Hence (4.5) shows that VH is $\tilde{\mathbf{M}}$-invariant, and thus T is a subscalar operator.

4.4.8 Corollary. Every hyponormal operator has property (β).

Proof. This follows from Th. 4.4.7 and the fact that property (β) is preserved under restriction.

The next corollary is immediate from Cor. 4.4.8 and Prop. 2.4.8.

4.4.9 Corollary. If T is hyponormal and quasisimilar to a decomposable operator, then T is quasi- decomposable.

Stampfli and Wadhwa [88, Cor.1] have proved a result along this line as follows.

4.4.10 Proposition. If T is hyponormal and quasisimilar to a normal operator, then T is normal.

On the other hand, both Radjabalipour [75] and Albrecht [4] have constructed examples of non-normal subnormal operators which are decomposable, so the hypothesis "normal" in Prop. 4.4.10 cannot be weakened to "decomposable" to obtain the same conclusion. But our next result shows that some relaxation is possible.

4.4.11 Proposition. Let T be hyponormal and quasisimilar to the C-scalar operator S on some Hilbert space. Then T is normal.

Proof. Since S is defined on Hilbert space, it is scalar-type spectral by Cor 1.4.6. But it is well-known that a scalar-type operator is similar to a normal operator, hence S is a fortiori quasisimilar to some normal N. Since T is thus quasisimilar to N, the conclusion follows by Prop. 4.4.10.

4.4.12 Remark. The results above have an immediate generalization to the case of M-hyponormal operators. Recall that $T \in L(H)$ is M-hyponormal if the exists $M > 0$ such that $\|T^*h\| \leq M\|Th\|$ for all $h \in H$. So Prop. 4.4.4 implies

4.4.13 Theorem. Every M-hyponormal operator is subscalar.

In the remainder of this section we study the relation between $\sigma(T)$ and $\sigma(\tilde{M})$. Suppose that Q_1 and Q_2 are two bounded open sets containing $\sigma(T)$. Then the corresponding spaces H_{Q_1} and H_{Q_2} coincide. Moreover, these latter spaces are isomorphic to the universal Frechet space

$$W^2{}_T(H) = W^2{}_{loc}(\mathbf{C};H)/[(z - T)W^2{}_{loc}(\mathbf{C};H)]^-,$$

which depends only on T and H, where here "loc" means those functions which are locally square-integrable. In fact, for $i = 1,2$, $W^2{}_{loc}(C;H)$ has the decomposition

$$W^2{}_{loc}(\mathbf{C};H) = W^2(Q_i,H) \oplus W^2{}_{loc}(C\backslash Q_i;H),$$

and since $z - T$ is bijective for $z \notin Q_i$,

$$[(z-T)W^2{}_{loc}(\mathbf{C};H)]^- = [(z - T)W^2(Q_i;H)]^- \oplus W^2{}_{loc}(C\backslash Q_i;H),$$

so we have for $i = 1,2$

$$W^2{}_T(H) \cong W^2(Q_i;H)/[(z - T)W^2(Q_i;H)]^-. \tag{4.11}$$

Now fix a disc D containing $\sigma(T)$ and endow $W^2{}_T(H)$ with the corresponding Hilbert space structure. For $T \in L(H)$, define the operator $I \otimes T$ on $W^2{}_{loc}(\mathbf{C};H)$ by

$$[(I \otimes T)f](z) = Tf(z) \qquad (f \in W^2{}_{loc}(\mathbf{C};H)).$$

Then the operators induced by M, resp. $I \otimes T$, coincide on $W^2{}_T(H)$.

Finally, use $\tilde{T}$ for $\tilde{M}$.

4.4.14 Theorem. With $W^2{}_T(H)$, $\tilde{T}$ defined above and with the natural imbedding $V: H \to W^2{}_T(H)$, we have

(a) $\tilde{T}$ is generalized scalar of order 2 with $\sigma(\tilde{T}) \subset \sigma(T)$;

(b) the linear space of the vectors $\{\tilde{U}(\phi)Vh: \phi \in C^\infty(\mathbf{C}), h \in H\}$ is dense in $W^2{}_T(H)$;

(c) if f is an analytic function in a neighborhood of $\sigma(T)$, then $Vf(T) = f(\tilde{T})V$.

Proof. (a). Let Q be an arbitrary neighborhood of $\sigma(T)$. From the fact that $W^2{}_T(H)$ has representation (4.11) and the inclusions

$$\sigma(\tilde{T}) = \sigma(\tilde{M}) \subset \sigma(M|W^2(Q;H)) \subset Q^-,$$

we conclude $\sigma(\tilde{T}) \subset \sigma(T)$. Assertion (b) is a result of the density of $C^\infty(D^-)$ in $W^2(D) = W^2(D;\mathbf{C})$, while (c) follows from the functional calculus.

4.4.15 Corollary. With the notation of Theorem 4.4.14, we have also $\partial\sigma(T) \subset \sigma(\tilde{T})$.

Proof. Since $\partial\sigma(T)$ is contained in the approximate point spectrum of $\sigma(T)$, it must also be contained in $\sigma(\tilde{T})$ by Lemma 4.4.6.

5. Invariant Subspaces for Hyponormal Operators

We show first that hyponormal operators whose spectra have nonempty interior necessarily have nontrivial invariant subspaces (Th. 4.5.6). If the spectrum of a hyponormal operator has empty interior, then a supplementary condition is required to infer existence of a nontrivial invariant subspace (Th. 4.5.7). (A proof independent of §3 is required because in general a scalar extension of a hyponormal operator is not clearly unconditionally decomposable.) We conclude the section with a remark on the general problem.

We continue to let T, $\tilde{M}$, V, H_Q, etc., be as in §4, and let $H^\infty(Q)$ be as in §2. We shall also use the following notations. For $\lambda \in \mathbf{C}$ and $r > 0$, let $D(\lambda,r)$ denote the open disc of radius r, center λ. For n

= 1, 2,..., write $D_n = D(0, 1-2^{-n})$. If $\lambda \in D(0,1)$, let $n(\lambda)$ be the smallest integer for which λ is in the closure of $D_n(\lambda)$. Put $D_\lambda = D(\lambda, 2^{-n(\lambda)-2})$. Without loss of generality we suppose $D(0,2) \subset \sigma(T)$. To simplify notation, write $S = \tilde{M}$. Define

$$H_\lambda = \{y \in H_Q: \sigma(y,S) \subset C\backslash D_\lambda\},$$

and let H^λ be the orthogonal complement of H_λ in H_Q. Denote by P_λ the orthogonal projection of H_Q onto H^λ. Finally , whenever we write $\hat{x} \in H_Q$, we mean $\hat{x} = Vx$ for some $x \in H$.

4.5.1 Lemma. Let n be a fixed positive integer. Then

(i) there exists $K_n > 0$ such that $\|(\lambda - S|H_\lambda)^{-1}\| \le K_n$ for all $\lambda \in D_n^-$;

(ii) there exists $\delta_n > 0$ such that for all $\lambda \in D_n^-$ and for any sequence of unit vectors $\{x_j\}$ in H with $(\lambda - T)x_j \to 0$ we have

$$\varliminf \|P_\lambda \hat{x}_j\| > \delta_n \quad \text{and} \tag{5.1}$$

$$\varliminf \langle \hat{x}_j, P_\lambda \hat{x}_j \rangle > \delta_n^2. \tag{5.2}$$

Proof. (i) Assume, to the contrary, that for some sequence $\{\lambda_j\}$ in D_n^- we have $\|(\lambda_j - S|H_{\lambda_j})^{-1}\| \to \infty$. We may clearly suppose $\lambda_j \to \lambda \in D_n^-$, so that for j sufficiently large

$$D(\lambda, 2^{-n-4}) \subset D(\lambda_j, 2^{-n-2}).$$

Now consider

$$H_c = \{y \in H_Q: \sigma(y,S) \subset C\backslash D(\lambda, 2^{-n-4})\},$$

and put $K = \|(\lambda - S|H_c)^{-1}\|$. Then $K \ge \|(\lambda - S|H_{\lambda_j})^{-1}\|$, contradicting the assumption. The existence of K_n is thus proved.

(ii) Let $\lambda \in D_n^-$ and let K_n be as in (i). Suppose $\hat{x} = Vx$ is a unit vector such that

$$\|(\lambda - S)\hat{x}\| < \frac{1}{4K_n} \quad \text{and} \quad \|P_\lambda \hat{x}\| < \frac{1}{8K_n\|S\|} < \frac{1}{4}.$$

The last inequality follows from the fact that $|\lambda| < \|S\|$, so that $1 \le \|\lambda - S\| \, \|(\lambda - S|H_\lambda)^{-1}\| < 2K_n\|S\|$. Hence

$$\|(\lambda - S)(I - P_\lambda)\hat{x}\| < \frac{1}{4K_n} + \frac{\|\lambda - S\|}{8K_n\|S\|} < \frac{1}{2K_n}.$$

On the other hand, clearly $\|\hat{x} - P_\lambda \hat{x}\| > 3/4$, while by part (i) $\|(\lambda - S)y\| \ge \|y\| K_{n-1}$ for all $y \in H_\lambda$. Hence

$$\|(\lambda - S)(I - P_\lambda)\hat{x}\| > \frac{3}{4K_n}.$$

This contradiction proves (5.1), while (5.2) follows from the identity $\langle x,Px\rangle = \|Px\|^2$ for any orthogonal projection P.

Now let S_λ be the compression $S_\lambda = P_\lambda S P_\lambda$. It follows from Lemma 4.3.7 that $\sigma(S/H_\lambda) \subset D_\lambda^-$. We claim $\sigma(S_\lambda) = \sigma(S/H_\lambda)$. Since $S_\lambda{}^* = P_\lambda S^* P_\lambda = P_\lambda S^*|H^\lambda$, it follows that $S_\lambda = (P_\lambda S^*|H^\lambda)^* = (SP_\lambda)/H_\lambda = S/H_\lambda$, and our claim is proved. Thus $\sigma(S_\lambda) \subset D_\lambda^-$.

For $f \in H^\infty(D_{n+1})$, $\hat{x} \in H_Q$, $\lambda \in D_n^-$ and $y \in H^\lambda$, define $\hat{x} \otimes y$ by the formula

$$\langle f, \hat{x} \otimes y\rangle = \langle \hat{x}, f(S_\lambda)^* y\rangle = \langle f(S_\lambda)\hat{x}, y\rangle. \tag{5.3}$$

If $f_n \to 0$ weak* in $H^\infty(D_{n+1})$, by the proof of Th. 4.3.3 $f_n \to 0$ uniformly on compact sets in D_{n+1}. In particular, $f_n(S_\lambda) \to 0$ since $D^- \subset D_{n+1}$. By (5.3), $\hat{x} \otimes y$ is weak* continuous. Note also that if $f \in H^\infty(D_{n+1})$ and $\lambda \in D_m \subset D_m^- \subset D_n$ $(m < n)$, then for $\hat{x}$ and y given above $\langle f, \hat{x} \otimes y\rangle$ has the same value for both n and m. We use the symbol $\|.\|_{D_n}$ for the norm of an element of $H^\infty(D_n)^*$ (e.g. $\|\hat{x} \otimes y\|_{D_n}$). For $m \le n$ we have $H^\infty(D_{n+1}) \subset H^\infty(D_m)$, hence clearly

$$\|\hat{x} \otimes y\|_{D_n} \le \|\hat{x} \otimes y\|_{D_m}.$$

Following §3 we write e_λ for point evaluation on $H^\infty(D_{n+1})$: $\varepsilon_\lambda(f) = f(\lambda)$. The main idea of the proof of Th. 4.5.6 below is to show that ε_0 has a representation $\hat{x} \otimes y$.

4.5.2 Lemma. Fix n and let $\lambda \in D_n^-$. Let $\{\hat{x}_j\}$ be a sequence of unit vectors in H_Q for which $(\lambda - T)x_j \to 0$ in H. Then

$$(5.4) \qquad \|\hat{x}_j \otimes P_\lambda \hat{x}_j - \langle \hat{x}_j, P_\lambda \hat{x}_j \rangle \varepsilon_\lambda\|_{D_{n+1}} \to 0.$$

Proof. Let $\varepsilon > 0$ and let $D = D(\lambda, 2^{-n(\lambda)}/3)$. Since $\sigma(S_\lambda) \subset D_\lambda^- \subset \Delta$ by the remark above and the definition of D_λ^-, we have for $f \in H^\infty(D_{n+1})$

$$\|f(S_\lambda)\| = \|(2\pi i)^{-1} \int_{\partial\Delta} f(\mu)(\mu - S_\lambda)^{-1} d\mu\| \leq a_\lambda \|f\|_D,$$

where $a_\lambda > 0$ depends only on λ. (Moreover, we may suppose $a_\lambda \geq 1$). Then if we take $\delta < \varepsilon/2a_\lambda$, the last inequality implies $\|f(S)\| < \varepsilon/3$ whenever $\|f\|_D < \delta$ and $f \in H^\infty(D_{n+1})$.

Now let $f \in H^\infty(D_{n+1})$ with $\|f\|_{D_{n+1}} = 1$. Denote the radius of D_{n+1} by r. If $f(z) = \Sigma_0^\infty c_j z^j$ is its power series, then by the Cauchy integral formula it is easy to show $|c_j| \leq r^{-j}$ (all j). Let R denote the distance from the origin in C to the farthest point on $\partial\Delta$ $(R = |\lambda| + 3^{-1}2^{-n(\lambda)})$. It follows that

$$(5.5) \qquad \|f\|_\Delta \leq \sum_{j=0}^{\infty} |c_j| R^j \leq \sum_{j=0}^{\infty} (R/r)^j = \frac{r}{r-R},$$

and, if $c_0, c_1, \ldots, c_N$ all vanish, then (5.5) becomes

$$(5.6) \qquad \|f\|_\Delta \leq \frac{r}{r-R}\left(\frac{R}{r}\right)^N \to 0 \quad (N \to \infty).$$

Thus, if $\|f\|_{D_{n+1}} = 1$ for $f \in H^\infty(D_{n+1})$, by (5.6) we can find N sufficiently large but independent of f such that $f = f_1 + f_2$ where f_1

is a polynomial of degree at most N and $\|f_2\|_D < \delta$. Note also that

$$\|f_1\|_{D_{n+1}} \leq \Sigma_0^N c_j r^j \leq N + 1;$$

that is, $\{f_1 : \|f\|_{D_{n+1}} = 1\}$ is uniformly bounded.

Suppose $f = f_1 + f_2$ as above. Since $S_\lambda = P_\lambda S P_\lambda = P_\lambda S$ and $(I - P_\lambda)H_Q$ is S-invariant, it follows from the hypothesis $(\lambda - T)x_j \to 0$ that

$$\|f_1(S_\lambda)P_\lambda \hat{x}_j - f_1(\lambda)P_\lambda \hat{x}_j\| \to 0 \ (j \to \infty) \tag{5.7}$$

uniformly for all $f \in H^\infty(D_{n+1})$ of unit norm. Choose j so large that (5.7) is less than $\varepsilon/3$. Then for $f \in H^\infty(D_{n+1})$ of norm 1 we estimate

$$|\langle f, \hat{x}_j \otimes P_\lambda \hat{x}_j - \langle \hat{x}_j, P_\lambda \hat{x}_j \rangle \varepsilon_\lambda \rangle|$$

$$= |\langle f, \hat{x}_j \otimes P_\lambda \hat{x}_j \rangle - \langle \hat{x}_j, P_\lambda \hat{x}_j \rangle f(\lambda)|$$

$$\leq |\langle \hat{x}_j, f(S_\lambda)^* P_\lambda \hat{x}_j \rangle - \langle \hat{x}_j, P_\lambda \hat{x}_j \rangle f_1(\lambda)| + \delta$$

$$\leq \delta + |\langle \hat{x}_j, f_2(S_\lambda)^* P_\lambda \hat{x}_j \rangle + \langle f_1(S_\lambda) P_\lambda \hat{x}_j, P_\lambda \hat{x}_j \rangle - \langle P_\lambda \hat{x}_j, P_\lambda \hat{x}_j \rangle f_1(\lambda)|$$

$$< \delta + \varepsilon/3 + \varepsilon/3 < \varepsilon,$$

where we have applied first the condition on f_2 and then (5.7) and condition $\|f_2\|_\Delta < \delta$. Hence (5.4) follows from the uniformity of weak convergence.

The next two results are similar to Lemma 4.3.9.

4.5.3 Lemma. Let $\{\hat{x}_j\}$ be a sequence in H_Q converging weakly to zero, and let $y \in P_\lambda H_Q$ for some $\lambda \in D_n^-$ and some n. Then $\|\hat{x}_j \otimes y\| \to 0$.

Proof. Let $\varepsilon > 0$ and decompose $f \in H^\infty(D_{n+1})$ as in the previous proof. Let B_1, B_2 denote those sets of f_1, f_2 resp. (for $\|f\| = 1$) and let

$B_i(S_\lambda)$ be the set of all $f_i(S_\lambda)$ $(i = 1,2)$. Then the set

$$K_\lambda = \{f(S_\lambda)^* y: f \in H^\infty(D_{n+1}), \|f\|_{D_{n+1}} = 1\}$$

$$\subset B_1(S_\lambda)^* y + B_2(S_\lambda)^* y$$

$$\subset B_1(S_\lambda)^* y + B_3,$$

where B_3 is the unit ball in H^λ. As $\varepsilon > 0$ is arbitrary and $B_1(S_\lambda)^* y$ is bounded in a space of dimension at most $N+1$, K_λ can be covered by finitely many ε-balls in H_Q. Since K_λ is thus compact and $\hat{x}_j \to 0$ weakly, we have

$$\|\hat{x}_j \otimes y\|_{D_{n+1}} = \sup\{|\langle f, \hat{x}_j \otimes y\rangle|: \|f\|_{D_{n+1}} = 1\}$$
$$= \sup\{|\langle \hat{x}_j, w\rangle|: w \in K_\lambda\} \to 0,$$

so the proof is complete.

4.5.4 Lemma. Let $\{\hat{x}_j\}$ converge weakly to zero in H_Q. Then for any $\hat{x} = Vx \in H_Q$ and for all $\lambda \in D_n$,

$$\|\hat{x} \otimes P_\lambda \hat{x}_j\|_{D_{n+1}} \to 0. \tag{5.8}$$

Proof. Conclusion (5.8) follows much as in the last argument by taking $y = P_\lambda \hat{x}$.

Our next crucial step shows how to iterate successive approximations to a point evaluation.

4.5.5 Lemma. Let $n = 2, 3, \ldots$ be fixed. Suppose $\hat{x}, y \in H_Q$ such that

$$\|\hat{x} \otimes y - \varepsilon_0\| < \delta_n^2 2^{-2n} M \tag{5.9}$$

where $M > 0$, δ_n are defined in Lemma 4.5.1 and y is a linear combination of vectors from H^λ for some λ's in D_{n-1}^-. Then there exist $x' \in H$ and $y' \in H_Q$ such that

(5.10) $$\|\hat{x}' \otimes y' - \varepsilon_0\| < \delta_{n+1}{}^2 2^{-2n-2} M$$

and

(5.11) $$\|\hat{x} - \hat{x}'\| < \delta_n 2^{-n} \text{ and } \|y - y'\| < 2^{-n},$$

where y' is a linear combination of vectors from H^{λ}.

Proof. Without loss of generality, let $M = 1$. Observe that for $f \in H^\infty(D)$ and $\lambda \in D$ we have $|\langle f, \varepsilon_\lambda \rangle| \le \|f\|_D$, hence $\|\varepsilon_\lambda\|_D \le 1$. But each such ε_λ is weak* continuous, again by application of the proof of Th. 4.3.3. Hence $\varepsilon_\lambda \in M(D)$. Moreover, $\|f\|_D = \sup\{|\langle f, \varepsilon_\lambda \rangle| : \lambda \in D\}$, and so by Lemma 4.3.2 the unit ball of $M(D)$ is the absolutely convex closed hull of $\{\varepsilon_\lambda : \lambda \in D\}$. Thus by (4.6) we may choose $\lambda_1, \lambda_2, \ldots, \lambda_m \in D$ and $c_1, c_2, \ldots, c_m \in C$ such that

(5.12) $$\|\hat{x} \otimes y - \varepsilon_0 - \Sigma_j c_j \varepsilon_{\lambda_j}\|_{D_n} < \delta_n{}^2 2^{-2n-2} \text{ and}$$

$$\Sigma_j |c_j| < \delta_n{}^2 2^{-2n}.$$

For each $j = 1,2,\ldots,m$ choose a sequence of unit vectors $\{x_{jk}\}$ so that $(\lambda_j - T)x_{jk} \to 0$ $(k \to \infty)$. Define

$$\beta_{jk} = [\langle \hat{x}_{jk}, P_{\lambda_j} \hat{x}_{jk} \rangle]^{-1};$$

by (5.2) β_{jk} is bounded (above) for all j,k. By Lemma 4.4.2, for each j,

(5.13) $$\|\beta_{jk} \hat{x}_{jk} \otimes P_{\lambda_j} \hat{x}_{jk} - \varepsilon_{\lambda_j}\| \to 0 \quad (k \to \infty).$$

It follows from (5.9) and (5.13) that for all $j = 1,\ldots,m$ and for k sufficiently large $(k > N)$

(5.14) $$\|\hat{x} \otimes y - \varepsilon_0 - \sum_{j=1}^{m} c_j \beta_{jk} \hat{x}_{jk} \otimes P_{\lambda_j} \hat{x}_{jk}\|_{D_{n+1}} < \delta_n{}^2 2^{-2n-2}.$$

For $j = 1,\ldots,m$ let $r_j \in C$ such that $r_j{}^2 = c_j$. Without loss of generality we may suppose $\hat{x}_{jk} \to 0$ weakly for each j, hence by Lemmas 4.5.3 and 4.5.4 we can choose $k_1, k_2, \ldots, k_m$ sequentially, each of which is larger than N, so as to make each of the following quantities arbitrarily small.

$$\sum_{j=1}^{m} \|r_j \hat{x}_{jk_j} \otimes y\|_{D_{n+1}} \tag{5.15}$$

$$\sum_{j \neq p} \|r_j \bar{r}_p \beta_{pk_p} \hat{x}_{jk_j} \otimes P_{\lambda_p} \hat{x}_{pk_p}\|_{D_{n+1}} \tag{5.16}$$

$$\sum_{j=1}^{m} \|r_j \beta_{jk_j} \hat{x} \otimes P_{\lambda_j} \hat{x}_{jk_j}\|_{D_{n+1}} \tag{5.17}$$

$$\sum_{j \neq p} |r_j \bar{r}_p \langle \hat{x}_{jk_j}, \hat{x}_{pk_p} \rangle| \tag{5.18}$$

$$\sum_{j \neq p} |\bar{r}_j r_p \bar{\beta}_{jk_j} \beta_{pk_p} \langle P_{\lambda_j} \hat{x}_{jk_j}, P_{\lambda_p} \hat{x}_{pk_p} \rangle|. \tag{5.19}$$

If (5.15)-(5.17) is each small enough, then by (5.14)

$$\|\{\hat{x} + \sum_{j=1}^{m} r_j \hat{x}_{jk_j}\} \otimes \{y + \sum_{j=1}^{m} \bar{r}_j \bar{\beta}_{jk_j} P_{\lambda_j} \hat{x}_{jk_j}\} - \varepsilon_0\|_{D_{n+1}} < \delta_{n+1}{}^2 2^{-2n-2}.$$

If (5.18), (5.19) are small, then $\|\Sigma\, r_j \hat{x}_{jk}\| < \delta_n 2^{-n}$ and by (5.2)

$$\|\sum_{j=1}^{m} r_j \hat{x}_{jk_j} P_{\lambda_j} \hat{x}_{jk_j}\| \leq (\sum_{j=1}^{m} |c_j| \delta_n{}^{-2})^{1/2} < 2^{-n}.$$

If we put

$$\hat{x}' = \hat{x} + \sum_j r_j \hat{x}_{jk_j} \quad \text{and} \quad y' = y + \sum_j \bar{r}_j \hat{x}_{jk_j} P_{\lambda_j} \hat{x}_{jk_j},$$

then (5.10) and (5.11) follow.

4.5.6 Theorem. If T is hyponormal and $\sigma(T)$ has nonempty interior, then T has a nontrivial rationally cyclic invariant subspace.

Proof. To initiate the recursive process of Lemma 4.5.5, we are required to find $\hat{x} = Vx$, $y \in H^{\lambda}$ (for example) such that

$$\|\hat{x} \otimes y - \varepsilon_0\|_{D_1} < M\,\delta_1^{\,2}/4, \tag{5.20}$$

but this follows by the arbitrariness of $M > 0$.

Applying now Lemma 4.5.5, we find Cauchy sequences $\{\hat{x}_n\}$ and $\{y_n\}$ in H_Q such that $\hat{x}_n \otimes y_n \to \varepsilon_0$ in the norm topology of $H^\infty(D(0,1))^*$. The limits of these sequences, $\hat{x}_0$, y_0, satisfy $\hat{x}_0 \otimes y_0 = \varepsilon_0$. If f is any rational function with poles off $\sigma(T)$, then $f \in H^\infty(D(0,1))$ and $\langle f(T), \hat{x}_0 \otimes y_0\rangle = f(0)$. Hence by the similarity of T and $S|VH$,

$$\langle f(S|VH)\hat{x}_0, y_0\rangle = f(0). \tag{5.21}$$

Since clearly $\langle \hat{x}_0, y_0\rangle = 1$, both $\hat{x}_0$ and y_0 are nonzero. Next let $P_1 y_0 = y_1$ where P_1 is the orthogonal projection of H_Q onto VH. Then clearly $\langle \hat{x}_0, y_1\rangle = 1$, hence $y_1 \neq 0$. But also (5.21) implies that if r is a rational function with poles off $\sigma(T)$ and $r(0) = 0$, then $\langle r(S|VH)\hat{x}_0, y_1\rangle = 0$. Thus the subspace W of VH defined by

$$W = \overline{\mathrm{sp}}\{r(S|VH)\hat{x}_0 : r \in R(\sigma(T)),\ r(0) = 0\}$$

is orthogonal to $y_1 \in VH$ and is clearly S-invariant. Hence T also has a nontrivial rationally cyclic invariant subspace.

We now consider the case where $\sigma(T)$ has empty interior. This case is easier, provided we assume that $\sigma(T) \cap G$ is dominating in G for some open G. The precise statement is

4.5.7 Theorem. If T is hyponormal such that $\sigma(T) \cap G$ is dominating in G for some open set G, then T has a nontrivial invariant subspace.

To prove Th. 4.5.7 we require several lemmas. The next lemma is well-known, but we give a proof for completeness.

4.5.8 Lemma. (a) The norm and spectral radius of a hyponormal operator coincide. (b) If a hyponormal operator is invertible, then its inverse is also hyponormal.

Proof. (a) For nonzero $x \in H$, note that if $n = 1,2,\dots$, then

$$\begin{aligned}\|T^n x\|^2 &= \langle T^n x, T^n x\rangle = \langle T^*T^n x, T^{n-1}x\rangle \\ &\le \|T^*T^n x\|\,\|T^{n-1}x\| \le \|T^{n+1}x\|\cdot\|T^{n-1}x\|,\end{aligned}$$

hence $\|T^n\|^2 \le \|T^{n+1}\|\,\|T^{n-1}\|$. The sequence defined by $a_n = \|T^{n+1}\|/\|T^n\|$ is thus nondecreasing, so

$$\|T\| = a_0 \le (a_0 a_1 \dots a_{n-1})^{1/n} = \|T^n\|^{1/n}.$$

Since the limit of the last term as $n \to \infty$ is the spectral radius $r(T)$ and $r(T) \le \|T\|$ always holds, (a) is proved.

(b) If T^{-1} exists, then write $y = Tx$ $(x \neq 0)$, and note that $\|T^*T^{-1}y\| \le \|y\|$, i.e. $\|T^*T^{-1}\| \le 1$. It follows that $\|(T^*)^{-1}T\| \le 1$, hence

$$\|(T^*)^{-1}y\| = \|(T^*)^{-1}Tx\| \le \|x\| = \|T^{-1}y\| \quad (y \in H),$$

and (b) is proved.

4.5.9 Lemma. If T is hyponormal and $\lambda \notin \sigma(T)$, then

$$\|(\lambda - S)^{-1}\| \le \frac{1}{\operatorname{dist}(\lambda, \sigma(T))}. \tag{5.22}$$

where S is defined above.

Proof. Let P be the orthogonal projection of $W^2(Q;H)$ onto $[(z-T)W^2(Q;H)]^\perp$ where Q is an open set containing $\sigma(T)$. Then S may be identified with PTP and hence we identify $(\lambda - S)^{-1}$ with $P(\lambda - T)^{-1}P$. Thus

$$\|(\lambda - S)^{-1}\| = \|P(\lambda - T)^{-1}P\| \le \|(\lambda - T)^{-1}\|. \tag{(5.23)}$$

By Lemma 4.5.8

$$\|(\lambda - T)^{-1}\| = r((\lambda - T)^{-1}) \tag{5.24}$$

$$= \sup\{|\mu|: \mu \in \sigma((\ - T)^{-1})\}$$
$$= \sup\{|\lambda - \eta|^{-1}: \eta \in \sigma(T)\}$$
$$= 1/\inf\{|\lambda - \eta|: \eta \in \sigma(T)\}$$
$$= 1/\mathrm{dist}(\lambda, \sigma(T)).$$

Now (5.22) follows from (5.23) and (5.24).

For $\beta \in \sigma(T)$ and $\rho > 0$, define $P_{\beta\rho}$ to be the orthogonal projection of H_Q onto the orthogonal complement of $H_Q(S, \mathbf{C}\backslash D(\beta,\rho))$. we say that $\beta \in \sigma(T)$ is <u>upright</u> if for each $\rho > 0$ and each sequence of unit vectors $\{\hat{x}_j\}$ in H_Q with $\|(\beta - T)x_j\| \to 0$ we have $\|P_{\beta\rho}\hat{x}_j\| \to 1$ (as $j \to \infty$).

4.5.10 Lemma. If T is hyponormal and $\sigma(T)$ has empty interior, then the set of upright elements in $\sigma(T)$ is dense.

Proof. Let $\beta' \in \sigma(T)$ and let $\varepsilon > 0$. Without loss of generality, we suppose β' is not isolated in $\sigma(T)$; otherwise, T clearly has a nontrivial invariant subspace. Choose $\beta \in D(\beta',\varepsilon) \cap \sigma(T)$ and $\alpha \notin \sigma(T)$ so that β is the point in $\sigma(T)$ nearest α. We prove β is upright. Let $\rho > 0$ be fixed, and let $\{\hat{x}_j\}$ be a sequence of unit vectors in H_Q such that $(\beta - T)x_j \to 0$. For $\delta > 0$ it suffices to show that $\|P_{\beta\rho}\hat{x}_j\| > \|\hat{x}_j\| - \delta$ for j sufficiently large.

Note that for λ in the line segment $[\beta,\alpha]$, it is clear that

$$|\lambda - \beta|^{-1} = \sup\{|\lambda - z|^{-1}: z \in \sigma(T)\}.$$

Hence as $\lambda \to \beta$ along this segment

$$\sup\{|(\lambda - \beta)(\lambda - z)^{-1}|: z \in \sigma(T)\backslash D(\beta,\rho)\} \to 0.$$

It follows that there exists $\lambda \in [\beta, \alpha]$ with

$$\|(\lambda - \beta)(\lambda - S|H_\rho)^{-1}\| < \delta/2, \tag{5.25}$$

where the subspace $H_\rho = \{y \in H_Q: \sigma(y,S) \cap D(\beta,\rho) = \varnothing\}$. By Lemma 4.5.9 $\|(\lambda - \beta)(\lambda - S)^{-1}\hat{x}_j\| \leq 1$, hence (5.25) implies

$$\begin{aligned} &\|(\lambda - \beta)(\lambda - S)^{-1}\hat{x}_j\| \\ &\leq \|(\lambda - \beta)(\lambda - S)^{-1}P_{\beta\rho}\hat{x}_j\| + \|(\lambda - \beta)(\lambda - S)^{-1}(I - P_{\beta\rho})\hat{x}_j\| \\ &\leq \|P_{\beta\rho}\hat{x}_j\| + \frac{\delta}{2}. \end{aligned} \tag{5.26}$$

But since $(\lambda-\beta)(\lambda - S)^{-1} - 1 = (\lambda - S)^{-1}(S - \beta)$ and the hypothesis on T shows that $\|(S - \beta)\hat{x}_j\| \to 0$, we have also

$$(5.27) \qquad \|(\lambda-\beta)(\lambda - S)^{-1}\hat{x}_j - \hat{x}_j\| \to 0.$$

It thus follows from (5.26), (5.27) that

$$\|\hat{x}_j\| - \frac{\delta}{2} \leq \|P_{\beta\rho}\hat{x}_j\| + \frac{\delta}{2}$$

for j sufficiently large, and this proves the lemma.

Proof of Th. 4.5.7. By Lemma 4.5.10 the set of upright elements is dense in $\sigma(T) \cap G$. In this case the proof of Lemma 4.5.5 can be repeated with G being substituted for D_2, D_3,... as long as upright λ are used because (5.2) holds for such λ.

Theorems 4.5.6, 4.5.7 and 4.2.9 now yield the following.

4.5.11 Theorem. If T is hyponormal and $R(\sigma(T)) \neq C(\sigma(T))$, then T has a nontrivial invariant subspace.

We close this section with some remarks on the general question of invariant subspaces for hyponormal operators. Our final result shows (via Th.2.4.10) that a hyponormal operator without proper invariant subspaces must be arbitrarily uniformly close to a decomposable operator in L(H).

4.5.12 Theorem. Let $T \in L(H)$ be hyponormal. Then either T has a nontrivial invariant subspace or T is a compact perturbation of a normal operator.

Proof. Suppose that T has no nontrivial invariant subspace. Then $\sigma_p(T^*) = \emptyset$ and hence T^* has SVEP. By Cor. 4.4.8 and Th. 2.3.21 T has a rationally cyclic vector. In this case it follows from [19, Th. 4] that $T^*T - TT^* \in K_1(H)$. In particular, T is a essentially normal. Let N be a normal operator on H with $\sigma_e(N) = \sigma_e(T)$. Then $\mathrm{ind}(\lambda - N) = \mathrm{ind}(\lambda - T) = 0$ for all $\lambda \notin \sigma_e(T)$. By Th. 2.3.18, $T = VNV^* + K$ where V is unitary and K is compact. Since VNV^* is normal, the proof is complete.

Theorem 4.5.12 reduces the invariant subspace problem for hyponormal operators to that of the case N + K. This appears to be a difficult question.

Chapter V - MULTIVARIATE THEORY

In this final chapter of the monograph, we give a survey of some elementary aspects of the "multivariate" theory of spectral decomposition. A separate table of references for this material appears at the end of this book.

1. Preliminaries

Assume that A is a complex algebra with unit 1, and let $\mathbf{Z}$ denote the set of integers. A sequence $(X,\alpha) = ((X^p,\alpha^p); p \in \mathbf{Z})$ of A-modules X^p and A-module homomorphisms α^p: $X^p \to X^{p+1}$ satisfying $\alpha^{p+1}\alpha^p = 0$ for all $p \in \mathbf{Z}$ is called a cochain-complex (or merely complex). The cohomology of the complex (X,α) is the sequence of quotient A-modules

$$H^p(X,\alpha) = N(\alpha^p)/R(\alpha^{p-1}) \qquad (p \in \mathbf{Z}),$$

where N and R denote kernel and range of the respective maps; each $H^p(X,\alpha)$, $p \in \mathbf{Z}$, is called a cohomology group for (X,α). The complex (X,α) is said to be exact if its cohomology is trivial: $H^p(X,\alpha) = 0$ for all p.

If (X,α) and (Y,β) are two complexes and for each $p \in \mathbf{Z}$ there is an A-module homomorphism f^p: $X^p \to Y^p$ such that $\beta^p f^p = f^{p+1}\alpha^p$, then $f = (f^p; p \in \mathbf{Z})$ is called a cochain homomorphism from X to Y. It is easy to verify that each cochain homomorphism f from X to Y induces a family of A-module homomorphisms $(f_*^p; p \in \mathbf{Z})$, where each f_*^p acts from $H^p(X,\alpha)$ to $H^p(Y,\beta)$ by the formula

$$f_*^p(x + R(\alpha^{p-1})) = f(x) + R(\beta^{p-1}) \qquad (x \in N(\alpha^p))$$

Let (X,α), (Y,β) and (Z,γ) be complexes with cochain homomorphisms $f{:}X \to Y$ and $g{:}Y \to Z$ such that $f = (f^p)$ and $g = (g^p)$ and

$$0 \to X^p \xrightarrow{f^p} Y^p \xrightarrow{g^p} Z^p \to 0$$

is exact, then the sequence

$$(1.1) \qquad 0 \to X \xrightarrow{f} Y \xrightarrow{g} Z \to 0$$

is said to be exact.

The proof of the following theorem can be found in [25'].

5.1.1. Theorem. If the sequence (1.1) is exact, then there is a family of homomorphisms $(\theta^p;\ p \in Z)$, called connecting homomorphism, such that the sequence

$$(1.2) \qquad \ldots \to H^p(X,\alpha) \xrightarrow{f_*^p} H^p(Y,\beta) \xrightarrow{g_*^p} H^p(Z,\gamma) \xrightarrow{\theta_*^p} H^{p+1}(X,\alpha) \to \ldots$$

is exact, where θ_* is defined by

$$\theta_*^p(z + R(\gamma^{p-1})) = x + R(\alpha^p)$$

with $z \in N(\gamma^p)$, $x \in N(\alpha^{p+1})$, $f^{p+1}(x) = \beta^p(y)$ and $g^p(y) = z$ for some $y \in Y^p$.

5.1.2. Corollary. If the sequence (1.1) is exact, then the exactness of any two of the complexes (X,α), (Y,β), (Z,γ) implies that of the third.

Proof. By Theorem 5.1.1, the exactness of (1.1) implies that of (1.2). Now assume that (X,α), (Y,β) are exact. Then $H^p(X,\alpha) = 0$ and $H^p(Y,\beta) = 0$ for all $p \in Z$, thus the exactness of the section

$$0 = H^p(Y,\beta) \xrightarrow{g_*^p} H^p(Z,\gamma) \xrightarrow{\theta_*^p} H^{p+1}(X,\alpha) = 0$$

implies $H^p(Z,\gamma) = 0$ for all p, so (Z,γ) is exact. The other two cases are treated in the same way

Let $\sigma = (\sigma_1,\sigma_2,\ldots,\sigma_n)$ be a family of indeterminates. We denote by $\Lambda[\sigma]$ the exterior algebra over C generated by σ. For $p \in Z$, $0 \le p \le n$, $\Lambda^p[\sigma]$ denotes the linear subspace of $\Lambda[\sigma]$ consisting of those elements of degree p in $\sigma_1,\sigma_2,\ldots,\sigma_n$. We set $\Lambda^p[\sigma] = 0$ for $p < 0$ or $p > n$. By definition, elements of $\Lambda^p[\sigma]$ have the form

$$f = \sum_{1 \le j_1 < \ldots < j_p \le n} k_{j_1 \ldots j_p} \, \sigma_{j_1} \wedge \ldots \wedge \sigma_{j_p}$$

where $\wedge$ denotes the exterior product which has the property that $\sigma_i \wedge \sigma_j = -\sigma_j \wedge \sigma_i$ for all i, j = 1,2,..,n, and $k_{j_1 \ldots j_p} \in \mathbb{C}$.

If X is an A-module, then $\Lambda[\sigma,X]$ [resp. $\Lambda^p[\sigma,X]$] is defined to be the tensor product $X \otimes \Lambda[\sigma]$ [resp. $X \otimes \Lambda^p[\sigma]$]. Elements in $\Lambda^p[\sigma,X]$ have the following form.

$$\xi = \sum_{1 \le j_1 < \ldots < j_p \le n} x_{j_1 \ldots j_p} \, \sigma_{j_1} \wedge \ldots \wedge \sigma_{j_p} \tag{1.3}$$

where the symbol $\otimes$ has been suppressed. The element ξ will be called an exterior form of degree p. Note that $\Lambda[\sigma,X]$ can be regarded as a $\Lambda[\sigma,A]$-module with the product induced by the equality

$$\alpha \wedge \xi = \sum_{1 \le k_1 < \ldots < k_q \le n} \sum_{1 \le j_1 < \ldots < j_p \le n} \alpha_{k_1 \ldots k_q} x_{j_1 \ldots j_p} \sigma_{k_1} \wedge \ldots \wedge \sigma_{k_q} \wedge \sigma_{j_1} \wedge \ldots \wedge \sigma_{j_p} \tag{1.4}$$

where ξ is given by (1.3) and

$$\alpha = \sum_{1 \le k_1 < \ldots < k_q \le n} \alpha_{k_1 \ldots k_q} \, \sigma_{k_1} \wedge \ldots \wedge \sigma_{k_q}$$

lies in $\Lambda^p[\sigma,A]$.

Assume that $a = (a_1, \ldots, a_n)$ is a commuting system in A. We define an endormorphism δ_a on $\Lambda[\sigma,X]$ by the relation

$$\delta_a \xi = (a_1 \sigma_1 + \ldots + a_n \sigma_n) \wedge \xi \quad (\xi \in \Lambda[\sigma,X]). \tag{1.5}$$

Let $\delta_a{}^p$ be the restriction of δ_a to $\Lambda^p[\sigma,X]$. Then $\delta_a{}^{p+1} \delta_a{}^p = 0$ for all $p \in \mathbb{Z}$ by the commutativity of the system $(a_1, \ldots, a_n)$. Thus we obtain the complex

$$((\Lambda^p[\sigma,X], \delta_a{}^p); p \in \mathbb{Z}), \tag{1.6}$$

which is clearly of finite length since $\Lambda^p[\sigma,X] = 0$ for $p < 0$ or $p > n$.

The cohomology of (1.6) will be denoted $(H^p(X,a);\ p \in Z)$.

5.1.3 Definition. The commuting system $a = (a_1,\dots,a_n)$ in A is said to be nonsingular on X if the complex (1.6) is exact, or its cohomology is trivial; otherwise, a is singular.

In this chapter our main concern will be the special case $A = L(X)$ where X is either a Banach or Frechet space. Let $a = (a_1,\dots,a_n)$ be a commuting system in $L(X)$. From the foregoing, we obtain the complex $((\Lambda^p[\sigma,X],\ \delta_a{}^p);\ p \in Z)$ of finite length, which is then called a complex of Banach or Frechet spaces, according to the nature of X. (We note, however, that we are still considering a special case.) In what follows, without special mention, we shall assume that X is a Banach or Frechet space over the complex field **C**..

5.1.4 Definition. The set $\sigma(a,X)$ defined by

$$(1.7) \qquad \{z \in \mathbf{C}^n:\ z - a = (z_1 - a_1,\dots,z_n - a_n) \text{ is singular on } X\}$$

is called the joint (Taylor) spectrum of system a over the complex field **C**.

Let Y be a closed subspace of X which is a_j-invariant for each $j = 1,\dots,n$. Then Y is called a-invariant.

5.1.5 Proposition. If Y is an a-invariant closed subspace of X for the system $a = (a_1,\dots,a_n)$, then each of $\sigma(a,X)$, $\sigma(a,Y)$ and $\sigma(a, X/Y)$ is contained in the union of the other two.

Proof. Since the sequence

$$0 \to Y \xrightarrow{i} X \xrightarrow{j} X/Y \to 0$$

is exact, where i [j] is the natural imbedding [surjection], the sequence

$$0 \to \Lambda^p[\sigma,Y] \xrightarrow{i^p} \Lambda^p[\sigma,X] \xrightarrow{j^p} \Lambda^p[\sigma,X/Y] \to 0$$

is thus exact for each p, where i^p, j^p are maps induced by i, j. By definition,

$$0 \to \Lambda[\sigma,Y] \xrightarrow{I} \Lambda[\sigma,X] \xrightarrow{J} \Lambda[\sigma,X/Y] \to 0$$

is exact, where $I = (i^p)$, $J = (j^p)$. For given $z \in C^n$, let $\alpha = z-a$ denote the n-fold translate in (1.7), $\alpha|Y$ its restriction to Y and α/Y the coinduced system on X/Y. Let δ_α, $\delta_{\alpha|Y}$ and $\delta_{\alpha/Y}$ be the endomorphisms defined by (1.5) with a,..., replaced by α,... . Since it is easy to see that $\delta_\alpha{}^p i^p = i^{p+1}(\delta_{\alpha|Y})^p$ and $(\delta_{\alpha/Y})^p j^p = j^{p+1}\delta_\alpha{}^p$, the correspondences I: $\Lambda[\sigma,Y] \to \Lambda[\sigma,X]$ and J: $\Lambda[\sigma,X] \to \Lambda[\sigma,X/Y]$ are cochain homomorphisms. Now Cor. 5.1.2 applies to give the conclusion of the proposition.

For later use we introduce two kinds of operators. Clearly $\Lambda[\sigma]$ has a natural Hilbert space structure in which the elements $1 \in C$ (= $\Lambda^0[\sigma]$) and

$$\sigma_{j_1}\wedge\ldots\wedge\sigma_{j_p} \quad (1 \le j_1 < \ldots < j_p \le n;\ p = 1,2,\ldots,n)$$

form an orthonormal basis. Define operators

$$s_j\xi = \sigma_j \wedge \xi \qquad (\xi \in \Lambda[\sigma]\ ;\ j = 1,2,\ldots,n). \tag{1.8}$$

Then their adjoints are defined by

$$s_j^*(\xi_j' + \sigma_j \wedge \xi_j'') = \xi_j'' \quad (j = 1,\ldots,n), \tag{1.9}$$

where $\xi = \xi_j' + \sigma_j \wedge \xi_j'' \in \Lambda[\sigma]$ with ξ_j', ξ_j'' not containing σ_j. The following anticommutative relations are evident:

$$s_j s_k + s_k s_j = 0, \tag{1.10}$$

$$s_j s_k^* + s_k^* s_j = \begin{cases} 1 & \text{if } j = k \\ 0 & \text{if } j \ne k \end{cases}, \tag{1.11}$$

where j,k = 1,...,n.

5.1.6 Proposition. Let $a = (a_1,\ldots,a_n)$ and $b = (b_1,\ldots, b_m)$ be commuting systems in L(X) such that (a,b) is also a commuting system. If a is nonsingular, then (a,b) is also.

Proof. It suffices to prove the assertion for the case m = 1. Let $\sigma' = (\sigma_1,\sigma_2,\ldots,\sigma_n,\sigma_{n+1})$, $a' = (a_1,\ldots,a_{n+1})$, $a_{n+1} = b_1$ and form the sequence

$$(1.12) \qquad 0 \to \Lambda^p[\sigma,X] \xrightarrow{\mu^p} \Lambda^{p+1}[\sigma',X] \xrightarrow{\nu^p} \Lambda^{p+1}[\sigma,X] \to 0$$

where μ^p is induced by s_{n+1} and ν^p by $s_{n+1}{}^*s_{n+1}$, resp. Since $R(\mu^p) = N(\nu^p)$, μ^p is injective and ν^p is surjective, (1.12) is exact. Moreover, it is easy to compute that

$$\mu^{p+1}\delta_\alpha{}^p = \delta_{\alpha'}{}^{p+1}\mu^p,$$
$$\nu^{p+1}\delta_{\alpha'}{}^{p+1} = \delta_\alpha{}^{p+1}\nu^p;$$

hence the exactness of the complex $(\Lambda[\sigma,X],\delta_a) = ((\Lambda^p[\sigma,X], \delta_a{}^p); p\in Z)$, from the hypothesis, implies that of the complex $(\Lambda[\sigma',X], \delta_{a'}) = ((\Lambda^p[\sigma',X],\delta_{a'}{}^p); p\in Z)$ by Cor. 5.1.2. Thus a' is nonsingular.

5.1.7 Lemma. Let $a = (a_1,\dots,a_n)$ be a commuting system in $L(X)$ and let v be a permutation on $\{1,2,\dots,n\}$. Define the permuted system $a_v = (a_{v(1)},\dots,a_{v(n)})$. Then we have

$$\sigma(a_v,X) = \{(z_{v(1)},\dots,z_{v(n)}): (z_1,\dots,z_n) \in \sigma(a,X)\}.$$

Proof. Define the automorphism $\tilde{v}$ on $\Lambda[\sigma,X]$ by the relation

$$\tilde{v}(x\sigma_{j_1}\wedge\dots\wedge\sigma_{j_p}) = x\sigma_{v^{-1}(j_1)}\wedge\dots\wedge\sigma_{v^{-1}(j_p)}$$

where $x\in X$ and $1 \le j_1 < \dots < j_p \le n$. Since $\tilde{v}\delta_\alpha = \delta_{\alpha_v}\tilde{v}$, the systems $\alpha = z - a$ and $\alpha_v = z_v - a_v$ are simultaneously nonsingular, where δ_α, δ_{α_v} are defined by (1.5) with α, α_v in place of a, resp.

5.1.8 Corollary. Let $a = (a_1,\dots,a_n)$ be a commuting system in $L(X)$. Then $\sigma(a,X)$ is a bounded subset of C^n.

Proof. Let Π_j be the projection of C^n onto its jth coordinate. Proposition 5.1.6 and Lemma 5.1.7 imply that $\Pi_j[\sigma(a,X)] \subset \sigma(a_j,X)$ for $j = 1,2,\dots,n$, and this gives the boundedness of $\sigma(a,X)$.

After introducing some preparatory material, we shall show that $\sigma(a,X)$ is in fact a nonempty compact subset of C^n when $X \ne \{0\}$.

The following theorem gives a sufficient condition for a commuting system to be nonsingular.

5.1.9 Theorem. Let $a = (a_1,\dots,a_n)$ be a commuting system in $L(X)$. If there exists $b_i \in L(X)$ $(i = 1,\dots,n)$ commuting with each a_j such that $a_1b_1 + \dots + a_nb_n = 1$, then a is nonsingular.

Proof. We define the operator on $\Lambda[\sigma,X]$ by $\gamma_b = b_1s_1^* + \dots + b_ns_n^*$ and note that for every $\xi \in \Lambda[\sigma,X]$

$$(1.13) \qquad (\delta_a\gamma_b + \gamma_b\delta_a)\xi = \sum_{j=1}^{n}\sum_{k=1}^{n} a_jb_k(s_js_k^* + s_k^*s_j)\xi = \xi$$

by (1.11). Now let $\xi \in N(\delta_a^p)$ and let $\eta = \gamma_b^p\xi$ where γ_b^p is the restriction of γ_b to $\Lambda^p[\sigma,X]$. Then (1.13) implies $\delta_a^{p-1}\eta = (\delta_a^{p-1}\gamma_b^p + \gamma_b^{p+1}\delta_a^p)\xi = \xi$. Consequently, $R(\delta_a^{p-1}) = N(\delta_a^p)$, and so a is nonsingular.

To consider the dual of a complex and further properties of commuting systems, we return to the general case of $(X,\alpha) = ((X^p,\alpha^p);\ p \in Z)$, where each X^p is a Banach space. Clearly $\alpha^{p*} \in L(X^{(p+1)*},X^{p*})$ and $\alpha^{p*}\alpha^{(p+1)*} = \xi$. Hence $(X^*,\alpha^*) = ((X^{-p*}, \alpha^{(-p-1)*});\ p \in Z)$ is a complex of Banach spaces which we call the dual of (X,α).

5.1.10 Proposition. The complex (X,α) is exact if and only if its dual is exact.

Proof. Let Y be any Banach space, and let $\alpha \in L(X,Y)$. Then the following well-known facts, (i) $R(\alpha)^\perp = N(\alpha^*)$, (ii) $R(\alpha)$ closed iff $R(\alpha^*)$ closed iff $R(\alpha^*) = N(\alpha)^\perp$, imply the equivalence of the sequences

$$\{R(\alpha^{p-1}) = N(\alpha^p):\ p \in Z\} \quad \text{and}$$
$$\{R(\alpha^{p*}) = N(\alpha^{(p-1)*}):\ p \in Z\}.$$

Hence (X,α) is exact if and only if (X^*,α^*) is exact.

5.1.11 Theorem. Let $a = (a_1,\dots,a_n)$ be a commuting system in $L(X)$. Then $\sigma(\alpha^*,X^*) = \sigma(\alpha,X)$.

Proof. By Prop. 5.1.10 the system a is nonsingular exactly if a^* is nonsingular. This implies the equality of their joint spectra.

5.1.12 Proposition. Let $(X,\alpha) = ((X^p,\alpha^p);p\in Z)$ be a complex of Banach spaces. If (X,α) is exact, then for each $p \in \mathbf{Z}$ there exists $\delta_p > 0$ such that every other complex $(X,\beta) = ((X^p,\beta^p);p\in \mathbf{Z})$ satisfies $R(\beta^{p-1}) = N(\beta^p)$ whenever

$$(1.14) \qquad \|\beta^{p-1} - \alpha^{p-1}\| < \delta_p \quad \text{and} \quad \|\beta^p - \alpha^p\| < \delta_p.$$

Proof. The hypotheses imply the equalities $R(\alpha^j) = N(\alpha^{j+1})$ for $j = p-1, p$. From the open mapping theorem there exists $r > 0$ such that for each $y \in N(\alpha^{j+1})$ we can find $x \in X^j$ satisfying

$$(1.15) \qquad \alpha^j x = y \quad \text{and} \quad \|x\| \le r\|y\| \qquad \text{for } j = p-1, p.$$

Choose $\delta = \delta_p$ in (1.14) so that $r\delta(2+r\delta) < 1$. For each $u \in N(\beta^p)$, (1.14) implies

$$(1.16) \qquad \|\alpha^p u\| = \|\alpha^p u - \beta^p u\| < \delta\|u\|.$$

Since $\alpha^p u \in N(\alpha^{p+1})$, we may choose $u' \in X^p$ such that

$$(1.17) \qquad \alpha^p u' = \alpha^p u, \qquad \|u'\| \le r\|\alpha^p u\| \le r\delta\|u\|$$

by (1.15), (1.16). The equality $\alpha^p(u - u') = 0$ combined with (1.15) implies the existence of $v_1 \in X^{p-1}$ such that

$$\alpha^{p-1}v_1 = u - u', \qquad \|v_1\| \le r\|u - u'\|.$$

Hence

$$(1.18) \qquad |v_1\| \le r(\|u\| + \|u'\|) \le r(1 + r\delta)\|u\|.$$

Put $u_1 = u - \beta^{p-1}v_1$. Then

$$(1.19) \qquad \begin{aligned} \|u_1\| &\le \|u - \alpha^{p-1}v_1\| + \|\alpha^{p-1}v_1 - \beta^{p-1}v_1\| \\ &\le \|u'\| + r\delta(1 + r\delta)\|u\| \\ &\le r\delta(2 + r\delta)\|u\| \end{aligned}$$

by (1.14), (1.17) and (1.18). Since $\beta^p u_1 = \beta^p u - \beta^p\beta^{p-1}v_1 = 0$, repeating the preceeding process shows existence of $v_2 \in X^{p-1}$ such that

$$\|v_2\| \le r(1 + r\delta)\|u_1\| \le r\varepsilon(1 + r\delta)\|u_1\|,$$

where $\varepsilon = r\delta(2 + r\delta)$. If we put $u_2 = u_1 - \beta^{p-1}v_2$, then an estimate similar to (1.19) gives

$$\|u_2\| \le \varepsilon\|u_1\| \le \varepsilon^2\|u\|.$$

Inductively, we can find sequences $\{v_k\}$ in X^{p-1} and $\{u_k\}$ in X^p satisfying

$$u_{k+1} = u_k - \beta^{p-1}v_{k+1}$$
$$\|v_k\| \le r\varepsilon^{k-1}(1 + r\delta)\|u\|$$
$$\|u_k\| \le \varepsilon^k\|u\|$$

for $k \ge 1$. Let $v = \Sigma_k v_k$. Then, since

$$\|u - \beta^{p-1}\Sigma_{j\le k} v_j\| = \|u_k\| \le \varepsilon^k\|u\| \to 0,$$

one obtains $u = \beta^{p-1}v$, or equivalently, $R(\beta^{p-1}) = N(\beta^p)$, and the proposition is proved.

5.1.13 Corollary. Let $(X,\alpha) = ((X^p,\alpha^p);p\in \mathbf{Z})$ be a complex of Banach spaces of finite length. If (X,α) is exact, then there exists $\delta > 0$ such that the complex (X,β) is also exact if it satisfies

(1.20) $$\|\beta^p - \alpha^p\| < \delta$$

for all p for which $X^p \ne \{0\}$.

Proof. We may assume that $X^p = \{0\}$ for $p < 0$ and $p > n$. By Prop 5.1.12 there exists δ_p for each p such that $R(\beta^{p+1}) = N(\beta^p)$ whenever $\|\beta^j - \alpha^j\| < \delta_p$ $(j = p-1,p)$. Let $\delta = \min\{\delta_p : 0 \le p \le n\}$. Then clearly (X,β) is exact if (1.20) holds.

5.1.14 Corollary. If $a = (a_1,\dots,a_n)$ is a commuting system in $L(X)$, then $\sigma(a,X)$ is compact.

Proof. By Cor. 5.1.8 $\sigma(a,X)$ is bounded, and by Cor. 5.1.13 it is closed. Hence the set is compact.

We also need some facts from the theory of vector-valued functions. Let Ω be open in $\mathbf{C}^n$. Then $C^k(\Omega,X)$ denotes the space of all X-valued functions having continuous derivatives through order k, where X is a Banach space, let $C_0{}^k(\Omega,X)$ be the set of those f in $C^k(\Omega,X)$ with compact support and let $A(\Omega,X)$ be the analytic X-valued functions on Ω. It is well-known that these three linear spaces are Frechet spaces in their usual topologies. We shall also use the following operators.

$$\partial = \sum_{j=1}^{n} \frac{\partial}{\partial z_j} dz_j \quad \text{and} \quad \bar{\partial} = \sum_{j=1}^{n} \frac{\partial}{\partial \bar{z}_j} d\bar{z}_j,$$

where $z \in \mathbf{C}^n$, and $dz = dz_1 \wedge \ldots \wedge dz_n$, $d\bar{z} = d\bar{z}_1 \wedge \ldots \wedge d\bar{z}_n$.

Let W denote one of the spaces defined above. We write $\Lambda^p[d\bar{z}, W]$ to denote the space spanned by the differential forms $f\, d\bar{z}_{j1} \wedge \ldots \wedge d\bar{z}_{jp}$ of degree p in $d\bar{z}_1, \ldots, d\bar{z}_n$, where $f \in W$

5.1.15 Theorem (Stokes' formula). Assume that $\Delta\,(\subset \Delta^-) \subset \Omega$ is an open relatively compact set with a piecewise smooth boundary $\partial\Delta$. Then for every $f \in \Lambda^{n-1}[d\bar{z}, C^\infty(\Omega)]$ we have

$$\int_{\partial\Delta} f \wedge dz = \int_{\Delta} \bar{\partial} f \wedge dz. \tag{1.21}$$

Proof. Since $\partial f \wedge dz = 0$, by Stokes formula for smooth functions we have

$$\int_{\Delta} \bar{\partial} f \wedge dz = \int_{\Delta} (\partial + \bar{\partial}) f \wedge dz = \int_{\Delta} df \wedge dz$$

$$= \int_{\Delta} d(f \wedge dz) = \int_{\partial\Delta} f \wedge dz,$$

where $d = \partial + \bar{\partial}$.

5.1.16 Corollary. If $f \in \Lambda^{n-1}[d\bar{z}, C_0{}^\infty(\Omega)]$, then

$$\int_{\Omega} f \wedge dz = 0. \tag{1.22}$$

Proof. Let us choose D satisfying the properties given in Th. 5.1.15 and containing supp f. Then (1.22) follows immediately from (1.21)

5.1.17 Definition. Let $K \subset \Omega \subset C^n$ with K compact, Ω open. Then the set

$$\hat{K} = \cap\{z \in \Omega: |f(z)| \leq ||f||_K, f \in A(\Omega)\}$$

is called the _holomorphically convex hull_ of K, where $||.||_K$ denotes the uniform norm on K. If $\hat{K}$ is compact in Ω for each compact $K \subset \Omega$, then Ω is said to be a _domain of holomorphy_.

As an example, let Ω be a convex open set in C^n. Since for each compact set K in Ω we know that $\hat{K}$ is contained in the convex hull of K and since the latter is compact in Ω, so also is $\hat{K}$. Thus Ω is a domain of holomorphy.

5.1.18 Theorem. For every domain of holomorphy $U \subset C^n$, the following sequence is exact.

$$0 \to A(U,X) \xrightarrow{i} C^\infty(U,X) \xrightarrow{\bar\partial} \Lambda^1[d\bar z, C^\infty(U,X)] \xrightarrow{\bar\partial} \dots \tag{1.23}$$

$$\dots \xrightarrow{\bar\partial} \Lambda^n[d\bar z, C^\infty(U,X)] \to 0.$$

Proof. By [1', Cor 4.2.6] the sequence

$$0 \to A(U) \xrightarrow{i} C^\infty(U) \xrightarrow{\bar\partial} \Lambda^1[d\bar z, C^\infty(U)] \xrightarrow{\bar\partial} \dots \tag{1.24}$$

$$\xrightarrow{\bar\partial} \Lambda^n[d\bar z, C^\infty(U)] \to 0$$

is exact. Since it is well-known that $C^\infty(U,X)$ is isomorphic to $C^\infty(U) \hat{\otimes} X$, an easy argument shows that (1.24) implies (1.23).

5.1.19 Theorem. If Ω is a domain of holomorphy in $\mathbf{C}^n$, then there exists $f \in A(\Omega)$ such that f cannot be analytically continued outside Ω.

Proof. Let D be an open polydisc in $\mathbf{C}^n$ centered at the origin. For each $\xi \in \Omega$, let $r = r(\xi) > 0$ be such that the polydisc $D_\xi = \xi + rD$ is the largest one contained in Ω. Assume that E is a countable dense set in Ω, and construct a sequence $\{\xi_j\}$ from E such that each element of E appears infinitely often in the sequence. Moreover, let $\{K_j\}$ be a sequence of compact sets in Ω with the property that $K_j \subset K_{j+1}$ ($j = 1,2,\ldots$). Since Ω is a domain of holomorphy, the holomorphically convex hull $\hat{K}_j$ of each K_j is compact in Ω, and thus for each ξ_j there is a $z_j \in D_{\xi_j}$ with $z_j \notin \hat{K}_j$. By the definition of $\hat{K}_j$, there exists $f_j \in A(\Omega)$ such that $|f_j(z_j)| > \|f_j\|_{K_j}$. Without loss of generality we may assume that

$$(1.25) \qquad f_j(z_j) = 1 \quad \text{and} \quad \|f_j\|_{K_j} < 2^{-j}.$$

(In fact, dividing f_j by $f_j(z_j)$, if necessary, gives the first relation in (1.25), while the second results by replacing f_j with a sufficiently high power of itself.) Now set

$$(1.26) \qquad f(z) = \prod_{j=1}^{\infty} [1 - f_j(z)].$$

Then $f \in A(\Omega)$ and it is not identically zero on each component of Ω. Construction of the sequence $\{\xi_j\}$ and f implies that f vanishes at infinitely many points of D_ξ for each $\xi \in E$. If f could be continued analytically to a neighborhood of D_ξ^-, then f would vanish on D_ξ^-. Hence this contradiction shows that f cannot be continued analytically outside D_ξ^-. Since E is dense in Ω, what was just proved shows that f cannot be continued analytically outside Ω.

In the sequel we shall restrict our attention to the case of Banach spaces without further mention.

Let $\{X^p;\ p \in \mathbf{Z}\}$ be a sequence of Banach spaces, and let Ω be open in C^n. For each $z \in \Omega$, suppose that $\alpha^p(z) \in L(X^p, X^{p+1})$. If $\alpha^{p+1}(z)\alpha^p(z) = 0$ for all $p \in Z$ and $z \in \Omega$, then the complex

$$(1.27) \qquad \cdots \to X^{p-1} \xrightarrow{\alpha^{p-1}(z)} X^p \xrightarrow{\alpha^p(z)} X^{p+1} \to \cdots$$

is called a parametrized complex of Banach spaces. If the maps are continuous [resp. analytic], then the complex (1.27) is said to be parametrized continuously [analytically]. If the α^p are continuous [analytic], then they induce maps

$$(\alpha^p f)(z) = \alpha^p(z)f(z) \qquad (f \in C(\Omega, X^p)$$
$$[\text{resp. } (\alpha^p f)(z) = \alpha^p(z)f(z) \qquad (f \in A(\Omega, X^p)].$$

Our main concern is the relationship between the exactness of (1.27) for each $z \in \Omega$ and the exactness of

$$(1.28) \qquad \cdots \to C(\Omega, X^{p-1}) \xrightarrow{\alpha^{p-1}} C(\Omega, X^p) \xrightarrow{\alpha^p} C(\Omega, X^{p+1}) \to \cdots$$

or that of

$$(1.29) \qquad \cdots \to A(\Omega, X^{p-1}) \xrightarrow{\alpha^{p-1}} A(\Omega, X^p) \xrightarrow{\alpha^p} A(\Omega, X^{p+1}) \to \cdots$$

if (1.27) is continuously or analytically parametrized. Also of interest is the intermediate case

$$(1.30) \qquad \cdots \to C^\infty(\Omega, X^{p-1}) \xrightarrow{\alpha^{p-1}} C^\infty(\Omega, X^p) \xrightarrow{\alpha^p} C^\infty(\Omega, X^{p+1}) \to \cdots$$

5.1.20 Theorem. Assume that the complex $(X^p, \alpha^p(z))$ of Banach spaces is continuously parametrized by $z \in \Omega$. If it is exact for each $z \in \Omega$, then the sequence (1.28) is also exact.

Proof. Let $g \in C(\Omega, X^p)$ be such that $\alpha^p(z)g(z) = 0$. We first assume that supp $g = K$ is compact. From the proof of Prop. 5.1.12 it follows that there exists $M_K > 0$ such that each $w \in K$ has a neighborhood U_w so that whenever $z \in U_w$ and $u \in N(\alpha^p(z))$ there corresponds $v \in X^{p-1}$ satisfying

$$(1.31) \qquad u = \alpha^{p-1}(z)v \quad \text{and} \quad \|v\| \le M_K\|u\|.$$

Let $\{U_j = U_{w_j}: j = 1,2,\ldots,m\}$ be a finite open cover of K in Ω such that

$$\|\alpha^{p-1}(z) - \alpha^{p-1}(w_j)\| \le \frac{1}{4M_K},$$

$$\|g(z) - g(w_j)\| \le \frac{1}{4}\|g\|_K$$

for fixed $w_j \in U_j \cap K$ and for all $z \in U_j$. Choose $\phi_j \in C_0(\Omega)$ with supp $\phi_j \subset U_j$ and $0 \le \phi_j \le 1$ $(1 \le j \le m)$ such that $\Sigma\, \phi_j = 1$ on some neighborhood of K. For each j, by (1.31) there is $x_j \in X^{p-1}$ such that $g(w_j) = \alpha^{p-1}(w_j)x_j$ and $\|x_j\| \le M_K\|g\|_K$. Set $f_1 = \Sigma_j\, \phi_j x_j$ and $g_1 = g - \alpha^{p-1}f_1$. Then for $z \in \Omega$ in any of the U_j, one has

$$\|g_1(z)\| \le \sum_{j=1}^{m} \phi_j(z)\|g(z) - \alpha^{p-1}(z)x_j\|$$

$$\le \sum_{j=1}^{m} \phi_j(z)\{\|g(z) - g(w_j)\| + \|\alpha^{p-1}(w_j)x_j - \alpha^{p-1}(z)x_j\|\}$$

$$\le \frac{1}{4}\|g\|_K + \frac{1}{4}\|g\|_K = \frac{1}{2}\|g\|_K$$

and

$$\|f_1(z)\| \le M_K\|g\|_K.$$

Since supp f_1 and supp g_1 are both contained in the union of the U_j, the estimates above remain valid for all $z \in \Omega$. Moreover, $\alpha^p(z)g_1(z) = \alpha^p(z)[g(z) - \alpha^{p-1}(z)f_1(z)] = 0$ for $z \in \Omega$. By repeating this procedure, one obtains two sequences $\{f_n\}$ in $C(\Omega, X^{p-1})$ and $\{g_n\}$ in $C(\Omega, X^p)$ such that

$$g_k = g_{k-1} - \alpha^{p-1}f_{k-1} = \cdots = g - \alpha^{p-1}\Sigma_1^{k-1} f_n;$$

$$\|g_k(z)\| \le 2^{-k}\|g\|_K;$$

$$\|f_k(z)\| \le 2^{-k+1}M_K\|g\|_K \quad (z \in \Omega).$$

Thus the series Σf_k converges uniformly on Ω to a function $f \in C(\Omega, X^{p-1})$ and $g = \alpha^{p-1}f$.

For the general case, we write Ω as the countable union of compact sets K_j in Ω such that $K_j \subset \text{Int } K_{j+1}$ for all j. Let $\{\psi_j\}$ be a corresponding sequence from $C_0(\Omega)$ satisfying the conditions that $\psi_j = 1$ on a neighborhood of K_j and supp $\psi_j \subset \text{Int } K_{j+1}$. By the first part of the proof, given $g \in C(\Omega, X^p)$ with $\alpha^p g = 0$, one can find $f_j \in C(\Omega, X^{p-1})$ such that $\alpha^{p-1}f_j = g\chi_j$, where $\chi_1 = \psi_1$ and $\chi_j = \psi_j - \psi_{j-1}$ for $j \ge 2$. Since supp $\chi_j \subset \text{Int } K_{j+1} \backslash K_{j-1}$ for $j \ge 2$, we can choose f_j so that supp $f_j \subset (\text{Int } K_{j+1}) \backslash K_{j-1}$. Moreover, f_1 may be chosen with supp $f_1 \subset \text{Int } K_2$. Then the series $\Sigma_j f_j$ converges uniformly on every compact set in Ω and its sum $f \in C(\Omega, X^{p-1})$ satisfies $\alpha^{p-1}f = g$ since $\Sigma_j \chi_j = 1$ on Ω. This completes the proof.

By means of a more complicated argument, one can also prove the following

5.1.21 Theorem. Assume that the following analytically parametrized complex of Banach spaces,

$$\cdots \rightarrow X^{p-1} \xrightarrow{\alpha^{p-1}(z)} X^p \xrightarrow{\alpha^p(z)} X^{p+1} \rightarrow \cdots,$$

is exact for $z \in \Omega$, an open set in C^n. Then

(i) the sequence (1.30) is exact;

(ii) if Ω is a domain of holomorphy, then the sequence (1.29) is also exact.

See [34', Th II.10.8] for the proof of (i) and [22', Ch. IV] for that of (ii).

Remark. Theorem 5.1.21 remains valid for the case of Frechet spaces; we use this fact later.

The Cauchy-Weil integral is also basic for the multivariate theory of operators. Let U be open in C^n, let $a_j \in A(U, L(X))$ ($j = 1, \ldots, n$), and let $\alpha = a(z) = (a_1(z), \ldots, a_n(z))$ be a commuting system for each $z \in U$, where X is a Frechet space. Put $\delta_\alpha = \delta_{a(z)} = a_1(z)\sigma_1 + \dots + a_n(z)\sigma_n$ and denote

$$\gamma_U(\alpha,X) = \{z \in U: ((\Lambda^p[\sigma,X],\delta_{a(z)}{}^p); p = 0,\dots,n) \text{ is exact}\}.$$

It follows from Th. 5.1.21(ii) that if V is a domain of holomorphy in $\gamma_U(\alpha,X)$, then $((A(V,\Lambda^p[\sigma,X]), \delta_\alpha{}^p); p = 0,\dots,n)$ is exact, where $\delta_\alpha{}^p$ is the restriction of δ_α to $A(V, \Lambda^p[\sigma,X])$. Since it is clear that $A(V, \Lambda^p[\sigma,X])$ and $\Lambda^p[\sigma, A(V,X)]$ are isomorphic, the complex

$$((\Lambda^p[\sigma,A(V,X)], \delta_\alpha{}^p); p = 0,\dots,n)$$

is also exact.

Consider the set of indeterminates $\sigma \cup d\bar{z} = (\sigma_1,\dots,\sigma_n,d\bar{z}_1,\dots,d\bar{z}_m)$, and note that a_j commutes with $\frac{\partial}{\partial \bar{z}_k}$ for $j = 1,\dots,n$ and $k = 1,\dots,m$. We then obtain a new commuting system

$$(a_1, \dots,a_n, \frac{\partial}{\partial \bar{z}_1},\dots, \frac{\partial}{\partial \bar{z}_m})$$

and write

$$\delta_\alpha + \bar{\partial} = a_1\sigma_1 + \dots + a_n\sigma_n + \frac{\partial}{\partial \bar{z}_1}d\bar{z}_1 + \dots + \frac{\partial}{\partial \bar{z}_m}d\bar{z}_m.$$

Then

$$((\Lambda^p[\sigma \cup d\bar{z}, C^\infty(U,X)], \delta_\alpha+\bar{\partial}); p = 0,\dots, n+m)$$

is a complex.

Using Th. 5.1.21(ii) one can prove the following theorem (see [34',Th. III.8.1] for details).

5.1.22 Theorem. Assume that U is open in C^m and that $\alpha = (a_1,\dots,a_n)$ is a commuting system in $A(U,L(X))$. Then $H^p(C^\infty(V,X), \delta_\alpha+ \bar{\partial}) = 0$ for $1 \le p \le m+n$ and every open $V \subset \gamma_U(\alpha,X)$.

Proof. We first prove that the theorem holds for every open polydisc $V = D$ and $1 \le p \le m+n$. Let $\eta \in \Lambda^p[\sigma \cup d\bar{z}, C^\infty(D,X)]$ satisfy $(\delta_\alpha+$

$\bar\partial)\eta = 0$. Then η has the decomposition $\eta = \eta_0 + \ldots + \eta_p$ with η_j $(0 \le j \le p)$ having degree j in $d\bar z_1, \ldots, d\bar z_m$. Suppose ξ satisfies the equation $(\delta_\alpha + \bar\partial)\xi = \eta$. Then ξ has the decomposition $\xi = \xi_0 + \ldots + \xi_{p-1}$ where each ξ_j $(0 \le j \le p-1)$ is of degree j in $d\bar z_1, \ldots, d\bar z_m$. These decompositions give a system of equations as follows:

$$\bar\partial\xi_{p-1} = \eta_p$$
$$\delta_\alpha\xi_{p-1} + \bar\partial\xi_{p-2} = \eta_{p-1}$$
$$\ldots\ldots\ldots\ldots$$
$$\delta_\alpha\xi_j + \bar\partial\xi_{j-1} = \eta_j \quad (1 \le j \le p-1)$$
$$\ldots\ldots\ldots\ldots$$
$$\delta_\alpha\xi_0 = \eta_0$$
$$\bar\partial\eta_p = 0$$
$$\ldots\ldots\ldots\ldots$$
$$\delta_\alpha\eta_j + \bar\partial\eta_{j-1} = 0 \quad (1 \le j \le p)$$
$$\ldots\ldots\ldots\ldots$$
$$\delta_\alpha\eta_0 = 0.$$

We solve this system step by step. Since $\bar\partial\eta_p = 0$, there is a solution ξ_{p-1} for $\bar\partial\xi_{p-1} = \eta_p$ by Th. 5.1.18. Because $\bar\partial(\eta_{p-1} - \delta_\alpha\xi_{p-1}) = \bar\partial\eta_{p-1} + \delta_\alpha\eta_p = 0$, the equation $\bar\partial\xi_{p-2} = \eta_{p-1} - \delta_\alpha\xi_{p-1}$ has a solution ξ_{p-2} by Th. 5.1.18 again. Now if the solution ξ_j has been found, then solution ξ_{j-1} of $\delta_\alpha\xi_j + \bar\partial\xi_{j-1} = \eta_j$ can also be found, since

$$\bar\partial(\eta_j - \delta_\alpha\xi_j) = \bar\partial\eta_j + \delta_\alpha(\eta_{j+1} - \delta_\alpha\xi_{j+1}) = 0$$

by the relation between η_j and η_{j+1}. We clarify the last step as follows. Let ξ_0^* be a solution of the equation $\bar\partial\xi_0^* = \eta_1 - \delta_\alpha\xi_1$. Then $\delta_\alpha\bar\partial\xi_0^* = \delta_\alpha\eta_1 = -\bar\partial\eta_0$, and hence $\bar\partial(\delta_\alpha\xi_0^* - \eta_0) = 0$. Since the degree of the form $\delta_\alpha\xi_0^* - \eta_0$ in $d\bar z_1, \ldots, d\bar z_m$ is zero, the coefficients of $\delta_\alpha\xi_0^* - \eta_0$ are analytic. By the equality $\delta_\alpha(\delta_\alpha\xi_0^* - \eta_0) = 0$ and the hypothesis $D \subset \gamma_U(\alpha, X)$, we may find an analytic solution θ_0 to the equation $\delta_\alpha\xi_0^* - \eta_0 = \delta_\alpha\theta_0$. Then $\xi_0 = \xi_0^* + \theta_0$ is the desired form and $\xi = \xi_0 + \ldots + \xi_{p-1}$ is a solution to the equation $(\delta_\alpha + \bar\partial)\xi = \eta$ on D.

In the general case, we want to solve $(\delta_\alpha+\bar\partial)\xi=\eta$ on V subject to $(\delta_\alpha+\bar\partial)\eta = 0$, where V is an arbitrary open set in $\gamma_U(\alpha,X)$. If p = 0, then $\eta \in A(V,X)$ and $\delta_\alpha\eta = 0$ by the condition $(\delta_\alpha+\bar\partial)\eta = 0$. Hence $\eta = 0$ by the inclusion $V \subset \gamma_U(\alpha,X)$. Next, assume that the conclusion holds for degree p-1. Let $\{D_j: j = 1,2,...\}$ be a sequence of open polydiscs in $\gamma_U(\alpha,X)$ whose union is $\gamma_U(\alpha,X)$. By the first part of the proof, we can find forms ξ_1 and ξ_2 such that $(\delta_\alpha+\bar\partial)\xi_j = \eta$ on D_j (j=1,2), so that $(\delta_\alpha+\bar\partial)(\xi_1-\xi_2) = 0$ on $D_1 \cap D_2$. By the induction hypothesis, there is a form $\theta \in \Lambda^{p-2}[\sigma \cup d\bar z, C^\infty(D_1 \cap D_2,X)]$ such that $\xi_1-\xi_2 = (\delta_\alpha+\bar\partial)\theta$ on $D_1 \cap D_2$. Put $\zeta_1 = \xi_1$ and

$$\zeta_2 = \left\{ \begin{array}{ll} \zeta_1 & \text{on } D_1 \\ \xi_2 + (\delta_\alpha + \bar\partial)\theta & \text{on } D_2 \end{array} \right\}.$$

Then $(\delta_\alpha+\bar\partial)\zeta_2 = \eta$ on $D_1 \cup D_2$. If necessary, by shrinking D_1, D_2 and multi- plying θ by a suitable C_0^∞-function, we may assume that ζ_2 is an element of $\Lambda^{p-1}[\sigma \cup d\bar z, C^\infty(D_1 \cup D_2,X)]$. Continuing successively in this way, we can find a sequence of forms $\{\zeta_j\}$ such that

$$\zeta_j \in \Lambda^{p-1}[\sigma \cup d\bar z, C^\infty(D_1 \cup \ldots \cup D_j,X)],$$
$$\zeta_{j+1} = \zeta_j \text{ and } (\delta_\alpha+\bar\partial)\zeta_j = \eta \text{ on } D_1 \cup \ldots \cup D_j.$$

If we define $\xi = \lim \zeta_j$, then ξ is the required solution, and the proof is complete.

For the commuting system $\alpha = (a_1,\ldots,a_n)$ in A(U,L(X)) and the open set $V \subset U$, we define the space

$$Z_\alpha{}^p(V,X) = \{\xi \in \Lambda^p[\sigma \cup d\bar z, C^\infty(V,X)]: (\delta_\alpha+\bar\partial)\xi = 0\}.$$

5.1.23 Corollary. With the hypotheses of Th. 5.1.22, for every $\xi \in Z_\alpha{}^p(V,X)$ there exists $\eta \in \Lambda^{p-1}[\sigma \cup d\bar z, C^\infty(V,X)]$ such that $(\delta_\alpha+\bar\partial)\eta = \xi$. Moreover, we may choose the form η so that supp η is contained in an arbitrary neighborhood of supp ξ.

Proof. By Th. 5.1.22 there exists $\eta_0 \in \Lambda^{p-1}[\sigma \cup d\bar{z}, C^\infty(V,X)]$ such that $(\delta_\alpha + \bar{\partial})\eta_0 = \xi$. Let W be an arbitrary neighborhood of supp ξ in V and put Ω = W\supp ξ. Then $(\delta_\alpha + \bar{\partial})\eta_0 = 0$ on Ω, and hence we have $\eta_0 = (\delta_\alpha + \bar{\partial})\theta_0$ for some $\theta_0 \in \Lambda^{p-2}[\sigma \cup d\bar{z}, C^\infty(\Omega,X)]$ again by the previous theorem. Choose a function $\phi \in C^\infty(V)$ such that $\phi = 0$ on a neighborhood of supp ξ and $\phi = 1$ outside another neighborhood in W. Then the support of the form $\eta = \eta_0 - (\delta_\alpha + \bar{\partial})\phi\theta_0$ is contained in W and η satisfies $(\delta_\alpha + \bar{\partial})\eta = \xi$.

5.1.24 Definition. Assume that U is open in $\mathbf{C}^m$ and that $\alpha = (a_1, \ldots, a_n)$ is a commuting system in A(U,L(X)) such that $k_U(\alpha,X) = U\backslash\gamma_U(\alpha,X)$ is compact. For every $\xi \in Z_\alpha{}^m(U,X)$, denote by ξ_α a solution to the equation $(\delta_\alpha + \bar{\delta})\eta = \xi$ in the open set $\gamma_U(\alpha,X)$. Let $\phi \in C^\infty(V)$ be such that $\phi = 0$ on a neighborhood of $k_U(\alpha,X)$ and that supp$(1 - \phi)$ is compact. We define the integral

$$(1.32) \qquad \mu_\alpha(\xi) = (-1)^{m+1}\int_U P_\sigma[(\delta_\alpha + \bar{\partial})\phi\xi_\alpha - \xi] \wedge dz,$$

where P_σ is a mapping on $\Lambda[\sigma \cup d\bar{z}, C^\infty(U,X)]$ which annihilates those terms that contain at least one of $\sigma_1, \ldots, \sigma_n$ and leaves all others invariant. The map μ_α: $Z_\alpha{}^m(U,X) \to X$ is called the <u>Cauchy-Weil integral</u> of the commuting system a, while $\mu_\alpha(\xi)$ is called the Cauchy-Weil integral of a relative to ξ.

5.1.25 Remark. From the definition of P_σ, we see that the integrand contains only those terms of degree m in $d\bar{z}_1, \ldots, d\bar{z}_m$. If we put $\chi = (\delta_\alpha + \bar{\partial})\phi\xi_\alpha - \xi$, then χ and ξ have the same cohomology class with supp χ compact in U.Therefore, to define the Cauchy-Weil integral, the crucial point is the utilization of a form with compact support in U having the same cohomology class as ξ. Moreover, the solution ξ_α of the equation $(\delta_\alpha + \bar{\partial})\eta = \xi$ and the function $\phi \in C^\infty(U)$ with the required properties stated in the definition may be arbitrary. In order to verify that μ_α is well-defined, we prove the following proposition.

5.1.26 Proposition. The mapping μ_α is independent of the choice of ξ_α and ϕ; hence it is well-defined.

Proof. Let ξ_j $(j = 1,2)$ be two solutions to the equation $(\delta_\alpha + \bar\partial)\eta = \xi$, and let $\phi_j \in C^\infty(U)$ be corresponding functions according to Def. 5.1.24. Then the forms $\theta_j = (\delta_\alpha + \bar\partial)\phi_j\xi_j$ have compact support and value zero on $k_U(a,X)$. Clearly, the difference $\theta_1 - \theta_2$ vanishes on a neighborhood of $k_U(\alpha,X)$. It follows from Corollary 5.1.23 that $\theta_1 - \theta_2 = (\delta_\alpha + \bar\partial)\lambda$ where λ is a form with compact support. By Stokes' formula we have

$$\int P_\sigma(\theta_1 - \theta_2) \wedge dz = \int P_\sigma(\delta_\alpha + \bar\partial)\lambda \wedge dz = \int P_\sigma \bar\partial\lambda \wedge dz$$

$$= \int \bar\partial(P_\sigma\lambda) \wedge dz = 0,$$

and this gives the required independence.

In order to understand better the Cauchy-Weil integral, we observe the following special case.

5.1.27 Proposition. Assume that U is open in C, let $\alpha = (a_1)$ lie in $A(U,L(X))$ and suppose that $k_U(\alpha,X)$ is compact. Let $\xi = \xi_1\sigma_1 + \xi_2 d\bar z$ be an element of $Z_\alpha^1(U,X)$. Then

$$\mu_\alpha(\xi) = \int_\Gamma a_1^{-1}(z)\xi_1(z)\,dz + \int_\Delta \xi_2(z)dz \wedge d\bar z, \tag{1.33}$$

where $\Delta\ (\subset \Delta^- \subset U)$ is open with piecewise smooth boundary Γ.

Proof. Since $(a_1\sigma_1 + \frac{\partial}{\partial \bar z}d\bar z)\xi = 0$, we have

$$a_1(z)\xi_2(z) - \frac{\partial}{\partial \bar z}d\bar z\xi_1(z) = 0.$$

Let ξ_α be a solution of the equation $(a_1\sigma_1 + \frac{\partial}{\partial \bar z}d\bar z)\eta = \xi$. Then

$$(1.34) \qquad a_1(z)\xi_\alpha(z) = \xi_1(z) \text{ and } \frac{\partial}{\partial \bar{z}_1}\xi_\alpha(z) = \xi_2(z),$$

for $z \in \gamma_U(\alpha,X)$. Let $\phi \in C^\infty(U)$ be such that $\phi = 0$ on a neighborhood of $k_U(\alpha,X)$ and $\mathrm{supp}(1-\phi) \subset \Delta$. Then

$$\begin{aligned}\mu_\alpha(\xi) &= (-1)^2\int P_\sigma[(\delta_\alpha + \bar{\partial})\phi\xi_\alpha - \xi] \wedge dz \\ &= \int_\Delta \bar{\partial}(\phi\xi_\alpha)\wedge dz - \int_\Delta P_\sigma \xi \wedge dz \\ &= \int_\Gamma a_1^{-1}(z)\xi_1(z)\, dz + \int_\Delta \xi_2(z) dz \wedge d\bar{z}\end{aligned}$$

by (1.34) and Stokes' formula.

In particular, if $f \in A(U)$ and $\alpha = z - a$ with $a \in L(X)$, then (1.33) becomes

$$\mu_\alpha(f x \sigma_1) = \int_\Gamma R(z;a) f(z)\, dz.$$

The Cauchy-Weil integral is thus an extension of the functional calculus for a single operator.

We next study some properties of Cauchy-Weil integral. Let U be open in C^{n+m}. The points of U will be denoted by pairs (z,w) with $z \in C^n$ and $w \in C^m$. The operator $\bar{\partial}$ defined on C^{n+m} will be associated with the set of indeterminates $(d\bar{z}, d\bar{w}) = (d\bar{z}_1,\dots,d\bar{z}_n, d\bar{w}_1,\dots,d\bar{w}_m)$. The symbols $\bar{\partial}_z$ and $\bar{\partial}_w$ will denote the operator on C^n and C^m resp. Besides the operator δ_α defined by the commuting system $\alpha = (\alpha_1,\dots,\alpha_n)$ in $A(U,L(X))$, let $\beta = (\beta_1,\dots,\beta_m)$ be a second commuting system in $A(U,L(X))$ with the corresponding operator

$$\delta_\beta = \beta_1\tau_1 + \dots + \beta_m\tau_m$$

where $\tau = (\tau_1,\dots,\tau_m)$ is another system of indeterminates.

5.1.28 Definition. Let $\alpha = (\alpha_1,\dots,\alpha_n)$ be a commuting system in

$A(U,L(X))$ such that $k_U(\alpha,X)$ is C^n-compact in U, i.e. for every compact set K in W, the projection of U onto the last m coordinates, the set $(C^n \times K) \cap k_U(\alpha,X)$ is compact. Let $\phi \in C^\infty(U)$ be such that $\phi = 0$ on a neighborhood of $k_U(\alpha,X)$ and $\mathrm{supp}(1-\phi)$ is C^n-compact. For every $\xi \in Z_\alpha{}^n(U,X)$, we define the integral

$$\mu_\alpha(\xi) = (-1)^{n+1}\int_{C^n} P_\sigma[(\delta_\alpha+\bar\partial)\phi\xi_\alpha - \xi] \wedge dz, \tag{1.35}$$

where ξ_α is a solution of the equation $(\delta_\alpha+\bar\partial)\eta = \xi$ in the open set $\gamma_U(\alpha,X)$ and P_σ is the projection given in Def. 5.1.24. The mapping μ_α is called the <u>Cauchy-Weil integral</u> of the commuting system α, while $\mu_\alpha(\xi)$ is also called the Cauchy-Weil.integral of α relative to ξ.

We observe that in integral (1.35) the only significant terms are those of degree n in $d\bar z_1,\ldots,d\bar z_n$

The next proposition is an analog of Prop. 5.1.26.

5.1.29 Proposition. The mapping defined by (1.35) is independent of the choice of ξ_α, ϕ and defines an X-valued analytic function on the open set W.

Proof. With notation similar to that of Prop. 5.1.26, we see that the supports of the forms

$$\theta_j = (\delta_\alpha+\bar\partial)\phi_j\xi_j - \xi \qquad (j = 1,2)$$

are C^n-compact and that the difference $\theta_1 - \theta_2$ is null in a neighborhood of the set $k_U(\alpha,X)$. By Corollary 5.1.23, we have $\theta_1 - \theta_2 = (\delta_\alpha+\bar\partial)\lambda$ where the support of λ is C^n-compact and its degree is n-1. By Stokes' formula,

$$\int P_\sigma(\theta_1-\theta_2) \wedge dz = \int P_\sigma(\delta_\alpha+\bar\partial)\lambda \wedge dz$$

$$= \int P_\sigma(\bar\partial_z\lambda)\lambda \wedge dz = \int \bar\partial_z(P_\sigma\lambda) \wedge dz = 0.$$

Therefore, the integral μ_α is independent of the choice of ξ_α and ϕ.

Next we note that

$$\bar\partial_w\mu_\alpha(\xi)(w) = \int_{C^n} \bar\partial_w[(\bar\partial\phi P_\sigma\xi_\alpha)(z,w) - P_\sigma\xi(z,w)] \wedge dz$$

and that

$$\int_{C^n} \bar\partial_z[(\bar\partial\phi P_\sigma\xi_\alpha)(z,w) - P_\sigma\xi(z,w)] \wedge dz = 0.$$

The latter follows from the Stokes' formula. Then

$$\begin{aligned}\bar\partial_w\mu_\alpha(\xi)(w) &= \int_{C^n} \bar\partial[(\bar\partial\phi P_\sigma\xi_\alpha)(z,w) - P_\sigma\xi(z,w)] \wedge dz\\ &= -\int_{C^n} P_\sigma(\bar\partial\xi)(z,w) \wedge dz\\ &= \int_{C^n} P_\sigma(\delta_\alpha\xi)(z,w) \wedge dz = 0.\end{aligned}$$

Thus $\mu_\alpha(\xi)(w)$ is analytic for $w \in W$.

5.1.30 Theorem. Assume that U is open in C^{n+m} and that W is the projection of U onto the last m coordinates. Let $\alpha = (\alpha_1,\dots,\alpha_n)$ and $\beta = (\beta_1,\dots,\beta_m)$ be systems in $A(U,L(X))$ and $A(W,L(X))$ resp. such that $(\alpha,\beta) = (\alpha_1,\dots,\alpha_n,\beta_1,\dots,\beta_m)$ is a commuting system. If $k_U(\alpha,X)$ is C^n-compact in U and $k_W(\beta,X)$ is compact, then for every $g \in Z_\alpha{}^n(U,X)$ and every $f \in A(W)$, we have

$$\mu_{(\alpha,\beta)}(f\tau_1 \wedge\dots\wedge \tau_m \wedge g) = \mu_\beta(f\mu_\alpha(g)\tau_1 \wedge\dots\wedge \tau_m).$$

We omit the proof of this theorem; see [34', Th. III.8.11].

5.1.31 Corollary. Assume that the following properties hold.

(i) $V \subset C^n$, $W \subset C$ are open and $U = V \times W$;

(ii) $\alpha = (\alpha_1,\dots, \alpha_n)$ and $\beta = (\beta_1)$ are systems in $A(U,L(X))$ and $A(W,L(X))$ resp. and $(\alpha,\beta) = (\alpha_1,\dots, \alpha_n, \beta_1)$ is a commuting system;

(iii) $k_U(\alpha,X)$ is C^n-compact, $k_W(\beta,X)$ is compact and $\beta_1{}^{-1} \in A(\gamma_W(\beta,X),L(X))$.

Then, for every $g \in A(U,X)$ and every $f \in A(W)$, we have

$$\mu_{(\alpha,\beta)}(fg\tau \wedge \sigma) = \int_\Gamma \beta_1{}^{-1}(w)f(w)\mu_\alpha(g\sigma)(w)\, dw, \tag{1.36}$$

where $\Gamma \subset W$ is a piecewise smooth oriented contour with $k_W(\beta,X)$

inside and $\sigma[\tau]$ denote $\sigma_1\wedge...\wedge\sigma_n$ $[\tau_1\wedge...\wedge\tau_m]$ for simplicity.

Proof. By Th. 5.1.30 we have the equality $\mu_{(\alpha,\beta)}(fg\tau\wedge\sigma) = \mu_\beta(f\mu_\alpha(g\sigma)\tau)$. Since $f\mu_\alpha(g\sigma)$ is independent of z and dz, application of Prop. 5.1.27 yields (1.36).

The following example gives a special case of Cor. 5.1.31. Suppose that U is open in C^n and $w = (w_1,...,w_n)$ is fixed in U. Then $z - w = (z_1-w_1, ..., z_n-w_n)$ is a commuting system in A(U,L(C)) with $\{(w_1,...,w_n)\} = k_U(z-w,C)$. Applying Cor. 5.1.31 repeatedly, we obtain

$$\mu_{z-w}(g\wedge\sigma) = \int_{\Gamma_1}...\int_{\Gamma_n}\Pi_j(z_j-w_j)^{-1}g(z)\,dz_1...dz_n$$

for every $g\in A(U)$, where Γ_j is a circle with center w_j, $j = 1,...,n$.

Theorem 5.1.30 provides a rule of calculation for the Cauchy-Weil integral. We shall now study the change-of-variable rule for this integral.

Let U be open in C^{n+m}, and let $\alpha=(\alpha_1,..., \alpha_n)$ be a commuting system in A(U,L(X)). Now let $u = (u_{jk};\ k,j = 1,...,n)$ be a matrix of commuting elements from A(U,L(X)) which also commute with all α_k, and define

$$\beta_j = \sum_{k=1}^{n} u_{jk}\alpha_k. \tag{1.37}$$

Then $\beta = (\beta_1,...,\beta_n)$ is a commuting system in A(U,L(X)). We seek a relation for the Cauchy-Weil integrals of α and β. We first define a transformation of the space $\Lambda[\sigma\cup d\bar{z}\cup d\bar{w}, C^\infty(U,X)]$ into itself associated with matrix u. Set

$$\hat{u}(\sigma_j) = \sum_{k=1}^{n} u_{kj}\sigma_k \quad (j = 1,...,n).$$

Each element $\xi\in\Lambda[\sigma\cup d\bar{z}\cup d\bar{w}, C^\infty(U,X)]$ of degree p in $\sigma_1,...,\sigma_n$ has the following form

$$\xi = \sum_{1\le j_1<...<j_p\le n} \sigma_{j1}\wedge...\wedge\sigma_{jp}\wedge\xi_{j1...jp},$$

where $\xi_{j1\ldots jp} \in \Lambda[d\bar{z} \cup d\bar{w}, C^\infty(U,X)]$. Now define the mapping

(1.38) $$\tilde{u}(\xi) = \sum_{1 \le j_1 < \ldots < j_p \le n} \hat{u}(\sigma_{j1}) \wedge \ldots \wedge \hat{u}(\sigma_{jp}) \wedge \xi_{j1\ldots jp}.$$

The mapping defined in (1.38), $\tilde{u}$, can be extended to the whole space by linearity and is then called the special transformation induced by the matrix u. From (1.38) one has for given indices $j_1, \ldots, j_p$

$$\tilde{u}\delta_\alpha(\sigma_{j1} \wedge \ldots \wedge \sigma_{jp} \wedge \xi_{j1\ldots jp})$$
$$= \tilde{u}\Big\{\sum_{k=1}^{n} \alpha_k \sigma_k \wedge \sigma_{j1} \wedge \ldots \wedge \sigma_{jp} \wedge \xi_{j1\ldots jp}\Big\}$$
$$= \sum_{k=1}^{n} \alpha_k \hat{u}(\sigma_k) \wedge \hat{u}(\sigma_{j1}) \wedge \ldots \wedge \hat{u}(\sigma_{jp}) \wedge \xi_{j1\ldots jp}$$
$$= \sum_{j=1}^{n}\Big\{\sum_{k=1}^{n} u_{jk}\alpha_k\Big\}\sigma_j \wedge \hat{u}(\sigma_{j1}) \wedge \ldots \wedge \hat{u}(\sigma_{jp}) \wedge \xi_{j1\ldots jp}$$
$$= \delta_\beta \tilde{u}(\sigma_{j1} \wedge \ldots \wedge \sigma_{jp} \wedge \xi_{j1\ldots jp}).$$

This, together with the evident equality $\tilde{u}\bar{\partial} = \bar{\partial}\tilde{u}$, gives

(1.39) $$\tilde{u}(\delta_\alpha + \bar{\partial}) = (\delta_\beta + \bar{\partial})\tilde{u}.$$

5.1.32 Proposition. Let U be open in C^{n+m}, and let $\alpha = (\alpha_1, \ldots, \alpha_n)$ be a commuting system in $A(U,L(X))$. Let $(u_{jk};\ j,k = 1,\ldots,n)$ be a matrix of commuting elements in $A(U,L(X))$ which also commute with each component of α. Let β be defined by (1.37), and let $\tilde{u}$ be the special transformation induced by u. If both sets $k_U(\alpha,X)$ and $k_U(\beta,X)$ are C^n-compact, then for every $\xi \in Z_\alpha{}^n(U,X)$ we have $\mu_\beta(\tilde{u}\xi) = \mu_\alpha(\xi)$.

Proof. Let $\phi \in C^\infty(U)$ be such that $\phi = 0$ in a neighborhood of $k_U(\alpha,X) \cup k_U(\beta,X)$ as in Def. 5.1.28 and let ξ_α be the solution of the equation $(\delta_\alpha + \bar{\partial})\theta = \xi$. By equality (1.39) has a solution of the equation $(\delta_\beta + \bar{\partial})\theta = \tilde{u}\xi$. From the definition of the Cauchy-Weil integral,

$$(-1)^{n+1}\mu_\beta(\tilde{u}\xi) = \int P_\sigma[(\delta_\beta + \bar{\partial})\phi\tilde{u}\xi_\alpha - \tilde{u}\xi] \wedge dz$$

$$= \int [\bar{\partial}(\phi P_\sigma \tilde{u}\xi_\alpha) - P_\sigma \tilde{u}\xi] \wedge dz$$

$$= \int \tilde{u}[\bar{\partial}(\phi P_\sigma \xi_\alpha) - P_\sigma \xi] \wedge dz = (-1)^{n+1}\mu_\alpha(\xi).$$

Here we have used the facts that $P_\sigma \tilde{u} = \tilde{u}P_\sigma$ and that $\tilde{u}$ leaves those forms invariant which do not contain $\sigma_1, \sigma_2, \ldots, \sigma_n$.

5.1.33 Proposition. Let U be open in C^m and let $\alpha = (\alpha_1, \ldots, \alpha_n)$ and $\beta = (\beta_1, \ldots, \beta_n)$ be commuting systems in $A(U,L(X))$ and $A(U,L(Y))$ resp. Assume that both $k_U(\alpha, X)$ and $k_U(\beta, Y)$ are compact. If $u \in L(X,Y)$ has the property that $u\alpha_j(z) = \beta_j(z)u$ for all $z \in U$ and $j = 1, \ldots, n$, then the equality

$$\mu_\beta(u\xi) = u\mu_\alpha(\xi) \tag{1.40}$$

holds for every $\xi \in Z_\alpha{}^m(U,X)$.

Proof. Since the equality $u(\delta_\alpha + \bar{\partial})\xi = (\delta_\beta + \bar{\partial})u\xi$ is clear, an argument like the proof of Th. 5.1.32 (with u replacing $\tilde{u}$) gives (1.40).

5.1.34 Proposition. Let U be open in C^m, and let $\alpha = (\alpha_1, \ldots, \alpha_n)$ be a commuting system in $A(U,L(X))$ such that $k_U(\alpha, X)$ is compact. Then the mapping $\mu_\alpha: Z_\alpha{}^m(V,X) \to X$ is continuous, where $V = U \backslash k_U(\alpha, X)$.

Proof. Because $Z_\alpha{}^m(V,X)$ is the null space of the operator $\delta_\alpha + \bar{\partial}$, it is closed and thus a Frechet space. By Cor. 5.1.23, the map

$$\delta_\alpha + \bar{\partial}: \Lambda^{m-1}[\sigma \cup d\bar{z}, C^\infty(V,X)] \to Z_\alpha{}^m(V,X)$$

is surjective. Let $\{\xi_k\}$ be a sequence in $Z_\alpha{}^m(V,X)$ converging to ξ. By the open mapping theorem, there exists a sequence of forms $\{\eta_k\}$ such that $\eta_k \to \eta$ $(k \to \infty)$, $(\delta_\alpha + \bar{\partial})\eta_k = \xi_k$ and $(\delta_\alpha + \bar{\partial})\eta = \xi$. Let $\phi \in C^\infty(U)$ be as in Def. 5.1.24. Since by the compactness of $k_U(\alpha, X)$ we may assume that all supp ξ_k are contained in some compact $K \subset V$ (passing to restrictions, if necessary) and P_σ is continuous, we have

$$\bar\partial(\phi P_\sigma \eta_k) - P_\sigma \xi_k \to \bar\partial(\phi P_\sigma \eta) - P_\sigma \xi \qquad (k \to \infty)$$

uniformly on every compact subset of V. Hence $\mu_\alpha(\xi_k) \to \mu_\alpha(\xi)$ as $k \to \infty$, and this is the required continuity.

We now return to the case of a commuting system $a = (a_1,\dots,a_n)$ in L(X), where X is a Banach space. If we put $\alpha = z-a = (z_1-a_1,\dots, z_n-a_n)$, then α is also a commuting system. In the next result we use δ_{z-a} instead of δ_α for ease of reading.

5.1.35 Proposition. For the given commuting system $a = (a_1,\dots,a_n)$ with $\alpha = (z_1-a_1,\dots,z_n-a_n)$ as above, we have the relation $K_{C^n}(\alpha,X) = \sigma(a,X)$.

Proof. If $w \notin \sigma(a,X)$, then the complex $((\Lambda^p[\sigma,X],(\delta_{z-a})^p);p=0,\dots,n)$ is exact for each z in a neighborhood of w. By Th. 5.1.21, $w \notin K_{C^n}(\alpha,X)$, hence $K_{C^n}(\alpha,X) \subset \sigma(a,X)$.

Conversely, assume that $w \notin K_{C^n}(\alpha,X)$. We show by reverse induction that $H^p(X,w-a) = 0$ for $p = 0,1,\dots,n$. Let $\xi \in \Lambda^n[\sigma,X]$. Then $\delta_{z-a}{}^n\xi = 0$ for all $z \in C^n$. By hypothesis this implies existence of a form $\eta(z)$ such that $\delta_{z-a}{}^{n-1}\eta(z) = \xi$ for z in some neighborhood of w. In particular, $\eta = \eta(w)$ satisfies $(\delta_{w-a})^{n-1}\eta = \xi$.

Now suppose that $H^k(X,w-a) = 0$ for $k = n, n-1,\dots,p+1$. Let $\xi \in \Lambda^p[\sigma,X]$ satisfy $(\delta_{w-a})^p\xi = 0$. Since $H^{p+1}(X, w-a) = 0$, by a method in [34', Lemma II.10.5] we find a form $\xi(z)$ with analytic coefficients in a neighborhood of w such that $(\delta_{z-a})^p\xi(z) = 0$ and $\xi(w) = \xi$. By hypothesis, there is a form in a neighborhood of w for which $(\delta_{z-a})^{p-1}\eta(z) = \xi(z)$. In particular, $(\delta_{w-a})^{p-1}\eta(w) = \xi$, and hence $H^p(X,w-a) = 0$. Since this completes the inductive step, it follows that $w \notin \sigma(a,X)$, and the proof is complete.

5.1.36 Theorem. Let $a' = (a_1,\dots,a_n,a_{n+1},\dots,a_{n+m})$ be a commuting system in L(X) and let $a = (a_1,\dots,a_n)$. Let Π: $C^{n+m} \to C^n$ be the projection onto the first n coordinates. Then $\Pi[\sigma(a',X)] = \sigma(a,X)$.

Proof. The inclusion $\Pi[\sigma(a',X)] \subset \sigma(a,X)$ follows from Th. 5.1.6. To obtain the reverse inclusion, it suffices to consider the case $m = 1$ and to show that if $\sigma(a,X)$ contains the origin in C^n, then $(0,w) \in$

$\sigma(a',X)$ for some $w \in \mathbf{C}$. If we assume the contrary, namely, $(0,w) \notin \sigma(a',X)$ for all complex w and there exists an open polydisc $U \subset \mathbf{C}^n$ such that U contains the origin of $\mathbf{C}^n$ and the system $(z-a, w-a_{n+1})$ is nonsingular on $A(U \times \mathbf{C}, X)$. From the identification of the spaces $A(U \times \mathbf{C}, X)$ and $A(\mathbf{C},X_U)$, where $X_U = A(U,X)$, we see that the system $\alpha = (\alpha_1,...\alpha_n)$ with $\alpha_j(z) = z_j - a_j$ is constant with respect to w. If we put $\alpha'(w) = (\alpha, w-a_{n+1})$, then α' has the property that $k_{\mathbf{C}}(\alpha',X_U) = \emptyset$.

Now let $\xi \in \Lambda^p[\sigma,X_U]$ be such that $\delta_\alpha\xi = 0$. Then the form $\xi_1 = \sigma_{n+1} \wedge \xi$ satisfies the equation $\delta_{\alpha'(w)}\xi_1 = 0$. Since ξ is independent of $w \in \mathbf{C}$, by [34', Lemma III.9.4] we may write $\xi_1 = \delta_{\alpha'(w)}\eta_1$ with η_1 independent of $w \in \mathbf{C}$. From the evident decomposition $\eta_1 = \eta_1' + \sigma_{n+1}\eta_1''$ with η_1', η_1'' not containing σ_{n+1}, the equality $\delta_{\alpha'(w)}\eta_1 = \xi_1$ implies $\eta_1' = 0$. In fact, the former equality implies that

$$\delta_\alpha(-\eta_1'') + (-1)^p(w - a_{n+1})\eta_1' = \xi. \tag{1.41}$$

Dividing both sides of (1.41) by w and letting $w \to \infty$, we obtain $\eta_1' = 0$, and hence $\delta_\alpha(-\eta_1'') = \xi$. Thus $(\Lambda^p[\sigma,X], \delta_\alpha)$ is exact, and this implies the contradiction $0 \notin \sigma(a,X)$. This completes the proof of the theorem.

5.1.37 Corollary. If $a = (a_1,...,a_n)$ is a commuting system in $L(X)$, $X \neq \{0\}$, then $\sigma(a,X)$ is nonempty compact subset of $\mathbf{C}^n$.

Proof. Let Π_j be the projection of $\mathbf{C}^n$ onto its jth coordinate. By Th. 5.1.36 and Lemma 5.1.7 $\Pi_j[\sigma(a,X)] = \sigma(a_j,X)$ for all j. But then the conclusion follows from Cor. 5.1.14 since $\sigma(a_j,X)$ is nonempty.

The next definition introduces the notion of functional calculus.

5.1.38 Definition. Assume that $a = (a_1,...,a_n)$ is a commuting system in $L(X)$, that U is open in $\mathbf{C}^n$ and that $\sigma(a,X) \subset U$. Let $\alpha = (z_1-a_1,...,z_n-a_n)$ for $z \in U$. For $f \in A(U)$, define the mapping

$$f(a)x = \frac{1}{(2\pi i)^n}\mu_\alpha(fx\sigma_1\wedge...\wedge\sigma_n), \tag{1.42}$$

where $x \in X$ is arbitrary and μ_α is given by Def. 5.1.24.

5.1.39 Theorem (Shilov, Arens-Calderon, Taylor). Let $a = (a_1,\ldots,a_n)$ be a commuting system in $L(X)$ and let U be open in C^n with $\sigma(a,X) \subset U$. Then the mapping Φ: $f \to f(a)$ defined by (1.42) is a continuous unital algebra homomorphism from $A(U)$ into $L(X)$ satisfying the conditions

(i) $\Phi(1) = I$ and $\Phi(z_j) = a_j$ for each j;

(ii) $\Phi(f)$ lies in the bicommutant of the system a for every $f \in A(U)$.

Proof. The facts that $f \to f(a)$ is continuous and $f(a) \in L(X)$ for each $f \in A(U)$ are simple consequences of Prop. 5.1.34.

Let $D = D_1 \times\ldots\times D_n$ be an open polydisc such that $\sigma(a_j,X) \subset D_j$ for each $j = 1,\ldots,n$, and let $\Gamma_j = \partial D_j$. By repeated application of Prop. 5.1.26 and Cor. 5.1.31 we have, with g replaced by x in (1.36), for each complex polynomial $p(z_1,\ldots,z_n)$,

$$\begin{aligned} p(a)x &= \frac{1}{(2\pi i)^n}\mu_\alpha(px\sigma_1\wedge\ldots\wedge\sigma_n) \\ &= \frac{1}{(2\pi i)^n}\int_{\Gamma_1}\ldots\int_{\Gamma_n}\Pi_j(z_j-a_j)^{-1}p(z)\,dz_1\ldots dz_n \\ &= p(a_1,\ldots,a_n)x. \end{aligned}$$

Hence both parts of (i) are now evident, and (ii) follows from Prop. 5.1.33.

It remains to show that Φ is multiplicative. Assume that $f, g \in A(U)$, and let $\tau = (\tau_1,\ldots,\tau_n)$ be a second system of indeterminates. From Th. 5.1.30 we have

$$\begin{aligned} (2\pi i)^{2n}f(a)g(a)x &= \mu_\alpha(f\mu_\alpha(gx\tau_1\wedge\ldots\wedge\tau_n)\sigma_1\wedge\ldots\wedge\sigma_n) \\ &= \mu_\gamma(fgx\sigma_1\wedge\ldots\wedge\sigma_n\wedge\tau_1\wedge\ldots\wedge\tau_n), \end{aligned} \tag{1.43}$$

where $\gamma(z,w) = (\alpha(z),\alpha(w))$. Now transform the system γ into a new system by means of the matrix $(u_{jk};\ j,k = 1,\ldots,2n)$, where $u_{jk} = 1$ on the diagonal $j = k$, $u_{jk} = -1$ for $k = n+j$ and zero at all other j,k. Then the new system is $(\rho,\alpha) = (z_1 - w_1,\ldots,z_n - w_n, w_1 - a_1,\ldots,w_n - a_n)$. Let $\tilde{u}$ be the special transformation associated with u. By Prop. 5.1.32

$$\begin{aligned} &\mu_{\gamma}(fgx\sigma_1\wedge\ldots\wedge\sigma_n\wedge\tau_1\wedge\ldots\wedge\tau_n) \\ (1.44) \qquad &= \mu_{(\rho,\alpha)}(\hat{u}fgx\sigma_1\wedge\ldots\wedge\sigma_n\wedge\tau_1\wedge\ldots\wedge\tau_n) \\ &= \mu_{(\rho,\alpha)}(fgx\sigma_1\wedge\ldots\wedge\sigma_n\wedge\tau_1\wedge\ldots\wedge\tau_n), \end{aligned}$$

where we have used the fact that $\sigma_1\wedge\ldots\wedge\sigma_n\wedge\tau_1\wedge\ldots\wedge\tau_n$ is left fixed by $\hat{u}$. By Th. 5.1.30

$$\begin{aligned} (1.45) \qquad &\mu_{(\rho,\alpha)}(fgx\sigma_1\wedge\ldots\wedge\sigma_n\wedge\tau_1\wedge\ldots\wedge\tau_n) \\ &= \mu_{\alpha}(f\mu_{\rho}(gx\tau_1\wedge\ldots\wedge\tau_n)\sigma_1\wedge\ldots\wedge\sigma_n) \\ &= (2\pi i)^n\mu_{\alpha}(fgx\sigma_1\wedge\ldots\wedge\sigma_n) \\ &= (2\pi i)^n(fg)(a)x, \end{aligned}$$

since $\mu_{\rho}(gx\tau_1\wedge\ldots\wedge\tau_n)(w) = (2\pi i)^n g(w)$ by the special case of Cor. 5.1.31. Now (1.43)-(1.45) complete the proof.

5.1.40 Corollary. Let $a = (a_1,\ldots,a_n)$ and $b = (b_1,\ldots,b_n)$ be commuting systems in L(X) and L(Y), resp. If $u \in L(X,Y)$ and satisfies $ua_j = b_j u$ for $j = 1,\ldots,n$ and if f is analytic on some neighborhood of $\sigma(a,X) \cup \sigma(b,X)$, then $uf(a) = f(b)u$.

Proof. By Prop. 5.1.33 we have

$$(1.46) \qquad \mu_{\beta}(fux\sigma_1\wedge\ldots\wedge\sigma_n) = u\mu_{\alpha}(fx\sigma_1\wedge\ldots\wedge\sigma_n)$$

where $\alpha = (z_1-a_1,\ldots,z_n-a_n)$ and $\beta = (z_1-b_1,\ldots,z_n-b_n)$. Hence the conclusion follows.

We conclude this section with a discussion of the spectral mapping theorem for commuting systems. We still assume that X is a Banach space and U is open in C^m. Let $\alpha = (\alpha_1,\ldots,\alpha_n)$ and $\beta = (\beta_1,\ldots,\beta_r)$ be commuting systems in A(U,L(X)) associated with $\sigma = (\sigma_1,\ldots,\sigma_n)$ and $\tau = (\tau_1,\ldots,\tau_r)$ resp. We further assume that α and β commute with each other. For $V = \gamma_U(\alpha,X)$, by Th. 5.1.22 the complex

$$((\Lambda^p[\sigma \cup d\bar{z}, C^\infty(V,X)],\ \delta_\alpha+\delta_\beta+\bar{\partial});\ p = 0,\ldots,n+m)$$

is exact, or equivalently, $(\alpha,\bar{\partial})$ is nonsingular on $C^\infty(V,X)$, and hence by Prop. 5.1.6 so is the complex

$$((\Lambda^p[\sigma \cup \tau \cup d\bar{z},\ C^\infty(V,X)]);\ p = 0,\dots,\ n + r + m).$$

Equivalently, $(\alpha,\beta,\bar\partial)$ is nonsingular on $C^\infty(V,X)$.

Let $\xi \in Z_\alpha{}^m(U,\Lambda^p[\tau,X])$ have the property $\delta_\beta\xi = 0$. Then $(\delta_\alpha + \delta_\beta + \bar\partial)\xi = 0$ and hence there exists a form η satisfying $(\delta_\alpha + \delta_\beta + \bar\partial)\eta = \xi$ on V. If $k_U(\alpha,X)$ is compact, then we define the integral

$$\tilde\mu_\alpha(\xi) = (-1)^{m+1}\int_U [(\delta_\beta + \bar\partial)\phi P_\sigma\eta - P_\sigma\xi] \wedge dz \tag{1.47}$$

where $\phi \in C^\infty(U)$ has the properties in Def. 5.1.24.

Remark. The integral $\tilde\mu_\alpha$ defined in (1.47) is clearly a form of degree p in $\tau_1,\dots,\tau_r$.

5.1.41 Lemma. The following properties hold for the integral $\tilde\mu_\alpha$.

(i) $\tilde\mu_\alpha$ is uniquely determined by the form ξ modulo an integral of the form $\int(\delta_\beta\nu) \wedge dz$ for some form ν with compact support;

(ii) $\tilde\mu_\alpha(\xi)$ is equal to $\mu_\alpha(\xi)$ modulo an integral of the form $\int(\delta_\beta\nu) \wedge dz$ for some form ν with compact support.

Proof. We prove (i) only and omit that of (ii). Let η_1, η_2 be solutions of $(\delta_\alpha + \delta_\beta + \bar\partial)\eta = \xi$ and let ϕ_1, ϕ_2 have the properties of ϕ defined in (1.47). Then the forms $\theta_j = (\delta_\alpha + \delta_\beta + \bar\partial)\phi_j\eta_j - \xi_j$ $(j=1,2)$ have compact support and $\theta_1 - \theta_2 = (\delta_\alpha + \delta_\beta + \bar\partial)\lambda$ where λ is a form whose support may be assumed compact also by Cor. 5.1.23. Thus

$$P_\sigma\int(\theta_1 - \theta_2) \wedge dz = \int(\delta_\beta + \bar\partial)P_\sigma\lambda \wedge dz = \int\delta_\beta P_\sigma\lambda \wedge dz$$
$$= \int(\delta_\beta\nu) \wedge dz.$$

Here we have used the equality $\int(\bar\partial P_\sigma\lambda) \wedge dz = 0$ by means of Stokes' formula.

5.1.42 Proposition. Let $a = (a_1,\dots,a_n)$ and $b = (b_1,\dots,b_n)$ be two commuting systems in $L(X)$, and let U be an open set in C^n containing $\sigma(a,X) \cup \sigma(b,X)$. Then the operators $f(a)$ and $f(b)$ induce the same action in $H^p(X,b-a)$ for $p = 0,1,\dots,n$. More precisely, $[f(b) - f(a)]\xi \in \delta_{b-a}(\Lambda^{p-1}[\tau,X])$ for every $\xi \in \Lambda^p[\tau,X]$ with $\delta_{b-a}\xi = 0$.

Proof. Let $\alpha = \alpha(z) = (z_1-a_1,\dots,z_n-a_n)$ and $\beta = \beta(z) = (z_1-b_1,\dots,z_n-b_n)$. We shall associate both systems α, β with the same system of indeterminates, $\sigma_1,\dots,\sigma_n$, whereas $\delta_{b-a} = (b_1-a_1)\tau_1 + \dots + (b_n-a_n)\tau_n$. Let u be the matrix as in the proof of Th. 5.1.39. Then clearly $\hat{u}(\sigma_j) = \sigma_j$ and $\hat{u}(\tau_j) = \tau_j - \sigma_j$ for $j = 1,2,\dots,n$. Then (1.39) can be rewritten as

$$\tilde{u}(\delta_\alpha + \delta_{b-a} + \bar{\partial}) = (\delta_\beta + \delta_{b-a} + \bar{\partial})\tilde{u}. \tag{1.48}$$

Now suppose that $\xi \in \Lambda^p[\tau,X]$ satisfies $\delta_{b-a}\xi = 0$ and $f \in A(U)$ is arbitrary. Let $\xi_0 = f\xi\sigma_1 \wedge \dots \wedge \sigma_n$. Then $\xi_0 \in \Lambda^n[\sigma \cup d\bar{z}, C^\infty(U,\Lambda^p[\tau,X])]$ and $P_\sigma(\delta_\alpha + \bar{\partial})\xi_0 = 0$ and $\delta_{b-a}\xi_0 = 0$. Since $(\alpha, b-a, \bar{\partial})$ is nonsingular in $C^\infty([U\backslash k_U(\alpha,X)],X)$, there exists $\eta_0 \in \Lambda^{n+p-1}[\sigma \cup \tau \cup d\bar{z}, C^\infty(V,X)]$ where $V = U\backslash k_U(\alpha,X)$ such that $(\delta_\alpha + \delta_{b-a} + \bar{\partial})\eta_0 = \xi_0$. By (1.48) $(\delta_\beta + \delta_{b-a} + \bar{\partial})\tilde{u}\eta_0 = \tilde{u}\xi_0$, and a simple calculation shows that

$$\tilde{u}\xi_0 = (\tilde{u}\xi)f\sigma_1 \wedge \dots \wedge \sigma_n = \xi f\sigma_1 \wedge \dots \wedge \sigma_n = \xi_0. \tag{1.49}$$

Let $\phi \in C^\infty(U)$ be such that ϕ vanishes on a neighborhood of $\sigma(a,X) \cup \sigma(b,X)$ and $\mathrm{supp}(1-\phi)$ is compact. Then (1.48) and (1.49) imply that

$$\begin{aligned}
\tilde{\mu}_\beta(\xi_0) &= \tilde{\mu}_\beta(\tilde{u}\xi_0) = (-1)^{n+1}\int P_\sigma(\delta_\beta + \delta_{b-a} + \bar{\partial})\phi\tilde{u}\eta_0 \wedge dz \\
&= (-1)^{n+1}\int P_\sigma\tilde{u}(\delta_\alpha + \delta_{b-a} + \bar{\partial})\phi\eta_0 \wedge dz \\
&= (-1)^{n+1}\int P_\sigma(\delta_\alpha + \delta_{b-a} + \bar{\partial})\phi\eta_0 \wedge dz \\
&= (-1)^{n+1}\int (\delta_{b-a} + \bar{\partial})\, P_\sigma\phi\eta_0 \wedge dz \\
&= \tilde{\mu}_\alpha(\xi_0),
\end{aligned}$$

where we have used the fact $P_\sigma\tilde{u} = P_\sigma$. Since $f(a)\xi = (2\pi i)^{-n}\mu_\alpha(\xi_0)$ and $f(b)\xi = (2\pi i)^{-n}\mu_\beta(\xi_0)$ and δ_{b-a} plays the role of δ_β in Lemma 5.1.41, we obtain $[f(b) - f(a)]\xi = \delta_{b-a}\lambda$ for some form $\lambda \in \Lambda^{p-1}[\tau,X]$ by that lemma and the fact that δ_{b-a} is independent of z. This completes the proof.

5.1.43 Lemma. Assume that $a = (a_1,...,a_n)$ is a commuting system in $L(X)$ and that U is open in $\mathbf{C}^n$ and contains $\sigma(a,X)$; let $f \in A(U)$. If $z_0 \in \sigma(a,X)$, then $(z_0,w_0) \in \sigma((a,f(a)),X)$ if and only if $w_0 = f(z_0)$.

See [34', Lemma III.10.5] for a proof of this lemma.

5.1.44 Theorem (Spectral mapping theorem). Let $a = (a_1,...,a_n)$ be a commuting system in $L(X)$, and let U be an open set in $\mathbf{C}^n$ containing $\sigma(a,X)$. Let $f = (f_1,...,f_m)$ where each $f_i \in A(U)$ for $i = 1,...,m$. Put $f(a) = (f_1(a),...,f_m(a))$. Then $\sigma(f(a),X) = f(\sigma(a,X))$.

Proof. By Th. 5.1.36 and Lemma 5.1.43 the theorem holds for $m = 1$. Now assume that it holds for $m \geq 1$. Set $g(z,w) = f_{m+1}(z)$. Then g is analytic on $U \times \mathbf{C}^m$ ($\supset \sigma((a,f(a)),X)$). By Th. 5.1.30 and a simple calculation, we have $g(a,f(a)) = f_{m+1}(a)$. if $z \in \sigma(a,X)$, then $(z,f(z)) \in \sigma((a,f(a)),X)$. But from Lemma 5.1.43 $(z,f(z),w_{m+1})$ lies in $\sigma((a,f(a)),g(a,f(a)),X) = \sigma((a,f(a)),f_{m+1}(a),X)$ if and only if $w_{m+1} = g(z,f(z)) = f_{m+1}(z)$. Thus for $z \in \sigma(a,X)$,

$$(z,w_1,...,w_{m+1}) \in \sigma((a,f(a),f_{m+1}(a)),X)$$

if and only if $w_j = f_j(z)$ for $j = 1,..., m+1$, so the theorem holds for $m+1$. Then by Th. 5.1.36 we obtain

$$\sigma(f(a),X) = \{f(z): z \in \sigma(a,X)\} = f(\sigma(a,X)),$$

and the theorem is proved.

5.1.45 Theorem (Shilov, Taylor). Let $a = (a_1,...,a_n)$ be a commuting system in $L(X)$. If $\sigma(a,X) = K_1 \cup K_2$, where K_1, K_2 are nonempty disjoint compact sets, then there are corresponding closed subspaces X_1, X_2 of X which are invariant under a and satisfy

(i) $X = X_1 \oplus X_2$;

(ii) $\sigma(a,X_j) = K_j$ for $j = 1,2$.

Proof. Let U_j be disjoint open neighborhoods of K_j, and let χ_1 be the characteristic function of U_1. Then $\chi_1 \in A(U_1 \cup U_2)$, $\chi_1(a)$ is defined and is a projection by Th. 5.1.39. Put $X_1 = \chi_1(a)X$ and $X_2 = (1-\chi_1(a))X$. Then X_1, X_2 are closed subspaces of X and are hyperinvariant under a because $\chi_1(a)$ is in the bicommutant of a. The relation (i) is obvious.

To verify (ii), we first prove that $\sigma(a,X_j) \subset \sigma(a,X)$ for $j = 1,2$. Let D be an open polydisc in $\mathbf{C}^n$ such that $\alpha(z) = (z_1-a_1,\ldots,z_n-a_n)$ is nonsingular on A(D,X). Assume that η is a form with coefficients in $A(D,X_1)$ such that $\delta_\alpha\eta = 0$. Then there exists a form ξ satisfying $\delta_\alpha\xi = \eta$ on D. The form $\chi_1\xi$ satisfies $\delta_\alpha(\chi_1\xi) = \eta$ and its coefficients are in $A(D,X_1)$. Therefore $\sigma(a,X_1) = k_{\mathbf{C}^n}(a,X_1) \subset \sigma(a,X)$ by Th. 5.1.35. Similarly, $\sigma(a,X_2) \subset \sigma(a,X)$.

From Cor 5.1.40, we have $\chi_1(a|X_1) = \chi_1(a)|X_1$. If $\sigma(a|X_1) \cap K_2 \neq \emptyset$, then $p = (1-\chi_1)(a|X_1)$ is a non-null idempotent in $L(X_1)$; in fact, $\sigma(p,X_1) = (1-\chi_1)(\sigma(a|X_1)) = \{0,1\}$. On the other hand, $p = \chi_1(a|X_1)(1-\chi_1(a|X_1)) = [\chi_1(1-\chi_1)](a|X_1) = 0$, a contradiction. Thus, by the last paragraph, $\sigma(a|X_1) \subset K_1$, and similarly $\sigma(a|X_2) \subset K_2$. These, together with the inclusion $\sigma(a,X) \subset \sigma(a|X_1) \cup \sigma(a|X_2)$ and the equality $\sigma(a,X) = K_1 \cup K_2$, give the result.

We also need the following results for later use.

5.1.46 Lemma. Let X be a Frechet space, and let $a = (a_1,\ldots,a_n)$ be a commuting system of continuous operators on X. Set $X_0 = \{0\}$ and $X_p = a_1X + \ldots + a_pX$. If the operator $\tilde{a}_p$ coinduced by a_p on X/X_{p-1} is injective for $1 \leq p \leq n$, then

$$H^{p-1}(\Lambda[\sigma,X],\alpha) = 0 \tag{1.50}$$

for $1 \leq p \leq n$. If we further have $X_n = X$, then a is nonsingular.

Proof. The lemma is evident for n = 1. We prove the general case by induction. Assume that the lemma holds for n = k. Let $a' = (a_1,\ldots,a_k,a_{k+1})$, $\sigma' = (\sigma_1,\ldots,\sigma_k,\sigma_{k+1})$. The sequence

$$0 \to \Lambda^p[\sigma,X] \xrightarrow{\mu^p} \Lambda^{p+1}[\sigma',X] \xrightarrow{\nu^p} \Lambda^{p+1}[\sigma,X] \to 0$$

is exact where μ^p, ν^p have the same meaning as in Prop. 5.1.6. By Th. 5.1.1 the sequence

$$(1.51)\qquad \rightarrow H^{p-1}(\Lambda[\sigma,X],\alpha) \xrightarrow{\mu_*} H^p(\Lambda[\sigma',X],\alpha') \xrightarrow{\nu_*} H^p(\Lambda[\sigma,X],\alpha)$$

$$\xrightarrow{\theta} H^p(\Lambda[\sigma,X],\alpha) \xrightarrow{\mu_*} H^{p+1}(\Lambda[\sigma',X],\alpha') \rightarrow \ldots.$$

is exact. By hypothesis one has that

$$(1.52)\qquad H^{p-1}(\Lambda[\sigma,X],\alpha) = 0 \qquad (p = 1,\ldots,k).$$

Now exactness of (1.51) implies $H^{p-1}(\Lambda[\sigma',X],\alpha') = 0$ $(p = 1,\ldots,k)$, so it remains to show $H^k(\Lambda[\sigma',X],\alpha') = 0$, and if $X_{k+1} = X$ then $H^{k+1}(\Lambda[\sigma',X],\alpha') = 0$.

Now assume that $\alpha'\xi = 0$ for $\xi \in \Lambda^k[\sigma',X]$. Since

$$\xi = \Sigma_{1\le j\le k+1} (-1)^{j-1} x_j \sigma_1\wedge\ldots\wedge\sigma_{j-1}\wedge\sigma_{j+1}\wedge\ldots\wedge\sigma_{k+1},$$

one has $a_1x_1 + \ldots + a_kx_k + a_{k+1}x_{k+1} = 0$, or equivalently, $\tilde{a}_{k+1}\tilde{x}_{k+1} = 0$. Since $\tilde{a}_{k+1}$ is injective, $\tilde{x}_{k+1} = 0$, i.e. $x_{k+1} = a_1y_1 + \ldots + a_ky_k$, for some $y_1,\ldots,y_k \in X$. Set

$$\eta_1 = \Sigma_{1\le j\le k}(-1)^{k+j-1} y_j \sigma_1\wedge\ldots\wedge\sigma_{j-1}\wedge\sigma_{j+1}\wedge\ldots\wedge\sigma_k .$$

Then $\alpha\eta_1 = (-1)^k x_{k+1}\sigma_1\wedge\ldots\wedge\sigma_k$. Therefore, we may write $\xi = \alpha\eta_1 + \xi_1\wedge\sigma_{k+1}$ with $\xi_1 \in \Lambda^{k-1}[\sigma,X]$. From $(\alpha + a_{k+1}\sigma_{k+1})\xi = \alpha'\xi = 0$ one obtains $\alpha[(-1)^k a_{k+1}\eta_1 + \xi_1] = 0$. Now (1.52) implies that there is a form η_2 such that $\alpha\eta_2 = (-1)^k a_{k+1}\eta_1 + \xi_1$. Set $\eta = \eta_1 + \eta_2\wedge\sigma_{k+1}$. Then $\alpha'\eta = \xi$, and hence $H^k(\Lambda[\sigma',X], \alpha') = 0$.

If we suppose that $X_{k+1} = X$, then $\alpha'\Lambda^k[\sigma',X]$ is clearly equal to $\Lambda^{k+1}[\sigma',X]$. Thus $H^{k+1}(\Lambda[\sigma',X], \alpha') = 0$ and the lemma is proved.

5.1.47 Proposition. Let $a = (a_1,\ldots,a_n)$ be a commuting system in $L(X)$, and let U be an open set in C^n such that $\sigma(a,X) \subset U$. If ξ is a form such that $\xi = (\delta_\alpha + \bar{\partial})\eta$ for some form η on U, then $\mu_\alpha(\xi) = 0$.

Proof. For $\alpha(z) = (z_1-a_1,\ldots,z_n-a_n)$, by Def. 5.1.24 and Prop. 5.1.26 we have

$$\begin{aligned}\mu_\alpha(\xi) &= (-1)^{n+1}\int_U P_\sigma[(\delta_\alpha + \bar\partial)\phi\eta - \xi] \wedge dz \\ &= (-1)^{n+1}\int_U P_\sigma[(\delta_\alpha + \bar\partial)(\phi\eta - \eta)] \wedge dz \\ &= (-1)^{n+1}\int_U \bar\partial(P_\sigma[\phi\eta - \eta]) \wedge dz.\end{aligned}$$

Since $\phi\eta - \eta$ has compact support, $\mu_\alpha(\xi) = 0$ by Stokes' formula.

5.1.48 Theorem. Let $a = (a_1,\ldots,a_n)$ be a commuting system in $L(X)$, where X is a Banach space. If $D = D_1 \times \ldots \times D_n$ is a polydisc in C^n containing $\sigma(a,X)$, then

$$H^{p-1}(\Lambda[\sigma, A(D,X)], \delta_\alpha) = 0$$

for $1 \leq p \leq n$.

Proof. Put $Z = A(D,X)$, hence Z is a Frechet space. Let $\alpha_j = z_j - a_j$ for each j, set $Z_0 = 0$ and $Z_p = \alpha_1 Z + \ldots + \alpha_p Z$. To prove the theorem, it suffices to prove that the coinduced operator $\tilde\alpha_p$ is injective on Z/Z_{p-1} for $1 \leq p \leq n$.

Suppose that $\alpha_1 f = 0$ for some $f \in Z$. By the projection property of $\sigma(a,X)$ one has $\sigma(a_1,X) \subset D_1$. Then $f = 0$ on $D_1 \backslash \sigma(a_1,X)$ and hence $f = 0$ on all D_1. So the theorem holds for $p = 1$. Assume now that the theorem holds for some $p \geq 1$. Let $f_1, \ldots, f_{p+1} \in Z$ be such that

$$\alpha_{p+1} f_{p+1} = \alpha_1 f_1 + \ldots + \alpha_p f_p \tag{1.53}$$

or equivalently, $\tilde\alpha_{p+1}\tilde f_{p+1} = 0$. Set

$$\theta = \Sigma_{1 \leq j \leq p}(-1)^{j-1} f_j \sigma_1 \wedge \ldots \wedge \sigma_{j-1} \wedge \sigma_{j+1} \wedge \ldots \wedge \sigma_p.$$

Then (1.53) shows that

$$(\delta_\alpha + \bar\partial)\theta = \alpha_{p+1} f_{p+1} \sigma_1 \wedge \ldots \wedge \sigma_p$$

on $D_1 \times \ldots \times D_p \supset \sigma((a_1,\ldots,a_p),X)$, where $\delta_\alpha = (a_1,\ldots,a_p)$. Therefore the Cauchy-Weil integral of the commuting system $(a_1,\ldots,a_p)$ relative to

$\alpha_{p+1}f_{p+1}$ is zero by Prop. 5.1.47. Hence for given $z_j \in D_j$, $j = p+1,\dots,n$,

$$(z_{p+1} - a_{p+1})\int_{\Gamma_1}\dots\int_{\Gamma_p}(\zeta_1 - a_1)^{-1}\dots(\zeta_p - a_p)^{-1}f_{p+1}\,d\zeta_1\dots d\zeta_p$$
$$= \int_{\Gamma_1}\dots\int_{\Gamma_p}(\zeta_1 - a_1)^{-1}\dots(\zeta_p - a_p)^{-1}(z_{p+1} - a_{p+1})f_{p+1}\,d\zeta_1\dots d\zeta_p$$
$$= \mu_\alpha(\alpha_{p+1}f_{p+1}) = 0$$

where the second equality follows by successive use of Cor. 5.1.31, and Γ_j is a circle in D_j surrounding $\sigma(a_j,X)$ $(j = 1,\dots,p)$. Hence

$$\int_{\Gamma_1}\dots\int_{\Gamma_p}(\zeta_1 - a_1)^{-1}\dots(\zeta_p - a_p)^{-1}f_{p+1}\,d\zeta_1\dots d\zeta_p = 0 \tag{1.54}$$

for any $z_j \in D_j$, $j = p+1,\dots,n$. On the other hand, the Cauchy integral formula gives

$$f_{p+1}(z) = \frac{1}{(2\pi i)^p}\int_{\Gamma_1}\dots\int_{\Gamma_p}\frac{f_{p+1}(\zeta_1,\dots,\zeta_p,z_{p+1},\dots,z_n)}{(\zeta_1 - z_1)\dots(\zeta_p - z_p)}d\zeta_1\dots d\zeta_p.$$

Using the identity

$$f_{p+1}(\zeta_1,\dots,\zeta_p,z_{p+1},\dots,z_n) \tag{1.55}$$
$$= (\zeta_1-a_1)\dots(\zeta_p-a_p)(\zeta_1- a_1)^{-1}\dots(\zeta_p-a_p)^{-1}f_{p+1}(\zeta_1,\dots,\zeta_p,z_{p+1},\dots,z_n)$$
$$= [(\zeta_1- z_1) - (z_1- a_1)]\dots[(\zeta_p- z_p) - (z_p- a_p)]$$
$$\times (\zeta_1- a_1)^{-1}\dots(\zeta_p- a_p)^{-1}f_{p+1}(\zeta_1,\dots,\zeta_p,z_{p+1},\dots,z_n)$$

and performing the product of the first factors, we obtain terms containing at least the factor $z_j- a_j$ except the term $(\zeta_1-z_1)\dots(\zeta_1-z_1)$; the integral corresponding to it is zero after substituting (1.55) into (1.54). All the factors z_j-a_j of the other terms may be put outside the corresponding integrals. Since the latter integrals are not modified as Γ_j varies, we obtain an equality of the form $f_{p+1}(z) = (z_1- a_1)g_1(z) + \dots + (z_p- a_p)g_p(z)$ for $z \in D$. Thus $\tilde{f}_{p+1} = 0$ and $\tilde{\alpha}_{p+1}$ is injective.

5.1.49 Proposition. Let U be an open polydisc in C^n containing $\sigma(a,X)$. Then an element $f\sigma_1\wedge\ldots\wedge\sigma_n$ belongs to the range of the map $\delta_\alpha : \Lambda^{n-1}[\sigma,A(U,X)] \to \Lambda^n[\sigma,A(U,X)]$ if and only if the Cauchy-Weil integral $\mu_\alpha(f)$ of a relative to f is equal to zero.

Proof. We first assume that $f\sigma_1\wedge\ldots\wedge\sigma_n$ belongs to the range of the operator δ_α. Then $f\sigma_1\wedge\ldots\wedge\sigma_n = \delta_\alpha\xi$ for some $\xi \in \Lambda^{n-1}[\sigma,A(U,X)]$. It follows from Prop. 5.1.47 that $\mu_\alpha(f) = 0$.

Conversely, assume that $\mu_\alpha(f) = 0$. Since $U = U_1 \times \ldots \times U_n$ is a polydisc, the projection property of the joint spectrum implies that $\sigma(a_j,X) \subset U_j$, $j = 1,\ldots,n$. Let Γ_j be any oriented contour in U_j surrounding $\sigma(a_j,X)$. From Cor. 5.1.31

$$\mu_\alpha(f) = \int_{\Gamma_1}\cdots\int_{\Gamma_n} (z_1 - a_1)^{-1}\ldots(z_n - a_n)^{-1}f(z)\, dz_1\ldots dz_n = 0. \tag{1.56}$$

On the other hand, by the Cauchy integral formula,

$$f(w) = \frac{1}{(2\pi i)^n}\int_{\Gamma_1}\cdots\int_{\Gamma_n}(z_1-w_1)^{-1}\ldots(z_n-w_n)^{-1}f(z)\, dz_1\ldots dz_n$$

where w_j is inside Γ_j, $j = 1,\ldots,n$. Rewriting f as

$$\begin{aligned} f(z) &= (z_1 - a_1)\ldots(z_n - a_n)(z_1 - a_1)^{-1}\ldots(z_n - a_n)^{-1}f(z) \\ &= [(z_1-w_1) - (w_1-a_1)] \ldots [(z_n-w_n) - (w_n-a_n)] \\ &\qquad \times (z_1-a_1)^{-1}\ldots(z_n-a_n)^{-1}f(z), \end{aligned}$$

performing in the last expression the product of the first n factors, and then passing to the integral, we obtain each term containing at least one factor w_j-a_j except the term containing $(z_1-w_1)\ldots(z_n-w_n)$. But (1.56) implies that this equals zero. Therefore,

$$f(w) = (w_1-a_1)f_1(w) + \ldots + (w_n-a_n)f_n(w)$$

where f_j are analytic functions on U by the arbitrariness of Γ_j in U_j.

5.1.50 Corollary. For every open polydisc U containing $\sigma(a,X)$, the map δ_α in the preceding proposition has closed range.

Proof. From Prop. 5.1.49, the range of δ_α consists of all forms $f\sigma_1\wedge\ldots\wedge\sigma_n$ having the Cauchy-Weil integral equal to zero. But the Cauchy-Weil integral, as a continuous mapping, has closed kernel.

2. Spectral Decomposition for Commuting Systems

In this section we study the spectral decomposition for commuting systems of operators on Banach space. Throughout this section we assume that X is a complex Banach space unless otherwise specified.

Assume that F is a closed set in $\mathbf{C}^n$; $A(F,X)$ will denote the set of all analytic X-valued functions defined on some neighborhood of F. Thus $A(\{z\},X)$ is the set of all analytic X-valued functions defined near z. Suppose that $a = (a_1,\ldots,a_n)$ is a commuting system in $L(X)$, and let $\delta_\alpha = (z_1-a_1)\sigma_1 + \ldots + (z_n-a_n)\sigma_n$, and let $\bar\partial$ be as above.

5.2.1 Definition. The commuting system $a = (a_1,\ldots,a_n)$ is said to have the <u>single-valued extension property</u> (SVEP) if $H^p(A(\{z\},X),\delta_\alpha) = 0$ for all $z \in \mathbf{C}$ and $p = 0,1,\ldots,n-1$.

Th. 5.1.21(ii) shows that the condition of the definition is automatically satisfied for all points of the <u>resolvent set</u> $\rho(a,X) = \mathbf{C}^n\backslash\sigma(a,X)$.

5.2.2 Lemma. If U is a domain of holomorphy in $\mathbf{C}^n$ and $p \geq 0$ is an integer, then $H^p(A(U,X),\delta_\alpha) = 0$ if and only if $H^p(C^\infty(U,X),\delta_\alpha+\bar\partial) = 0$.

Proof. We first assume that $H^p(A(U,X),\delta_\alpha) = 0$. Let $\xi \in \Lambda^p[\sigma \cup d\bar z, C^\infty(U,X)]$ be such that $(\delta_\alpha + \bar\partial)\xi = 0$. If $p = 0$ then ξ satisfies $\delta_\alpha\xi = 0$ and $\bar\partial\xi = 0$. The latter equality shows that ξ is analytic; from this, the former equality and the hypothesis $H^0(A(U,X),\delta_\alpha) = 0$, it follows that $\xi = 0$.

Next, assuming $p > 0$, we may write

$$\xi = \xi_{0,p} + \xi_{1,p-1} + \ldots + \xi_{p-1,1} + \xi_{p,0}, \tag{2.1}$$

where $\xi_{j,p-j}$ $(j = 0,1,\ldots,p)$ has degree j in σ and degree p-j in $d\bar z$. The equality $(\delta_\alpha + \bar\partial)\xi = 0$ implies

$$(2.2) \qquad \bar\partial\xi_{0,p} = 0, \qquad \delta_\alpha\xi_{0,p} + \bar\partial\xi_{1,p-1} = 0, \qquad \ldots,$$

$$\delta_\alpha\xi_{p-1,1} + \bar\partial\xi_{p,0} = 0, \qquad \delta_\alpha\xi_{p,0} = 0.$$

Since $\xi_{0,p}$ is a form of degree p in $d\bar z$, by Th. 5.1.18 there exists $\eta_{0,p-1} \in \Lambda^{p-1}[d\bar z, C^\infty(U,X)]$ such that $\bar\partial\eta_{0,p-1} = \xi_{0,p}$. Replacing this expression in the second equation in (2.2), we obtain

$$(2.3) \qquad \bar\partial[\xi_{1,p-1} - \delta_\alpha\eta_{0,p-1}] = 0.$$

The coefficients of $\xi_{1,p-1} - \delta_\alpha\eta_{0,p-1}$, regarded as a form in σ, are forms of degree p-1 in $d\bar z$. By (2.3) every such coefficient satisfies the same equality in (2.3), and by Th. 5.1.18 there exists a form $\eta_{1,p-2}$ such that $\xi_{1,p-1} - \delta_\alpha\eta_{0,p-1} = \bar\partial\eta_{1,p-2}$. Continuing in this way, we obtain a set of forms $\{\eta_{j,p-j-1}\}$ $(j = 0,1,\ldots,p-1)$ such that

$$(2.4) \qquad \xi_{j,p-j} - \delta_\alpha\eta_{j-1,p-j} = \bar\partial\eta_{j,p-j-1}.$$

Setting j = p-1 in (2.4) and replacing the result in the equality $\delta_\alpha\xi_{p-1,1} + \bar\partial\xi_{p,0} = 0$, one has

$$\bar\partial[\xi_{p,0} - \delta_\alpha\eta_{p-1,0}] = 0.$$

Therefore $\xi_{p,0} - \delta_\alpha\eta_{p-1,0}$, as a form in σ, has analytic coefficients. From the hypothesis and the equality $\delta_\alpha[\xi_{p,0} - \delta_\alpha\eta_{p-1,0}] = 0$, there exists another form ϕ such that $\delta_\alpha\phi = \xi_{p,0} - \delta_\alpha\eta_{p-1,0}$. Rewriting $\phi + \eta_{p-1,0}$ as $\eta_{p-1,0}$, one has $\xi_{p,0} = \delta_\alpha\eta_{p-1,0}$. Then $\xi = (\delta_\alpha + \bar\partial)\eta$ with $\eta = \Sigma_j \eta_{j,p-j-1}$.

The "if" part of the theorem can be proved in a similar way.

5.2.3 Lemma. For every open V in C^n, assume that

$$(2.5) \qquad H^{p-1}(C^\infty(V,X), \delta_\alpha + \bar\partial) = 0.$$

Let U_1, U_2 be two open sets in C^n, let $V = U_1 \cap U_2$, and let ξ be a form in

$\Lambda^p[\sigma \cup d\bar{z}, C^\infty(U_1 \cup U_2, X)]$. If there exist $\eta_j \in \Lambda^{p-1}[\sigma \cup d\bar{z}, C^\infty(U_j,X)]$ such that $(\delta_\alpha + \bar{\partial})\eta_j = \xi$ on U_j for $j = 1,2$, then there exists $\eta \in \Lambda^{p-1}[\sigma \cup d\bar{z}, C^\infty(U_1 \cup U_2, X)]$ such that $\xi = (\delta_\alpha + \bar{\partial})\eta$ on $U_1 \cup U_2$.

Proof. We first prove the following assertion. For every $\phi \in C^\infty(U_1 \cap U_2, X)$, there exists $\phi_j \in C^\infty(U_j,X)$ $(j = 1,2)$ such that $\phi = \phi_1 + \phi_2$ on $U_1 \cap U_2$. Let $U = U_1 \cup U_2$ and let $\{G_k\}$ be a locally finite countable cover of U subordinate to $\{U_1, U_2\}$. Corresponding to $\{G_k\}$ there exists a C^∞ partition of unity $\{h_k\}$. For all k such that $G_k \subset U_1$, define ϕ_{k2} on U_2 by

$$\phi_{k2}(z) = \left\{\begin{array}{ll} h_k(z)\phi(z) & \text{for } z \in G_k \cap U_2; \\ 0 & \text{for } z \in U_2 \setminus (G_k \cap U_2) \end{array}\right\}.$$

Since supp h_k is compact and contained in G_k, we have $\phi_{k2} \in C^\infty(U_2,X)$. For all other indices k, $G_k \subset U_2$ and we define ϕ_{k1} on U_1 by

$$\phi_{k1}(z) = \left\{\begin{array}{ll} h_k(z)\phi(z) & \text{for } z \in G_k \cap U_1 \\ 0 & \text{for } z \in U_1 \setminus (G_k \cap U_1) \end{array}\right\}.$$

Then $\phi_{k1} \in C^\infty(U_1,X)$. Since the cover $\{G_k\}$ is locally finite, the functions $\phi_j = \Sigma_k \phi_{kj}$ $(j=1,2)$ are well-defined and belong to $C^\infty(U_j,X)$. If $z \in U_1 \cap U_2$, then $\phi(z) = \Sigma\, h_k(z)\phi(z) = \phi_1(z) + \phi_2(z)$.

We can now prove the lemma. If $U_1 \cap U_2 = \emptyset$, we may define $\eta(z) = \eta_j(z)$ for $z \in U_j$. Otherwise, we have $(\delta_\alpha + \bar{\partial})[\eta_1 - \eta_2] = 0$ on $U_1 \cap U_2$. By (2.5) there exists a form $\phi \in \Lambda^{p-2}[\sigma \cup d\bar{z}, C^\infty(U_1 \cap U_2, X)]$ such that $\eta_1 - \eta_2 = (\delta_\alpha + \bar{\partial})\phi$. Applying the previous result to the coefficients of ϕ, we may write $\phi = \phi_2 - \phi_1$ with $\phi_j \in \Lambda^{p-2}[\sigma \cup d\bar{z}, C^\infty(U_j,X)]$ $(j=1,2)$. Define

$$\eta(z) = [\eta_j + (\delta_\alpha + \bar{\partial})\phi_j](z) \quad \text{for } z \in U_j,\ j = 1,2.$$

Then η is well-defined on $U_1 \cup U_2$ and $\eta \in \Lambda^{p-1}[\sigma \cup d\bar{z}, C^\infty(U_1 \cup U_2, X)]$.

Moreover, we have $(\delta_\alpha + \bar\partial)\eta = \xi$ on $U_1 \cup U_2$, hence the lemma is proved.

The following theorem gives several equivalent conditions for a commuting system having SVEP; it shows that SVEP is a local property.

5.2.4 Theorem. Let $a = (a_1, \ldots a_n)$ be a commuting system in $L(X)$. Then the following are equivalent:

(i) a has SVEP;

(ii) $H^p(A(U,X),\delta_\alpha) = 0$ for every domain of holomorphy U in C^n and $p = 0, \ldots, n-1$;

(iii) $H^p(C^\infty(U,X),\delta_\alpha + \bar\partial) = 0$ for every open set U in C^n and $p = 0, \ldots, n-1$;

(iv) $H^p(C^\infty(\{z\},X),\delta_\alpha + \bar\partial) = 0$ for every $z \in C^n$ and $p = 0, \ldots, n-1$.

Proof. The implication (iii) $\Rightarrow$ (iv) is evident, and (i) and (iv) are equivalent by Lemma 5.2.2. It remains to prove (ii) $\Rightarrow$ (iii) and (i) $\Rightarrow$ (ii).

We prove (iii) from (ii) by induction and the assumption that $H^p(A(D,X),\delta_\alpha) = 0$ for every open polydisc D in C^n and $p = 0, \ldots, n-1$. Let $\xi \in C^\infty(U,X)$ be such that $(\delta_\alpha + \bar\partial)\xi = 0$. Then $\bar\partial\xi = 0$ and $\delta_\alpha\xi = 0$, and the former implies that ξ is an analytic function. By hypothesis ξ is thus equal to zero on every polydisc contained in U. Therefore $\xi = 0$ on U, so $H^0(C^\infty(U,X),\delta_\alpha + \bar\partial) = 0$.

Now assume that $H^{p-1}(C^\infty(U,X),\delta_\alpha + \bar\partial) = 0$. Let $\xi \in \Lambda^p[\sigma \cup d\bar z, C^\infty(U,X)]$ be such that $(\delta_\alpha + \bar\partial)\xi = 0$. Let $\{K_j\}$ be an increasing sequence of compact sets which exhausts U. By Lemma 5.2.2, for each $z \in K_j$ (j fixed) and each polydisc D_z in U centered at z, we have $H^p(C^\infty(D_z,X), \delta_\alpha + \bar\partial) = 0$. Lemma 5.2.3 applied finitely many times shows that there exist a neighborhood V_j' of K_j and a form η_j on V_j' such that

$$\xi = (\delta_\alpha + \bar\partial)\eta_j \quad \text{on } V_j'. \tag{2.6}$$

Clearly we may assume that $\eta_j \in \Lambda^{p-1}[\sigma \cup d\bar z, C^\infty(U,X)]$ and that (2.6) holds on a neighborhood V_j of K_j with the property $V_j^- \subset V_j'$.

We claim that the sequence $\{\eta_j\}$ can be chosen so that $\eta_{j+1} = \eta_j$ on a neighborhood of K_j. Let us suppose that $\eta_1, \ldots, \eta_j$ have been determined to satisfy this condition. Assume that $(\eta_{j+1})^*$ satisfies the property $\xi = (\delta_\alpha + \bar\partial)(\eta_{j+1})^*$ on some neighborhood of K_{j+1}. Then $(\delta_\alpha + \bar\partial)[(\eta_{j+1})^* - \eta_j] = 0$ on a neighborhood of K_j. Since $H^{p-1}(C^\infty(U,X), \delta_\alpha + \bar\delta) = 0$ by assumption, there exists a form ϕ such that $(\eta_{j+1})^* - \eta_j = (\delta_\alpha + \bar\partial)\phi$ on a neighborhood of K_j. We may also suppose that $\phi \in \Lambda^{p-2}[\sigma \cup d\bar z, C^\infty(U,X)]$. Set $\eta_{j+1} = (\eta_{j+1})^* - (\delta_\alpha + \bar\partial)\phi$. Then $\xi = (\delta_\alpha + \bar\partial)(\eta_{j+1})^* = (\delta_\alpha + \bar\partial)\eta_{j+1}$ on a neighborhood of K_{j+1} and $\eta_{j+1} = \eta_j$ on a neighborhood of K_j.

Define $\eta = \eta_j$ on K_j for $j = 1,2,\ldots$. Then $\eta \in \Lambda^{p-1}[\sigma \cup d\bar z, C^\infty(U,X)]$ and $\xi = (\delta_\alpha + \bar\partial)\eta$. Therefore, $H^p(C^\infty(U,X), \delta_\alpha + \bar\partial) = 0$. This proves (ii) $\Rightarrow$ (iii).

(i) $\Rightarrow$ (ii). We use some elementary sheaf theory as found in [3']. Assume that U is open in C^n and Z is a Banach space. We let

$$F^p(U,Z) = \cup_{\zeta \in U}\{(\psi)_\zeta\colon \psi \in \Lambda^p[d\bar z, C^\infty(W,Z)] \text{ for some neighborhood } W \text{ of } \zeta\}$$

be the sheaf of all germs of C^∞-forms of degree p on U. Since $F^p(U,Z)$ is a fine sheaf of C_0^∞-modules, it is acyclic. Let $\mathcal{A}(U,Z)$ be the sheaf of all germs of analytic Z-valued functions on U. Then

$$0 \to \mathcal{A}(U,Z) \xrightarrow{i} F^0(U,Z) \xrightarrow{\bar\partial} F^1(U,Z) \xrightarrow{\bar\partial} \cdots \xrightarrow{\bar\partial} F^n(U,Z) \to 0$$

is an exact sequence of sheaves. If U is a domain of holomorphy, the following induced sequence of global sections is also exact.

$$0 \to A(U,Z) \xrightarrow{i} C^\infty(U,Z) \xrightarrow{\bar\partial} \Lambda^1[d\bar z, C^\infty(U,X)] \xrightarrow{\bar\partial} \cdots \xrightarrow{\bar\partial} \Lambda^n[d\bar z, C^\infty(U,X)] \to 0.$$

(see Th. 5.1.18). Hence $H^p(U, \mathcal{A}(U,Z)) \cong H^p(\Gamma(F(U,Z))) = 0$ for $1 \le p \le n$, i.e. $\mathcal{A}(U,Z)$ is acyclic. Since a satisfies condition (i),

$$0 \to \mathcal{A}(U,X) \xrightarrow{\delta_\alpha} \mathcal{A}(U,X^{n_1}) \xrightarrow{\delta_\alpha} \ldots$$

$$\to \mathcal{A}(U,X^{n_p}) \xrightarrow{\delta_\alpha} \ldots \xrightarrow{\delta_\alpha} \mathcal{A}(U,X)$$

is an exact sequence of acyclic sheaves, where $n_p = C_p{}^n$ and X^{n_p} is the n_p-fold direct sum of X. Therefore the induced sequence of global sections

$$0 \to A(U,Z) \xrightarrow{\delta_\alpha} [\sigma,A(U,X)] \xrightarrow{\delta_\alpha} \ldots \xrightarrow{\delta_\alpha} \Lambda^n[\sigma,A(U,X)]$$

is also exact, so (ii) holds. This completes the proof of the theorem.

5.2.5 Definition. Let $a = (a_1,\ldots,a_n)$ be a commuting system in L(X), X a Banach space, and let $x \in X$. Then the local spectrum $\sigma_a(x)$ of a at x is the complement in $\mathbf{C}^n$ of the local resolvent set

$$\rho_a(x) = \cup\{U \subset \mathbf{C}^n\text{: there is a } C^\infty\text{-form } \xi \in \Lambda^{n-1}[\sigma\cup d\bar{z}, C^\infty(U,X)] \text{ such that } x\,\sigma_1\wedge\ldots\wedge\sigma_n = (\delta_\alpha+\bar\partial)\xi\}.$$

The local analytic spectrum $sp_a(x)$ is the complement of the set

$$\cup\{U \subset \mathbf{C}^n\text{: there exist analytic functions } f_j\colon U \to X \text{ satisfying } x = \Sigma_j\,(z_j - a_j)f_j(z),\ 1 \le j \le n\}.$$

The manifold

$$X(a,F) = \{x \in X\colon\ \sigma_a(x) \subset F\} \tag{2.7}$$

is called a spectral subspace of a whenever F is closed in $\mathbf{C}^n$.

5.2.6 Theorem. For every commuting system $a = (a_1,\ldots,a_n)$ in

$L(X)$ and every $x \in X$, we have $\sigma_a(x) = sp_a(x)$.

Proof. The argument of this theorem is similar to Lemma 5.2.2; we give the details for completeness. To prove the inclusion $\sigma_a(x) \subset sp_a(x)$, let $z \notin sp_a(x)$. Then there exist a polydisc D_z with center z and n analytic functions $f_j: D_z \to X$ such that $x = \Sigma_j (\zeta_j - a_j)f_j(\zeta)$ on D_z. Put

$$\xi = \Sigma_j (-1)^{j+1} f_j \sigma_1 \wedge \ldots \sigma_{j-1} \wedge \sigma_{j+1} \wedge \ldots \wedge \sigma_n.$$

Then $(\delta_\alpha + \bar{\partial})\xi = x\sigma_1 \wedge \ldots \wedge \sigma_n$, so $z \notin \sigma_a(x)$.

Conversely, assume that $z \notin \sigma_a(x)$. Then there exists $\xi \in \Lambda^{n-1}[\sigma \cup d\bar{z}, C^\infty(D_z, X)]$ such that

$$(\delta_\alpha + \bar{\partial})\xi = x\sigma_1 \wedge \ldots \wedge \sigma_n. \tag{2.8}$$

Write $\xi = \Sigma_{j=0}^{n-1} \xi_{j,n-j-1}$ with $\xi_{j,n-j-1}$ of degree j in $d\bar{z}$ and degree $n-j-1$ in σ. By (2.8)

$$\begin{aligned} x\sigma_1 \wedge \ldots \wedge \sigma_n &= \delta_\alpha \xi_{0,n-1} \\ 0 &= \bar{\partial}\xi_{0,n-1} + \delta_\alpha \xi_{1,n-2} \\ &\ldots\ldots\ldots\ldots \\ 0 &= \bar{\partial}\xi_{j,n-j-1} + \delta_\alpha \xi_{n-1,0} \\ 0 &= \bar{\partial}\xi_{n-1,0}. \end{aligned} \tag{2.9}$$

By Th. 5.1.18 the last equality in (2.9) implies the existence of a form $\eta_{n-2,0}$ such that $\xi_{n-1,0} = \bar{\partial}\eta_{n-2,0}$. Hence $\bar{\partial}(\xi_{n-2,1} - \delta_\alpha \eta_{n-2,0}) = 0$. Again from Th 5.1.18 we obtain a form $\eta_{n-3,1}$ such that $\xi_{n-2,1} - \delta_\alpha \eta_{n-2,0} = \bar{\partial}\eta_{n-3,1}$. Proceeding successively, we obtain a family of forms $\{\eta_{n-2,0}, \ldots, \eta_{0,n-2}\}$ such that

$$\xi_{n-j,j-1} - \delta_\alpha \eta_{n-j,j-2} = \bar{\partial}\eta_{n-j-1,j-1} \quad (j = 1, \ldots, n-1), \tag{2.10}$$

where $\eta_{n-1,-1} = 0$. Setting $j = n-1$ in (2.10) and replacing the result in the second equality in (2.9), we have

$$\bar{\partial}(\xi_{0,n-1} - \delta_\alpha \eta_{0,n-2}) = 0.$$

Hence the coefficients of the form $\chi = \xi_{0,n-1} - \delta_\alpha \eta_{0,n-2}$ are analytic on D_z and $\delta_\alpha \chi = \delta_\alpha \xi_{0,n-1} = x\sigma_1 \wedge \ldots \wedge \sigma_n$, so $z \notin sp_a(x)$. The proof is complete.

Because of Th. 5.2.6 it is obvious that the spectral subspace $X(a,F)$ can also be written

$$X(a,F) = \{x \in X: sp_a(x) \subset F\}. \tag{2.11}$$

If a has SVEP, then by Lemma 5.2.3 each $x \in X$ has an associated global solution ξ of $(\delta_\alpha + \bar{\partial})\xi = x\sigma_1 \wedge \ldots \wedge \sigma_n$ on $\rho_a(x)$, hence by Theorems 5.2.4 and 5.2.6 for each domain of holomorphism U in $\rho_a(x)$ there are $f_1, \ldots, f_n \in A(U,X)$ satisfying $x = (z_1 - a_1)f_1(z) + \ldots + (z_n - a_n)f_n(z)$ for $z \in U$.

The next proposition is a consequence of Prop. 5.1.47.

5.2.7 Proposition. Let U be an open set in $\mathbf{C}^n$ containing $\sigma(a,X)$, and let $f \in \Lambda^n[\sigma, A(U,X)]$. If there is a form $g \in \Lambda^{n-1}[\sigma, A(U,X)]$ such that $f = \delta_\alpha g$, then the Cauchy-Weil integral $\mu_\alpha(f)$ of a relative to f is zero.

As a special case of Prop. 5.2.7, we conclude that if a has SVEP, then $\sigma_a(x) = 0$ implies $x = 0$.

5.2.8 Definition. Let $a = (a_1, \ldots, a_n)$ be a commuting system, and let $m \geq 2$ be given. If for every m-open cover $\{G_1, \ldots, G_m\}$ of $\sigma(a,X)$ there exists a corresponding system of a-invariant subspaces $\{X_1, \ldots, X_m\}$ such that

(i) $\sigma(a, X_j) \subset G_j \qquad (j = 1, \ldots, m)$,

(ii) $X = X_1 + \ldots + X_m$,

then a is said to have the m-<u>spectral decomposition property</u> (m-SDP); a is said to have SDP if it has m-SDP for every $m \geq 2$.

5.2.9 Lemma. Assume that the commuting system $a = (a_1, \ldots, a_n)$ has 2-SDP, D is a polydisc in $\mathbf{C}^n$ and p is some integer with $1 \leq p \leq n-1$. If $\delta_\alpha \xi = 0$ for $\xi \in \Lambda^p[\sigma, A(D,X)]$, then for every polydisc D_1 with $D_1^- \subset D$, there exists $\eta \in$ $\Lambda^{p-1}[\sigma, A(D_1,X)]$ such that $\xi = \delta_\alpha \eta$ on D_1.

Proof. Let G_1 be a polydisc in $\mathbf{C}^n$ such that $D_1^- \subset G_1 \subset G_1^- \subset D$, and let G_2 be open in $\mathbf{C}^n$ such that $\{G_1,G_2\}$ covers $\mathbf{C}^n$ and that $G_2^- \cap D_1^- = \emptyset$. By the 2-SDP, there exist a-invariant subspaces X_1, X_2 such that $X_1 + X_2 = X$ and $\sigma(a,X_j) \subset G_j$ $(j= 1,2)$. Since X/X_2 is topologically isomorphic to $X_1/X_1 \cap X_2$, we have $\sigma(a,X/X_2) = \sigma(a,X_1/X_1 \cap X_2)$, and since G_1 is a polydisc, it follows from Th. 5.1.36 that $\sigma(a,X_1 \cap X_2) \subset G_1$, therefore

$$\begin{aligned}\sigma(a,X/X_2) &= \sigma(a,X_1/X_1 \cap X_2)\\ &\subset \sigma(a,X_1) \cup \sigma(a,X_1 \cap X_2) \subset G_1\end{aligned}$$

by Prop. 5.1.5. So Th. 5.1.48 and the exactness of the sequence

$$0 \to X_2 \to X \to X/X_2 \to 0$$

show that $\xi/X_2 = \delta_\alpha(\eta_1/X_2)$ for some $\eta_1 \in \Lambda^{p-1}[\sigma,A(D,X)]$. On the other hand, the inclusions $\sigma(a,X_2) \subset G_2 \subset \mathbf{C}^n \backslash D_1$ imply that there is a form $\eta_2 \in \Lambda^{p-1}[\sigma,A(D_1,X_2)]$ such that $\xi - \delta_\alpha\eta_1 = \delta_\alpha\eta_2$. Set $\eta = \eta_1 + \eta_2$. Then $\xi = \delta_\alpha\eta$ with $\eta \in \Lambda^{p-1}[\sigma,A(D_1,X)]$, and the lemma is proved.

5.2.10 Theorem. If the commuting system $a = (a_1,\dots,a_n)$ has the 2-SDP, then a has SVEP.

Proof. By Th. 5.2.4 SVEP is a local property, so it suffices to prove that

$$H^p(A(D,X),\delta_\alpha) = 0 \tag{2.12}$$

for every polydisc D in $\mathbf{C}^n$ and $0 \leq p \leq n-1$. Let $p = 0$, and let $f \in H^0(A(D,X),\delta_\alpha) = A(D,X)$ such that $\delta_\alpha f = 0$. Then $f = 0$ on D by Lemma 5.2.9.

For the inductive step, suppose that $H^{p-1}(A(D,X),\delta_\alpha) = 0$ for $p \leq n-1$. Let $\{D_j\}$ be a sequence of open polydiscs with $D_j^- \subset D_{j+1} \subset D_{j+1}^- \subset D$ and let $\xi \in \Lambda^p[\sigma,A(D,X)]$ such that $\delta_\alpha\xi = 0$. By Lemma 5.2.9 there is a form η_1 with $\xi = \delta_\alpha\eta_1$ on D_2. Similarly there is a form η_2' with $\xi = \delta_\alpha\eta_2'$ on D_3. Since $\delta_\alpha(\eta_1 - \eta_2') = 0$ on D_2, there is a form $\chi \in \Lambda^{p-2}[\sigma,A(D_2,X)]$ such that $\eta_1 - \eta_2' = \delta_\alpha\chi$ on D_2. Let χ' be the sum of finitely many terms in the expansion of χ with the property that $\|\delta_\alpha\chi - \delta_\alpha\chi'\| \le 1/2$ on D_1^-. Set $\eta_2 = \eta_2' + \delta_\alpha\chi'$. Then $\delta_\alpha\eta_2 = \delta_\alpha\eta_2' = \xi$ on D_3 and

$$\|\eta_1 - \eta_2\| = \|(\eta_2' + \delta_\alpha\chi) - (\eta_2' + \delta_\alpha\chi')\| \le 1/2$$

on D_1^-. Continuing in this way, we obtain a sequence of forms $\{\eta_j\}$ such that

$$\xi = \delta_\alpha\eta_j \text{ on } D_{j+1}, \text{ for } j = 1,2,\dots;$$
$$\|\eta_{j+1} - \eta_j\| \le 1/2^j \text{ on } D_{j-1}^- \text{ for } j = 2,3,\dots.$$

Putting $\eta = \lim_j \eta_j$, we have $\eta \in \Lambda^{p-1}[\sigma, A(D,X)]$ and $\xi = \delta_\alpha\eta$. Thus $H^p(A(D,X),\delta_\alpha) = 0$, and the theorem is proved.

5.2.11 Theorem. If the commuting system $a = (a_1,\dots,a_n)$ has the 2-SDP, then $X(a,F)$ is closed for every closed set F in C^n.

Proof. Let $z \notin F$ and let G_1 be an open polydisc with $z \in G_1$ and $G_1^- \cap F = \emptyset$. Also let G_2 be the complement of a closed polydisc with $z \notin G_2^-$ such that $\{G_1, G_2\}$ covers C^n. By the 2-SDP for a, there exist a-invariant subspaces X_1, X_2 such that

$$X = X_1 + X_2 \text{ and } \sigma(a,X_j) \subset G_j \ (j = 1,2). \tag{2.13}$$

Since G_1 is a polydisc, it follows from Th. 5.1.36 that $\sigma(a, X_1 \cap X_2) \subset G_1$, hence

$$\sigma(a,X/X_2) = \sigma(a,X_1/X_1 \cap X_2) \subset \sigma(a,X_1) \cup \sigma(a,X_1 \cap X_2) \subset G_1$$

by Prop. 5.1.5. On the other hand, as x is in $X(a,F)$, there exists a form $\xi \in \Lambda^{n-1}[\sigma, A(C^n \setminus F, X)]$ such that

$$x\sigma_1 \wedge \dots \wedge \sigma_n = \delta_\alpha\xi = (\delta_\alpha + \bar{\delta})\xi$$

on $C^n \setminus F$. For every $y \in X$ let $\tilde{y}$ be the coset in X/X_2 containing y. Then the equality above implies $\tilde{x}\sigma_1\wedge\ldots\wedge\sigma_n = (\delta_\alpha + \bar{\partial})\tilde{\xi}$. Prop. 5.2.7 applies to the present case to show that $\tilde{x} = 0$, i.e. $x \in X_2$. Hence $X(a,F) \subset X_2$. If for each $z \notin F$ we fix $X_2 = X_z$ in (2.13), the foregoing shows that

$$X(a,F) \subset \cap_{z \notin F} X_z. \tag{2.14}$$

Since the reverse inclusion of (2.14) is evident, $X(a,F)$ is closed as the intersection of closed subspaces.

5.2.12 Remark. Clearly Ths. 5.2.10 and 5.2.11 remain valid if, in Def. 5.2.8, $m = 2$, G_1 is an open polydisc and G_2 is the complement of a closed polydisc such that $\{G_1,G_2\}$ covers C^n. But it still an open question whether $\sigma(a,X(a,F))$ is contained in F if a has 2-SDP.

The following defines another kind of spectral decomposition which is stronger than SDP.

5.2.13 Definition. Let $a = (a_1,\ldots,a_n)$ be a commuting system. Use $F(C^n)$ [resp. Inv(a)] to denote the set of all closed sets in C^n [resp. all a-invariant closed subspaces in X]. If there is a map $E\colon F(C^n) \to \mathrm{Inv}(a)$ such that

(i) $\sigma(a,E(F)) \subset F$,

(ii) $\sigma(a,X/E(F)) \subset C^n\setminus \mathrm{Int}\, F$,

we say that a is predecomposable.

We now observe some properties of predecomposable operators.

5.2.14 Lemma. If the commuting system a is predecomposable, then

(i) a has SVEP;

(ii) the spectral subspace $X(a,F)$ is closed for all closed F in C^n.

Proof. (i) Assume that $\xi \in \Lambda^p[\sigma,A(D,X)]$, where $0 \le p \le n-1$, $\delta_\alpha\xi = 0$ and D is an open polydisc in C^n. By Def 5.2.13(ii), every open polydisc G whose closure lies in D satisfies $\sigma(a,X/E(C^n\setminus G)) \subset G^-$. Hence the argument of Lemma 5.2.9 shows that

$$\xi/E(C^n\setminus G) \;=\; \delta_\alpha(\eta_1/E(C^n\setminus G))$$

for some $\eta_1 \in \Lambda^{p-1}[\sigma,A(D,X)]$. But by Def. 5.2.13(i), there exists $\eta_2 \in \Lambda^{p-1}[\sigma,A((G,E(C^n\setminus G)]$ such that $\xi - \delta_\alpha\eta_1 = \delta_\alpha\eta_2$, which, together with the argument of Th. 5.2.10, yields the conclusion.

(ii) Since $X(a,F)$ evidently satisfies

(2.15) $$X(a,F) \supset \cap\{E(H): H \text{ closed}, F \subset \operatorname{Int} H\},$$

it suffices to prove the opposite inclusion. But Th. 5.2.11 applies to the present case with $E(F)$ in place of X_z, hence we obtain the inclusion

(2.15.5) $$X(a,F) \subset \cap\{E(H): H \text{ closed}, F \subset \operatorname{Int} H\}.$$

This and (2.15) show that $X(a,F)$ is closed.

5.2.15 Lemma. For the given commuting system $a = (a_1,...,a_n)$, $x \in X$ and open set $U \subset C^n$, if there is a form $\xi \in \Lambda^{n-1}[\sigma \cup d\bar{z}, C^\infty(U,X)]$ such that

(2.16) $$x\sigma_1 \wedge .... \wedge \sigma_n = (\delta_\alpha + \bar{\partial})\xi,$$

then there is a form $\zeta \in \Lambda^{n-1}[\sigma \cup d\bar{z}, C^\infty(U \cup \rho(a,X),X)]$ satisfying (2.16) on $U \cup \rho(a,X)$.

Proof. We note that

$$H^p(C^\infty(U \cap \rho(a,X),X),\delta_\alpha + \bar{\partial}) = 0$$

for $p = 0,1,...,2n$ by Th. 5.1.22 and Prop. 5.1.35. Since this holds for the specific case $p = n-1$, Lemma 5.2.3 yields the conclusion of the present lemma.

5.2.16 Lemma. Assume that the following conditions hold:
(i) the commuting system $(a_1,...,a_n)$ is predecomposable;

(ii) $(\delta_\alpha + \bar{\partial})\xi \in E(F)$ for some $\xi \in \Lambda^{n-1}[\sigma \cup d\bar{z}, C^\infty(V,X)]$ where F [V] is closed [open] in C^n, resp.

Then for each $n > 1$ and each open neighborhood U of F, there is a form $\eta \in \Lambda^{n-2}[\sigma \cup d\bar{z}, C^\infty(V,X)]$ such that the coefficients of the difference $\xi - (\delta_\alpha + \bar{\partial})\eta$ belong to $E(U^-)$. For $n = 1$ we have the more precise result $\xi \in E(F)$.

Proof. We first assume that $n > 2$. Let $V_1 = V \cap (C^n \backslash F)$ and $V_2 = V \cap U$. Since V_1 is disjoint from $\sigma(a,E(F))$, there is a form $\xi_0 \in \Lambda^{n-1}[\sigma \cup$

$d\bar{z}, C^\infty(V_1,E(F))]$ such that $(\delta_\alpha + \bar\partial)\xi = (\delta_\alpha + \bar\partial)\xi_0$ on V_1. By the SVEP of a, the equation $\xi - \xi_0 = (\delta_\alpha + \bar\partial)\eta_1$ has a solution $\eta_1 \in \Lambda^{n-2}[\sigma \cup d\bar{z}, C^\infty(V_1,X)]$. On the other hand, as $V_2 \cap \sigma(a,X/E(U^-)) = \emptyset$, there is also a form $\eta_2 \in \Lambda^{n-2}[\sigma \cup d\bar{z}, C^\infty(V_2,X)]$ such that $\xi/E(U^-) = (\delta_\alpha + \bar\partial)(\eta_2/E(U^-))$. Here, as usual, we use the exactness of the sequence

$$0 \to E(U^-) \to X \to X/E(U^-) \to 0$$

and Th. 5.1.21(i). Since $V_1 \cap V_2 \cap \sigma(a,X/E(U)) = \emptyset$ and $(\delta_\alpha + \bar\partial)[(\eta_1 - \eta_2)/E(U^-)] = (\xi - \xi_0 - \xi)/E(U^-) = 0$, the equation $(\eta_1 - \eta_2)/E(U^-) = (\delta_\alpha + \bar\partial)\chi$ has a solution $\chi \in \Lambda^{n-3}[\sigma \cup d\bar{z}, C^\infty(V_1 \cap V_2, X/E(U^-))]$. From Lemma 5.2.3, we may decompose χ into the difference $\chi = \chi_2 - \chi_1$ with $\chi_j \in \Lambda^{n-3}[\sigma \cup d\bar{z}, C^\infty(V_j, X/E(U^-))]$. Set

$$\tilde\eta(z) = [\eta_j(z)/E(U^-)] + (\delta_\alpha + \bar\partial)\chi_j(z)$$

$(z \in V_j;\ j = 1,2)$. Then $\tilde\eta$ satisfies $\xi/E(U^-) = (\delta_\alpha + \bar\partial)\tilde\eta$ on V. If $\eta \in \Lambda^{n-2}[\sigma \cup d\bar{z}, C^\infty(V, X)]$ represents $\tilde\eta$, then $\xi - (\delta_\alpha + \bar\partial)\eta$ has values in $E(U^-)$ on V.

Next assume that $n = 1$. By Th. 1.2.1(iii) $a = (a_1)$ is decomposable, hence [32, Ch. 1, Prop. 4.16(i)] implies that E(F) is analytically invariant under a and hence a/E(F) has SVEP. The condition (ii) on ξ means that $(\partial/\partial\bar{z})(\xi/E(F)) = 0$ and $(z - a/E(F))(\xi/E(F)) = 0$. The first equality proves analyticity of $\xi/E(F)$ on V and the second shows $\xi/E(F) = 0$, or $\xi \in E(F)$.

For the case $n = 2$, we infer $\eta_1 \in C^\infty(V_1,E(F))$ and $\eta_2 \in C^\infty(V_2, X)$, hence $\eta_1 - \eta_2 \in C^\infty(V_1 \cap V_2, X)$. The relation $(\delta_\alpha + \bar\partial)[(\eta_1 - \eta_2)/E(U^-)] = 0$ implies that $\eta_1/E(F) = \eta_2/E(F)$ on $V_1 \cap V_2$. If we put $\tilde\eta(z) = \eta_j/E(F)$ for $z \in V_j$ $(j = 1,2)$, then the representation η of $\tilde\eta$ in $\Lambda^{n-2}[\sigma \cup d\bar{z}, C^\infty(V, X)]$ shows that $\xi - (\delta_\alpha + \bar\partial)\eta$ has values in $E(U^-)$ on V, and the lemma is proved.

Let B denote a Banach space or a function space such as A(U,X), etc., where U is open in $\mathbb{C}^n$, let $\alpha = (\alpha_1,\dots,\alpha_n)$ be a commuting system of endomorphisms in B and let $\delta_\alpha = \alpha_1\sigma_1 + \dots + \alpha_n\sigma_n$. We simplify the notation by using $F(B,\delta_\alpha)$ for the complex

$$0 \to B\ (=\Lambda^0[\sigma,B]) \xrightarrow{\delta_\alpha} \Lambda^1[\sigma,B] \xrightarrow{\delta_\alpha} \dots \xrightarrow{\delta_\alpha} \Lambda^n[\sigma,B] \to 0.$$

Let $F(B',\delta_{\alpha'})$ be another complex, and let $f = (f^p)$ be a cochain-homomorphism with $f^p\colon \Lambda^p[\sigma,B] \to \Lambda^{p+k}[\sigma,B']$ for each p, i.e. $f^{p+1}\delta_\alpha = \delta_{\alpha'} f^p$ for each p. We use $f_* = (f_*^p)$ to denote the induced cohomology maps (see §5.1)

$$f_*^p\colon H^p(B,\delta_\alpha) \to H^p(B',\delta_{\alpha'}), \qquad \tilde{\xi} \to f_*^p\tilde{\xi}.$$

In order to prove Theorem 5.2.17, we need some parts of a commuting diagram of cochain maps which has been used to advantage in [6',8',33']. We construct the whole diagram for completeness. To do so we need the function spaces

$$B = C^\infty(\mathbf{C}^{2n},X), \quad B_z = \{f \in B\colon \operatorname{supp} f \subset K \times \mathbf{C}^n,\ K \text{ compact in } \mathbf{C}^n\}$$

$$B_0 = C_0^\infty(\mathbf{C}^{2n},X), \qquad C = C^\infty(\mathbf{C}^n,X), \qquad C_0 = C_0^\infty(\mathbf{C}^n,X).$$

We write the elements of $\mathbf{C}^{2n}$ in the form $(z,w) = (z_1,\dots,z_n,w_1,\dots,w_n)$; these are used to construct the forms

$$\delta_\alpha = (z_1 - a_1)\sigma_1 + \dots + (z_n - a_n)\sigma_n$$

$$\delta_\beta = (w_1 - a_1)\tau_1 + \dots + (w_n - a_n)\tau_n\bar{\partial}$$

$$\delta_\alpha + \delta_\beta = (z_1 - a_1)\sigma_1 + \dots + (z_n - a_n)\sigma_n + (w_1 - a_1)\tau_1 + \dots + (w_n - a_n)\tau_n$$

$$\bar{\partial} = \bar{\partial}_z + \bar{\partial}_w = \frac{\partial}{\partial \bar{z}_1}d\bar{z}_1 + \dots + \frac{\partial}{\partial \bar{z}_n}d\bar{z}_n + \frac{\partial}{\partial \bar{w}_1}d\bar{w}_1 + \dots + \frac{\partial}{\partial \bar{w}_n}d\bar{w}_n.$$

With these notations we obtain the following commuting diagram of cochain maps.

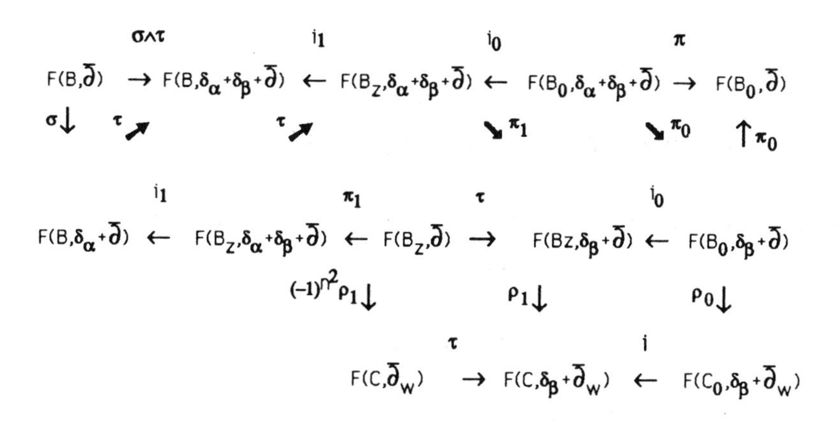

All maps denoted by i are imbeddings from a space of forms into another with coefficients belonging to some larger function space and induce isomorphisms of cohomology (see [32', Lemma 3.4.1]). The map $\sigma\wedge\tau$ is defined by

$$\sigma\wedge\tau\xi = \xi\wedge\sigma_1\wedge\ldots\wedge\sigma_n\wedge\tau_1\wedge\ldots\wedge\tau_n$$

for $\xi\in\Lambda^p[d\bar{z}\cup d\bar{w},B]$. Thus $\sigma\wedge\tau\xi$ is a form of degree $2n+p$ in $\sigma_1,\ldots,\sigma_n$, $\tau_1,\ldots,\tau_n,d\bar{z}_1,\ldots,d\bar{z}_n,d\bar{w}_1,\ldots,d\bar{w}_n$. Maps σ,τ behave similarly. The cochain maps denoted by π are projections. For instance,

$$\pi\colon \Lambda^p[\sigma\cup\tau\cup d\bar{z}\cup d\bar{w}, B_0] \to \Lambda^p[d\bar{z}\cup d\bar{w}, B_0]$$

is the projection annihilating those terms containing at least one of $\sigma_1,\ldots,\sigma_n,\tau_1,\ldots,\tau_n$ and leaving other terms invariant. The maps ρ are cochain maps of degree $-n$, e.g. $\rho\colon \Lambda^{p+n}[d\bar{z}\cup d\bar{w},B_z]\to\Lambda^p[d\bar{w},C]$ is defined by

$$\rho(\xi d\bar{z}_{j_1}\wedge\ldots\wedge d\bar{w}_{j_{p+n}-n}) = 0 \quad \text{if } j_n > n$$

and

$$\rho(\xi d\bar{z}_1\wedge\ldots\wedge d\bar{z}_n\wedge d\bar{w}_{j_1}\wedge\ldots\wedge d\bar{w}_{j_p}) = k(p,n)[\int_{C^n}\xi(z,w)\, d\bar{z}\wedge dz]d\bar{w}_{j_1}\wedge\ldots\wedge d\bar{w}_{j_p}$$

for $\xi\in B_z$, where $k(p,n) = (-1)^{(p+1)n}\pi^{-n}$.

We describe briefly how this diagram is used. Assume that $x\in X$ and that U, W are open in $\mathbf{C}^n$. If there exist forms χ_0,χ with $\operatorname{supp}\chi_0\subset U$, $\operatorname{supp}\chi\subset U\times W$ such that

$$\begin{aligned}(2.17)\qquad \chi_0 - x\sigma_1\wedge\ldots\wedge\sigma_n &= (\delta_\alpha + \bar\partial_z)\xi_1\\ \chi - x\sigma_1\wedge\ldots\wedge\sigma_n\wedge\tau_1\wedge\ldots\wedge\tau_n &= (\delta_\alpha + \delta_\beta + \bar\partial)\xi_2\end{aligned}$$

for some $\xi_1 \in \Lambda^{n-1}[\sigma \cup d\bar z, C^\infty(C^n,X)]$ and $\xi_2 \in \Lambda^{2n-1}[\sigma \cup \tau \cup d\bar z \cup d\bar w, C^\infty(C^{2n},X)]$, resp., then we have

$$(2.18)\qquad \tau\chi_0 - \chi = (\delta_\alpha + \delta_\beta + \bar\partial)\xi$$

for some $\xi \in \Lambda^{2n-1}[\sigma \cup \tau \cup d\bar z \cup d\bar w, B_z]$. In fact, since χ_0 and χ are independent of w, we may regard them as forms in $d\bar z_1,\ldots,d\bar z_n, d\bar w_1,\ldots,d\bar w_n,\sigma_1,\ldots,\sigma_n$ and rewrite (2.17) as

$$\chi_0 - x\sigma_1\wedge\ldots\wedge\sigma_n = (\delta_\alpha + \bar\partial)\xi_1.$$

But as x is independent of (z,w), we regard it as an element of $H^0(B,\bar\partial)$. Then

$$(2.19)\qquad (i_{1*})^{-1}\sigma_*(x) = [\chi_0],$$

$$(2.20)\qquad (i_{0*})^{-1}(i_{1*})^{-1}(\sigma\wedge\tau)_*(x) = [\chi];$$

here we use the fact that i_{0*}, i_{1*} are isomorphisms of cohomology. Now (2.19) implies

$$\tau_*(i_{1*})^{-1}\sigma_*(x) = [\tau\chi_0],$$

which, with (2.20), yields (2.18) for some $\xi \in \Lambda^{2n-1}[\sigma\cup\tau\cup d\bar z \cup d\bar w, B_z]$ by the commutativity of the diagram.

Now assume that $f \in A(U)$. We identify it with the function $U \times C^n \to C$, given by $(z,w) \to f(z)$, and consider $f\chi_0$, $f\chi$ to be forms with coefficients in $C^\infty(C^{2n},X)$. Since $f\chi_0$ is independent of w, one concludes that

$$\rho_{1*}\pi_{1*}\tau_*[f\chi_0] \in H^0(C,\delta_\beta + \bar\partial_w) \subset C$$

does not depend on w either, and hence defines an element $x_f \in X$. If $f = 1$ on U, then the definition of the Cauchy-Weil integral shows that $x_f = x$. If $(i_{0*})^{-1}\tau_*[f\chi_0] = [f\chi]$, then the equality $\rho_{1*}\pi_{1*}i_{0*} = i_*\rho_{0*}\pi_{0*}$ shows that

$$\tau x_f - \rho_0\pi_0((-1)^{n^2}f\chi) = (\delta_\beta + \bar{\partial}_w)\xi \tag{2.21}$$

for some $\xi \in \Lambda^{n-1}[\tau \cup d\bar{w}, C^\infty(C^n,X)]$; in particular, we have

$$\tau x_f = (\delta_\beta + \bar{\partial}_w)\xi \tag{2.22}$$

outside $\mathrm{supp}(\rho_0\pi_0(f\chi))$.

We can now prove the following theorem.

5.2.17 Theorem. Let the commuting system $a = (a_1,\ldots,a_n)$ be predecomposable, and let $\{U_1,U_2\}$ be an open cover of the closed set F in C^n. Then $E(F) \subset E(U_1^-) + E(U_2^-)$.

Proof. Let $V \subset C^n$ be compact such that $F \cap \sigma(a,X) \subset \mathrm{Int}\ V$ and such that $V \subset U_1 \cup U_2$. It follows from (2.15.5) and the evident inclusion $E(F) \subset X(a,F) = X(a, F\cap\sigma(a,X))$ that $E(F) \subset E(V)$ (the validity of $X(a,F) = X(a, F\cap\sigma(a,X))$ follows from Lemma 5.2.15). Thus we may assume that F is compact.

Let G_1, G_2, and G be open with compact closure satisfying inclusions

$$F \subset G_1 \cup G_2,$$
$$G_1^- \cap G_2^- \subset G \subset U_1 \cap U_2 \text{ and}$$
$$G_j^- \subset U_j \quad (j = 1,2).$$

Denote $Y = E(G_1^-\cap G_2^-)$, and assume $x \in E(F)$. The complement K of $(C^n\backslash F) \cup (G_1\cap G_2)$ is the disjoint union of two compact sets $K_1 = F\backslash G_2$ and $K_2 = F\backslash G_1$. Since $G_1\cap G_2 \subset \rho(a,X/Y)$ by Def 5.2.13(ii), there is, by Lemma 5.2.15, a form $\eta_0 \in \Lambda^{n-1}[\sigma \cup d\bar{z}, C^\infty(C^n\backslash K,X)]$ such that

$$(x/Y)\sigma_1\wedge\ldots\wedge\sigma_n = (\delta_\alpha + \bar{\partial}_z)(\eta_0/Y).$$

For each closed H in C^{2n}, define $E_{(a,a)}(H) = E(\{z \in C: (z,z) \in H\})$. Then it is easy to verify that $E_{(a,a)}$ satisfies (i) and (ii) of Def. 5.2.13.

Hence (a,a) is predecomposable. Since $x \in E(F) = E_{(a,a)}(\{(z,z)\colon z\in F\})$ and $(G_1\cap G_2)\times(G_1\cap G_2) \subset \rho((a,a),X/Y)$, there is a form η of degree 2n-1 on $C^{2n}\setminus\{(z,z)\colon z\in K\}$ such that

$$(x/Y)\sigma_1\wedge\ldots\wedge\sigma_n\wedge\tau_1\wedge\ldots\wedge\tau_n = (\delta_\alpha+\delta_\beta+\bar\partial)(\eta/Y).$$

Let V_j (j=1,2) be disjoint open sets with $K_j \subset V_j \subset U_j$, and let $V = V_1\cup V_2$. Multiplying η_0 and η by suitable C^∞-functions θ_0 and θ, resp., we may suppose that

$$\chi_0/Y = (x/Y)\sigma_1\wedge\ldots\wedge\sigma_n - (\delta_\alpha+\bar\partial_z)(\theta_0\eta_0/Y),$$
$$\chi/Y = (x/Y)\sigma_1\wedge\ldots\wedge\sigma_n\wedge\tau_1\wedge\ldots\wedge\tau_n - (\delta_\alpha+\delta_\beta+\bar\partial)(\theta\eta/Y)$$

are C^∞-forms on C^n, C^{2n} with $\mathrm{supp}(\chi_0/Y) \subset V$, $\mathrm{supp}(\chi/Y) \subset (V_1\times V_1)\cup(V_2\times V_2)$. It follows from (2.18) that

$$\tau(\chi_0/Y) - (\chi/Y) = (\delta_\alpha+\delta_\beta+\bar\partial)(\xi/Y) \tag{2.23}$$

for some $\xi \in \Lambda^{2n-1}[\sigma\cup\tau\cup d\bar z\cup d\bar w, C^\infty(C^{2n},X)]$. Considering the supports of the forms on the left side of (2.23) and observing that $Y = E_{(a,a)}(\{(z,z)\colon z\in G_1^-\cap G_2^-\})$, we infer from Lemma 5.2.16 that there exists a compact $M \subset V$ and a form $\zeta \in \Lambda^{2n-2}[\sigma\cup\tau\cup d\bar z\cup d\bar w, C^\infty(C^{2n},X)]$ such that $\xi' = \xi - (\delta_\alpha+\delta_\beta+\bar\partial)\zeta$ has values in $E(G) = E_{(a,a)}(G^-\times G^-)$ off $M \times C^n$. Define the function f on V by $f(z) = 1$ for $z \in V_1$, $f(z) = 0$ for $z \in V_2$. Then

$$\tau(f\chi_0/E(G^-)) - (f\chi)/E(G^-) = (\delta_\alpha+\delta_\beta+\bar\partial)(f\xi'/E(G^-))$$

makes sense on C^{2n}. By an analogous argument from (2.21), we obtain

$$\tau(\rho_1\pi_1(f\chi_0)/E(G^-)) = (\delta_\beta+\bar\partial_w)(\phi/E(G^-))$$

outside $\mathrm{supp}(\rho_0\pi_0(f\chi)/E(G^-))$ for some ϕ. Let $x_1, x_2 \in X$ such that

$$x_1/E(G^-) = \rho_1\pi_1(f\chi_0)/E(G^-)$$

$$x_2/E(G^-) = \rho_1\pi_1(1-f)\chi_0/E(G^-).$$

Then

$$x/E(G^-) = x_1/E(G^-) + x_2/E(G^-), \text{ and}$$
$$\tau(x_j/E(G^-)) = (\delta_\beta + \bar\partial_w)(\phi_j/E(G^-)) \text{ off a compact set in } U_j \ (j=1,2).$$

Since $\sigma(a,E(G^-)) \subset G^- \subset U_1 \cap U_2$, for $j = 1,2$ there is a solution to the equation $\tau x_j - (\delta_\beta + \bar\partial_w)\phi_j = (\delta_\beta + \bar\partial_w)\psi_j$ outside some compact subset of U_j. Therefore $x_j \in E(U_j)$, and hence

$$x = x - (x_1 + x_2) + x_1 + x_2 \in E(G^-) + E(U_1^-) + E(U_2^-).$$

From the evident inclusions $E(G^-) \subset X(a,G^-) \subset E(U_1^-)$, we obtain $x \in E(U_1^-) + E(U_2^-)$, and the proof is complete.

5.2.18 Corollary. If the commuting system $a = (a_1,\dots a_n)$ is predecomposable, then for every finite open cover $\{U_1,\dots,U_m\}$ of the closed set F in $\mathbf{C}^n$ we have

$$E(F) \subset E(U_1^-) + \dots + E(U_m^-). \tag{2.24}$$

Hence a has the SDP.

Proof. Conclusion (2.24) is reduced by easy induction to the case $m = 2$, so the predecomposable a has SDP.

5.2.19 Definition. Let $a = (a_1,\dots,a_n)$ be a commuting system, and let E be the mapping given by Def. 5.2.13. Then we define the <u>support</u> of E to be the set

$$\operatorname{supp} E = \cap\{F \subset \mathbf{C}^n\colon F \text{ closed and } X = E(F)\}.$$

5.2.20 Proposition. If a is predecomposable and E is the map given by Def. 5.2.13, then

$$\operatorname{supp} E = \sigma(a,X). \tag{2.25}$$

Proof. Since for every closed set F satisfying $E(F) = X$ we have

$\sigma(a,X) = \sigma(a,E(F)) \subset F$, the inclusion $\sigma(a,X) \subset \text{supp } E$ follows. To prove the opposite inclusion, let $\lambda \notin \sigma(a,X)$ and let V, F be closed sets with $\lambda \in \text{Int } V$, $V \cap \sigma(a,X) = \emptyset$ and $\sigma(a,X) \subset \text{Int } F$. Then $X = E(V) + E(F)$ by Th. 5.2.17. If $x \in E(V)$, then $\sigma_a(x) \subset V \cap \sigma(a,X) = \emptyset$. Hence $x = 0$ by Prop. 5.2.7. Therefore $E(V) = \{0\}$ and $X = E(F)$, and so supp $E \subset F$. Since F is arbitrary with the property $\sigma(a,X) \subset \text{Int } F$, we have supp $E \subset \sigma(a,X)$. Thus (2.25) is proved.

5.2.21 Definition. Let $m \geq 2$ be fixed. An m-<u>spectral capacity</u> $\mathcal{E}$ is a map from the family $F(\mathbf{C}^n)$ of all closed sets in $\mathbf{C}^n$ into the family of all closed subspaces in X satisfying the following properties:

(i) $\mathcal{E}(\emptyset) = \{0\}$, $\mathcal{E}(\mathbf{C}^n) = X$;

(ii) $\mathcal{E}(\cap_j F_j) = \cap_j \mathcal{E}(F_j)$ for each family of closed sets $\{F_j\}$ in $F(\mathbf{C}^n)$;

(iii) $X = \mathcal{E}(U_1^-) + \dots + \mathcal{E}(U_m^-)$ for each open m-cover of $\mathbf{C}^n$.

The commuting system a is said to be m-<u>decomposable</u> if there is an m-spectral capacity $\mathcal{E}$ such that

(iv) for each $F \in F(\mathbf{C}^n)$ the subspace $\mathcal{E}(F)$ is a-invariant and $\sigma(a,\mathcal{E}(F)) \subset F$.

The map $\mathcal{E}$ satisfying (i)-(iv) is called an m-spectral capacity for a. Moreover, if a is m-decomposable for each $m \geq 2$, then a is said to be <u>decomposable.</u>

5.2.22 Theorem. Assume that the commuting system a is 2-decomposable. Then the 2-spectral capacity $\mathcal{E}$ for a is uniquely determined and for each $F \in F(\mathbf{C}^n)$

$$\mathcal{E}(F) = X(a,F). \tag{2.26}$$

Proof. If a is 2-decomposable, then clearly it has 2-SDP. Hence a has SVEP and X(a,F) is closed for each $F \in F(\mathbf{C}^n)$ by Ths. 5.2.10 and 5.2.11. Repeating the argument of Th. 5.2.11, we have

$$X(a,F) = \cap\{\mathcal{E}(H): H \in F(\mathbf{C}^n) \text{ and } F \subset \text{Int } H\}. \tag{2.27}$$

Since $\cap\{H: H \in F(\mathbf{C}^n), F \subset \text{Int } H\} = F$, by Def. 5.2.21(ii) the right-hand side of (2.27) is equal to $\mathcal{E}(F)$, and hence (2.26) holds.

5.2.23 Proposition. If the commuting system a is 2-

decomposable, then

$$\sigma(a/\mathcal{E}(F)) \subset \mathbf{C}^n \setminus \mathrm{Int}\, F \tag{2.28}$$

for every $F \in F(\mathbf{C}^n)$.

Proof. For $\lambda \in \mathrm{Int}\, F$, choose an open set U in $\mathbf{C}^n$ such that $\{\mathrm{Int}\, F, U\}$ covers $\mathbf{C}^n$ and $\lambda \notin U^-$. Since $X = \mathcal{E}(F) + \mathcal{E}(U^-)$ by Def. 5.2.21(iii), one has that $X/\mathcal{E}(F)$ is topologically isomorphic to $\mathcal{E}(U^-)/\mathcal{E}(F) \cap \mathcal{E}(U^-) = \mathcal{E}(U^-)/\mathcal{E}(F \cap U^-)$. Thus Prop. 5.1.5 implies that

$$\sigma(a, \mathcal{E}(U^-)/\mathcal{E}(F \cap U^-)) \subset U^-.$$

Therefore $\lambda \notin \sigma(a, X/\mathcal{E}(F))$, (2.28) holds and the conditions of Def. 5.2.13 are satisfied with $\mathcal{E} = E$.

5.2.24 Proposition. Every 2-decomposable commuting system is decomposable.

Proof. By Prop. 5.2.23, if commuting system a is 2-decomposable, then it is predecomposable. Corollary 5.2.18 and equality $\mathcal{E} = E$ imply that $X = \mathcal{E}(U_1^-) + \dots + \mathcal{E}(U_m^-)$ for every $m \geq 2$, as required.

Using the notion of support in Def. 5.2.19, we have supp $\mathcal{E} = \sigma(a,X)$ by Prop. 5.2.22.

5.2.25 Proposition. Let $a = (a_1,\dots,a_n)$ be a commuting system and let $b = (a_1,\dots,a_k)$ for $k \leq n$. If a has SDP [resp. is predecomposable, decomposable], then b also has SDP [resp. is predecomposable, decomposable].

Proof. We consider only the case of SDP. Let $\{G_1,\dots,G_m\}$ be an open cover of $\sigma(b,X)$ where G_j is open in $\mathbf{C}^k$ for $j = 1,2,\dots,m$. Then $\{G_j \times \mathbf{C}^{n-k}: 1 \leq j \leq m\}$ is an open cover of $\sigma(a,X)$. Since a has SDP by hypothesis, there are a-invariant subspaces X_j $(1 \leq j \leq m)$ such that

$$\sigma(a,X_j) \subset G_j \times \mathbf{C}^{n-k} \quad (1 \leq j \leq m), \tag{2.29}$$

$$X = X_1 + \dots + X_m. \tag{2.30}$$

By the projection property of the Taylor spectrum, (2.29) implies

$\sigma(b,X_j) \subset G_j$ $(1 \le j \le m)$. This and (2.30) show that b has SDP.

In [9'] J. Eschmeier gave an example of a commuting pair (a_1,a_2) which is not decomposable but whose components a_1,a_2 are decomposable in the sense of Chap. 1.

In this final part of the section we introduce another kind of spectrum for a commuting system and study its spectral decomposition properties.

5.2.26 Definition. For the commuting system $a = (a_1,\dots,a_n)$, let

$$\rho'(a,X) = \{z \in \mathbf{C}^n: \text{there is a commuting system } (b_1,\dots,b_n) \text{ commuting with } a \text{ and } \textstyle\sum_j (z_j - a_j)b_j = I\}$$

and

$$\sigma'(a,X) = \mathbf{C}^n \backslash \rho'(a,X).$$

Theorem 5.1.9 shows that $\sigma(a,X) \subset \sigma'(a,X)$ always holds, and hence $\sigma'(a,X)$ is nonempty if $X \neq 0$. On the other hand, if we put $b_1 = (w-a_1)^{-1}$ for $w \notin \sigma(a_1,X)$ and $b_j = 0$ for $j > 1$, then $b = (b_1,\dots,b_n)$ is a commuting system commuting with a such that $\sum_j (z_j - a_j)b_j = I$. So

$$\sigma'(a,X) \subset \sigma(a_1,X) \times \dots \times \sigma(a_n,X),$$

which proves that $\sigma'(a,X)$ is bounded. In fact, we have

5.2.27 Theorem. If $a = (a_1,\dots,a_n)$ is a commuting system, then $\sigma'(a,X)$ is compact.

Proof. By the previous remark, we need only show that $\sigma'(a,X)$ is closed. Suppose $b = (b_1,\dots,b_n)$ is a commuting system in the commutant of a such that $\sum_j(z_j - a_j)b_j = I$. Then there exists a neighborhood U of $z = (z_1,\dots,z_n)$ such that

$$\|\textstyle\sum_j(\zeta_j - z_j)b_j\| < 1$$

for $\zeta = (\zeta_1,\dots,\zeta_n) \in U$. Hence $c = \sum_j(\zeta_j - a_j)b_j$ is invertible. If we let $b_j' = c^{-1}b_j$, then $(b_1',\dots,b_n')$ lies in the commutant of a, and $\sum_j(\zeta_j - a_j)b_j' = I$. Thus $\rho'(a,X)$ is open and $\sigma'(a,X)$ is compact.

5.2.28 Definition. Let $a = (a_1,...,a_n)$ be a commuting system and let $m \geq 2$ be given. If, for every open cover $\{G_1,...,G_m\}$ of $\sigma'(a,X)$ by m open sets, there exists a system of a-invariant subspaces $\{X_j\}$ such that

(i) $\sigma'(a,X_j) \subset G_j \quad (1 \leq j \leq m)$,

(ii) $X = X_1 + ... + X_m$,

then a is said to have the σ'-m-SDP. If a has σ'-m-SDP for each $m \geq 2$, then a has σ'-SDP.

In the same way, we define a to be σ'-m-decomposable by replacing $\sigma(a,\mathcal{E}(F)) \subset F$ with $\sigma'(a,\mathcal{E}(F)) \subset F$ in Def. 5.2.21(iv). If a is σ'-m-decomposable for each $m \geq 2$, then we say that a is σ'-decomposable.

5.2.29 Theorem. If the commuting system a has the σ'-m-SDP, then a is σ'-m-decomposable.

Proof. Since $\sigma(a,X) \subset \sigma'(a,X)$, a having σ'-m-SDP implies that a has σ-m-SDP. With $\sigma_a(x)$ defined as in Def. 5.2.5 and $X(a,F)$ by (2.7), it follows from Ths. 5.2.10 and 5.2.11 that if a has σ'-2-SDP, then it has SVEP and $X(a,F)$ is closed for closed F. Moreover, we have

$$X(a,F) = \cap_{z \notin F} X_z, \tag{2.31}$$

where X_z is a-invariant and defined as in the proof of Th. 5.2.11 with $z \notin \sigma'(a,X_z)$. In this way, there exists a commuting system $b = (b_1,...,b_n)$ in $L(X_z)$ such that b commutes with $a|X_z$ and

$$\Sigma_j (z_j - a_j)b_j = I|X_z.$$

Since it is easy to verify that $X(a,F)$ is b_j-invariant for each $j = 1,...,n$, we obtain

$$[\Sigma_j(z_j - a_j)b_j]|X(a,F) = I|X(a,F).$$

Therefore $z \notin \sigma'(a,X(a,F))$. As $z \notin F$ is arbitrary, we obtain the inclusion $\sigma'(a,X(a,F)) \subset F$. Now define

$$\mathcal{E}(F) = X(a,F) \quad (F \in F(\mathbf{C}^n)). \tag{2.32}$$

Then $\mathcal{E}$ is a σ'-m-spectral capacity for a, and a is thus σ'-m-decomposable. This completes the proof.

If $\mathcal{E}$ is a σ'-m-spectral capacity for a in Def 5.2.28, we have an analogy with (2.31):

$$X(a,F) = \cap\{\mathcal{E}(H): \text{ H closed, } F \subset \text{Int } H\}.$$

By the intersection property of $\mathcal{E}$ one has $X(a,F) = \mathcal{E}(F)$. Therefore, a σ'-m-spectral capacity for a is unique (if it exists).

5.2.30 Proposition. If the commuting system a is σ'-2-decomposable, then

$$\sigma'(a, X/X(a,F)) \subset \mathbf{C}^n \backslash \text{Int } F \quad (F \in F(\mathbf{C}^n)). \tag{2.33}$$

Proof. The argument is exactly the same as the proof of Prop. 5.2.23.

5.2.31 Corollary. Assume that the commuting system a is σ'-2-decomposable. If $\{U_1,U_2\}$ is an open cover of $F \in F(\mathbf{C}^n)$, then

$$X(a,F) \subset X(a,U_1^-) + X(a,U_2^-). \tag{2.34}$$

Proof. From Prop. 5.2.30 we have $\sigma'(a,X/X(a,F)) \subset \mathbf{C}^n \backslash \text{Int} F$. But σ'-2-decomposability implies 2-decomposability and hence predecomposability. Now the argument of Th. 5.2.17 applies to the present to yield (2.34).

The following theorem is an easy consequence of the foregoing discussion.

5.2.32 Theorem. For the commuting system a the following assertions are equivalent:

(i) a has the σ'-2-SDP;

(ii) a is σ'-2-decomposable;

(iii) a is σ'-decomposable;

(iv) a has the σ'-SDP.

3. Duality Theory for Commuting Systems

In this section we discuss the duality theory for commuting systems of bounded linear operators on the Banach space X. If $a = (a_1,\dots,a_n)$ is a commuting system in L(X), then the adjoint $a' = (a_1',\dots,a_n')$ is also a commuting system of operators on the dual space X'. (Throughout this section we use ' to denote duals rather than *.) Our first duality theorem is

5.3.1 Theorem. For the commuting system $a = (a_1,\dots,a_n)$, the following are equivalent:

(i) a is predecomposable;

(ii) a' is predecomposable having an associated map E' satisfying Def. 5.2.13 such that E'(F) is $\sigma(X',X)$-closed for each closed F in $\mathbf{C}^n$.

Proof. (i) ⇒ (ii). Let E be the map associated with predecomposable system a. Then E' defined by

$$E'(F) = E(\mathbf{C}^n \setminus \operatorname{Int} F)^{\perp} \quad (F \text{ closed in } \mathbf{C}^n), \tag{3.1}$$

satisfies conditions (i) and (ii) of Def. 5.2.13 for a' since E does. Moreover, (3.1) implies that E'(F) is $\sigma(X',X)$-closed, and so (ii) follows.

(ii) ⇒ (i). If E' satisfies the conditions of Def. 5.2.13 for a' and if E'(F) is $\sigma(X',X)$-closed for each closed F in $\mathbf{C}^n$, then E defined by

$$E(F) = {}^{\perp}E'(\mathbf{C}^n \setminus \operatorname{Int} F) \tag{3.2}$$

satisfies the same conditions for a, hence a is predecomposable. In fact, (3.2) implies that for each closed F

$$E(F)^{\perp} = E'(\mathbf{C}^n \setminus \operatorname{Int} F)$$

since $E'(\mathbf{C}^n \setminus \operatorname{Int} F)$ is $\sigma(X',X)$-closed. This completes the proof.

It is worth mentioning that if (i) [or (ii)] of Th 5.3.1 holds then

$$X'(a',F) = X(a,\mathbf{C}^n \setminus F)^{\perp} \tag{3.3}$$

$$X(a,F) = {}^{\perp}X'(a',\mathbf{C}^n \setminus F) \tag{3.4}$$

for every closed F in $\mathbf{C}^n$. To prove (3.3) and (3.4) we first observe that X(a,F) and X'(a',F) are closed in the the norm topology for each closed set F by Lemma 5.2.14, and that the maps A and B defined by

$$A(F) = {}^{\perp}X'(a',\mathbb{C}^n\setminus F)$$
$$B(F) = X(a,\mathbb{C}^n\setminus F)^{\perp}$$

preserve intersections. Let $\{F_\alpha\}$ be any family of closed sets in $\mathbb{C}^n$, and let $x \in \cap_\alpha A(F_\alpha)$, $x' \in X'(a',\mathbb{C}^n\setminus\cap_\alpha F_\alpha)$. By the compactness of $\sigma_{a'}(x')$ and Cor. 5.2.18, there are indices α_j ($j = 1,2,\ldots,m$ for some m) such that

$$x' \in X'(a',\sigma_{a'}(x')) \subset X'(a',\mathbb{C}^n\setminus F_{\alpha_1}) + \ \ldots\ + X'(a',\mathbb{C}^n\setminus F_{\alpha_m}).$$

Therefore, $\langle x,x'\rangle = 0$. Since x is arbitrary, it follows that $\cap_\alpha A(F_\alpha) \subset {}^{\perp}X'(a',\mathbb{C}^n\setminus\cap_\alpha F_\alpha) = A(\cap_\alpha F_\alpha)$. The opposite inclusion is evident since A is monotone. Hence we have

$$\cap_\alpha A(F_\alpha) = A(\cap_\alpha F_\alpha).$$

The equality $\cap_\alpha B(F_\alpha) = B(\cap_\alpha F_\alpha)$ can be proved in a similar way.

Now let F be an arbitrary, fixed closed set in $\mathbb{C}^n$, and let H be closed in $\mathbb{C}^n$ such that $F \subset \operatorname{Int} H$. Then

$$X(a,\mathbb{C}^n\setminus F)^{\perp} \subset E(\mathbb{C}^n\setminus \operatorname{Int} H)^{\perp} \subset X(a,\mathbb{C}^n\setminus H)^{\perp} = B(H),$$
$${}^{\perp}X'(a',\mathbb{C}^n\setminus F) \subset {}^{\perp}E'(\mathbb{C}^n\setminus \operatorname{Int} H) \subset {}^{\perp}X'(a',\mathbb{C}^n\setminus H) = A(H).$$

Since H is arbitrary and A, B preserve intersections, we have

$$X(a,\mathbb{C}^n\setminus F)^{\perp} \subset \cap\{E(\mathbb{C}^n\setminus \operatorname{Int} H)^{\perp}\colon H \text{ closed}, F \subset \operatorname{Int} H\}$$
$$\subset B(F) = X(a,\mathbb{C}^n\setminus F)^{\perp};$$

$${}^{\perp}X'(a',\mathbb{C}^n\setminus F) \subset \cap\{{}^{\perp}E'(a',\mathbb{C}^n\setminus \operatorname{Int} H)\colon H \text{ closed}, F \subset \operatorname{Int} H\}$$
$$\subset A(F) = {}^{\perp}X'(a',\mathbb{C}^n\setminus F).$$

By equality (2.15.5) each second term is $X'(a',F)$, $X(a,F)$, resp., and hence (3.3) and (3.4) follow.

The following result is crucial to the next duality theorem.

5.3.2 Theorem. Assume that a' has SVEP, Let F be closed in $\mathbb{C}^n$.

If $X'(a',F)$ is closed in the norm topology, then it is closed in the $\sigma(X',X)$ topology.

Proof. Let $\{x_\alpha'\}$ be a net in $X'(a',F)$ which converges to $x' \in X'$ in the $\sigma(X',X)$ topology. By the Krein-Smulyan Theorem we may assume that $\|x_\alpha'\| \leq 1$ for all α. For each open polydisc D with its closure contained in $\mathbf{C}^n \backslash F$, let $A_b(D,X')$ be the Banach space of all bounded analytic X'-valued functions on D with sup-norm. Clearly

$$\tau': \Sigma\oplus A_b(D,X') \to A_b(D,X'); \quad f = (f_j) \to \Sigma_j(z_j - a_j')f_j(z)$$

(sum of n copies) defines a continuous linear map between Banach spaces. By Th. 5.2.6, the image of τ' contains the closed subspace of $A_b(D,X')$ which consists of all constants belonging to the space $X'(a',F)$. Then we may choose a bounded net $\{f_\alpha\}$ in $\Sigma\oplus A_b(D,X')$ such that $\tau' f_\alpha = x_\alpha'$ for all α. Since $A(D)$ is a nuclear reflexive Frechet space and X is a DF-space, we have canonical topological isomorphisms

$$A(D,X') = A(D) \hat{\otimes} X' = (A(D)_\beta' \hat{\otimes} X)_\beta' \tag{3.5}$$

by [20',Th. 12 of II.3.2]. We infer from (3.5) that $A(D,X')$ is the strong dual of the locally convex space $A(D)_\beta' \hat{\otimes} X$. Since $A(D)_\beta'$ and X are barreled, so is $A(D)_\beta' \hat{\otimes} X$ by [20', Cor. 1 of I.1.4]. Thus a bounded set in $A(D,X')$ corresponds to a strongly bounded set in $(A(D)_\beta' \hat{\otimes} X)'$, which is equicontinuous due to the principle of uniform boundedness on barreled spaces. By the Alaoglu-Bourbaki theorem, the net $\{f_{\alpha,j}\}$ is relatively compact in the $\sigma(A(D,X'),A(D)_\beta'\hat{\otimes}X)$ topology, where $f_{\alpha,j}$ is the jth component of f_α, $j = 1,\ldots,n$. Without loss of generality, we may assume that $\{f_{\alpha,j}\}$ is weak* convergent to some $f_j \in A(D,X')$ for each j as explained above.

Let $x \in X$, let g be an element of $A(D,X')$ and let $\tilde{\lambda}$ denote point evaluation at $\lambda \in D$. Then g, regarded as a continuous linear functional, acts as

$$\langle \tilde{\lambda} \otimes x, g\rangle = \langle x, g(\lambda)\rangle.$$

Thus for $\lambda \in D$, $x \in X$, it follows that

$$\langle x, \Sigma_j (\lambda_j - a_j')f_j(\lambda)\rangle = \lim_\alpha \Sigma_j \langle(\lambda_j - a_j)x, f_{\alpha,j}(\lambda)\rangle$$
$$= \lim_\alpha (x, x_\alpha') = \langle x, x'\rangle,$$

and hence for $\lambda \in D$, $x' = \Sigma_j (\lambda_j - a_j')f_j(\lambda)$, which means that $\sigma_{a'}(x') \cap D = \emptyset$. Since D is arbitrary, we have $x' \in X'(a',F)$, and the theorem is proved.

5.3.3 Theorem. (i) If a is decomposable, then a' is predecomposable and (3.3) and (3.4) hold for each closed F in $\mathbf{C}^n$.

(ii) If a' is decomposable, then a is predecomposable and (3.3) and (3.4) hold for each closed F in $\mathbf{C}^n$.

Proof. Assertion (i) is an easy consequence of Th. 5.3.1, while (ii) follows from Ths. 5.3.1 and 5.3.2.

In the following second part of this section we discuss the duality theory of commuting systems with a local decomposition property.

5.3.4 Definition. If the given commuting system $a = (a_1,\ldots,a_n)$ has the SVEP and for every domain of holomorphy U in $\mathbf{C}^n$ the space

$$F(U) = H^n(A(U,X),\delta_\alpha) = \Lambda^n[\sigma, A(U,X)]/\delta_\alpha\Lambda^{n-1}[\sigma, A(U,X)]$$
$$\cong A(U,X)/\Sigma_j(z_j - a_j)A(U,X)$$

is separated (hence Frechet) in its natural quotient topology, then a is said to have property (β).

We regard X as an $A(\mathbf{C}^n)$-module with the module structure given by

$$A(\mathbf{C}^n) \times X \to X; \qquad (f,x) \to f(a)x.$$

5.3.5 Proposition. For the commuting system $a = (a_1,\ldots,a_n)$, the spaces $F(\mathbf{C}^n) = H^n(A(\mathbf{C}^n,X),\delta_\alpha)$ and X are isomorphic as $A(\mathbf{C}^n)$-modules.

Proof. Let $\rho : A(\mathbf{C}^n,X) \to X$ be the map defined by

$$\Sigma_{p\geq 0}\, c_p z^p \to \Sigma_{p\geq 0}\, a^p c_p \qquad (c_p \in X)$$

where p runs over multi-indices. Clearly, ρ is continuous and onto. The kernel of ρ consists of elements of the form $\Sigma_{p\geq 0} (z^p - a^p)c_p =$

$\Sigma_j(z_j - a_j)f_j(z)$ by factorization. This proves the claimed isomorphism.

The sheaf $\mathcal{F}$ associated with the presheaf

$$U \to F(U) = H^n(A(U,X),\delta_\alpha) \qquad (U \text{ open in } \mathbf{C}^n)$$

is called the canonical sheaf model of a. More generally, a Frechet sheaf $\mathcal{G}$ of A-modules on $\mathbf{C}^n$ is said to be a sheaf model for a if X and $\mathcal{G}(\mathbf{C}^n)$ are isomorphic as topological $A(\mathbf{C}^n)$-modules. We say that a has a Frechet soft sheaf model if a has a Frechet sheaf model which is soft.

Assume that a has the SVEP. Let $\mathcal{A}^p$ be the sheaf given by the presheaf

$$U \to \Lambda^p[\sigma,A(U,X)] \qquad (U \text{ open in } \mathbf{C}^n).$$

Then the sequence

$$(3.6) \qquad 0 \to \Lambda^0[\sigma,A(U,X)] \xrightarrow{\delta_\alpha} \Lambda^1[\sigma,A(U,X)] \to \dots$$
$$\to \Lambda^n[\sigma,A(U,X)] \to F(U) \to 0$$

induces a sequence of sheaves

$$(3.7) \qquad 0 \to \mathcal{A}^0 \to \mathcal{A}^1 \to \dots \to \mathcal{A}^n \to \mathcal{F} \to 0$$

which is clearly a resolution of $\mathcal{F}$. If a has SVEP, then $\mathcal{A}^p$ $(0 \leq p \leq n-1)$ is acyclic on every domain of holomorphy by the proof of Th. 5.2.4. Thus the canonical sheaf model $\mathcal{F}$ of a is acyclic on every domain of holomorphy. This remark enables us to prove the following

5.3.6 Proposition. If the commuting system a has the SVEP, then F(U) and $\mathcal{F}(U)$ are isomorphic for every domain of holomorphy U, where $\mathcal{F}(U)$ is the abelian group of sections of $\mathcal{F}$ over U.

Proof. If a has the SVEP, then the sequence (3.6) is exact by Th. 5.2.4 and the exactness of the sequence

$$(3.8)\qquad \Lambda^{n-1}[\sigma,A(U,X)] \xrightarrow{\delta_\alpha} \Lambda^{n}[\sigma,A(U,X)] \to F(U) \to 0.$$

Hence (3.7) is an exact sequence of sheaves which are acyclic on every domain of holomorphy. Thus

$$0 \to \Lambda^{0}[\sigma,A(U,X)] \to \Lambda^{1}[\sigma,A(U,X)] \to \dots$$

$$\dots \to \Lambda^{n}[\sigma,A(U,X)] \to \mathcal{F}(U) \to 0$$

is exact for every domain of holomorphy U. This together with the exactness of (3.6) proves that F(U) and $\mathcal{F}(U)$ are isomorphic.

If a has SVEP, then it follows from Prop. 5.3.5 and the definition of local joint spectrum that one has

$$(3.9)\qquad \sigma_a(x) = \operatorname{supp} x \qquad (x \in X),$$

where x on the right-hand side is regarded as a section of $\mathcal{F}$ over $\mathbf{C}^n$. Then via (3.9) we may identify X(a,F) with the space $\Gamma_F(\mathbf{C}^n, \mathcal{F})$ of all sections of $\mathcal{F}$ over $\mathbf{C}^n$ with support contained in F, that is,

$$(3.10)\qquad X(a,F) = \Gamma_F(\mathbf{C}^n, \mathcal{F}).$$

5.3.7 Definition. If K is compact in $\mathbf{C}^n$, then the spectral hull K^s of K is defined to be the set of all $w \in \mathbf{C}^n$ failing to satisfy the equation

$$1 = (w_1 - z_1)f_1(z) + \dots + (w_n - z_n)f_n(z)$$

for any $f_1,\dots,f_n$ analytic on K.

In the following 5.3.8-5.3.11, except 5.3.10, we assume that the the commuting system a has property (β).

5.3.8 Proposition. For each closed F in $\mathbf{C}^n$, one has

$$(3.11)\qquad \sigma(a,X(a,F)) \subset [F \cap \sigma(a,X)]^s.$$

Proof. It is easy to see that $X(a,F) = X(a,F\cap \sigma(a,X))$ by Lemma 5.2.15. Let b be the restriction of a to $X(a,F \cap \sigma(a,X))$, and let $A(F\cap\sigma(a,X))$ be the algebra of all functions analytic on some neighborhood of $F \cap \sigma(a,X)$. For each $f \in A(F \cap \sigma(a,X))$, we define

$$f(b)x = \frac{1}{(2\pi i)^n}\mu_\alpha(fx\sigma_1\wedge\ldots\wedge\sigma_n)$$

by means of the Cauchy-Weil integral μ_α, where $x \in X(a,F\cap\sigma(a,X))$. Thus the commuting system b has an $A(F\cap\sigma(a,X))$-functional calculus. Now assume that $w \notin [F\cap\sigma(a,X)]^s$. Then there exist $f_j \in A(F\cap\sigma(a,X))$ for $j = 1,\ldots,n$ such that

$$1 = (w_1 - z_1)f_1(z) + \ldots + (w_n - z_n)f_n(z),$$

so that

$$I = (w_1 - b_1)f_1(b) + \ldots + (w_n - b_n)f_n(b).$$

Now Th. 5.1.9 implies (3.11).

5.3.9 Corollary. For each closed F in C^n, the space X(a,F) has an $A(F\cap\sigma(a,X))$-module structure which is an extension of the $A(C^n)$-module structure given by the commuting system a.

Proof. For each $f \in A(F\cap\sigma(a,X))$ and for each $x \in X(a,F)$ define the map $(f,x) \to f(b)x$, where b is the restricted system of Prop. 5.3.8. Clearly this is an $A(F\cap\sigma(a,X))$-module structure for X(a,F) and extends the $A(C^n)$-module structure for X.

5.3.10 Lemma. Assume that the commuting system a has SVEP, and assume that $x \in X$ has its local spectrum $\sigma_a(x)$ contained in the disjoint union of two closed sets F_1 and F_2. Then there exists a unique decomposition $x = x_1 + x_2$ with $\sigma_a(x_j) \subset F_j$, $j = 1,2$.

Proof. Let U_1, U_2 be disjoint respective neighborhoods of F_1, F_2. Let $f(z) = 1$ for $z \in U_1$ and $f(z) = 0$ for $z \in U_2$. Then $f \in A(\sigma_a(x))$. If we define $x_1 = f(b)x$ and $x_2 = x - x_1$, where b is the restriction of a on $X(a,\sigma_a(x))$, then $x_1 + x_2$ is the desired decomposition.

(Further details for results 5.3.8-5.3.10 can be found in [6',26']).

It is easy to see that the isomorphism in Prop. 5.3.6 induces a Frechet A(U)-module structure on $\mathcal{F}(U)$ for each domain of holomorphy U. If V is an arbitrary open set in C^n, we choose a countable family of domains of holomorphy $\{U_r\}$ covering V and define the topology on $\mathcal{F}(V)$ as the restriction of the product topology induced by the inclusion i in the following exact sequence of continuous maps:

$$(3.12) \qquad 0 \to \mathcal{F}(V) \xrightarrow{i} \Pi_r\, \mathcal{F}(U_r) \to \Pi_{r,s}\, \mathcal{F}(U_r \cap U_s).$$

Thus the whole sheaf $\mathcal{F}$ becomes a Frechet sheaf of A-modules. The topology is independent of the covering $\{U_r\}$ as a result of the closed graph theorem. The exactness of (3.12) yields the following

5.3.11 Proposition. (β) is a local property.

5.3.12 Definition. The commuting system $a = (a_1,\dots,a_n)$ is said to have the local decomposition property (LDP) if it satisfies the conditions

(i) a has SVEP and $X(a,F)$ is closed for every closed F in $\mathbf{C}^n$;

(ii) for every pair of open sets U_1, U_2 in $\mathbf{C}^n$

$$X(a,U_1 \cup U_2) = X(a,U_1) + X(a,U_2).$$

5.3.13 Theorem. If the commuting system a has the LDP, then a' has a Frechet soft sheaf model.

Proof. We divide the proof into several steps.

1) For the open set U in $\mathbf{C}^n$, we regard $X(a,U)$ as the inductive limit

$$X(a,U) = \varinjlim_{U \supset K} X(a,K),$$

where K runs over all compact sets in U and each $X(a,K)$ is endowed with its norm topology. Together with the inductive topology of the imbeddings, $X(a,U)$ is a strict (LF)-space. The strong dual $\mathcal{F}'(U)$ of this (LF)-space is the projective limit of the strong dual spectrum:

$$\mathcal{F}'(U) = X(a,U)_\beta' = \varprojlim_{U \supset K} X(a,K)_\beta'.$$

In particular, $\mathcal{F}'(U)$ is a Frechet space. Since we may identify $X(a,K)_\beta'$ with $X'/X(a,K)^\perp$, $\mathcal{F}'(U)$ consists of all families

(3.13) $$u = (u_K / X(a,K)^{\perp})_{U \supset K}$$

with the property that $u_K - u_L \in X(a,K)^{\perp}$ for all compact sets K, L in U with $K \subset L$. The duality between X(a,U) and $\mathcal{F}(U)$ is given by

(3.14) $$\langle x,u \rangle = \langle x,u_K \rangle,$$

where u and K are as above and $x \in X(a,K)$. For open sets $V \subset U$, the canonical restriction map

$$r_V^U: \mathcal{F}(U) \to \mathcal{F}(V), \quad u \to (u_K/X(a,K)^{\perp})_{V \supset K}$$

is the adjoint of the continuous embedding $X(a,V) \to X(a,U)$. We write $u|V$ instead of $r_V^U(u)$. Clearly $\mathcal{F}$ together with the restrictions r_V^U is a presheaf.

2) We now claim that $\mathcal{F}$ is a sheaf. To this end we must verify two conditions. Let $\{U_\alpha\}$ be an open cover of U, and let $u \in \mathcal{F}(U)$ such that $u|U_\alpha = 0$ for each α. Then $u = 0$. For, since every $x \in X(a,U)$ can be written as $x = x_1 + \ldots + x_k$ with $x_j \in X(a,U_{\alpha_j})$ for suitable $\alpha_1,\ldots,\alpha_k$, it follows by (3.14) that

$$\langle x,u \rangle = \Sigma_j \langle x,u|U_{\alpha_j} \rangle = 0.$$

For each α let $u_\alpha \in \mathcal{F}(U_\alpha)$ be such that $u_\alpha = u_\beta$ on $U_\alpha \cap U_\beta$ for each α,β. Fix $\alpha_1,\ldots,\alpha_k$ and let $x_j \in X(a,U_{\alpha_j})$ be arbitrary with $x_1 + \ldots + x_k = 0$. Then we have

(3.15) $$\sum_{j=1}^{k} \langle x_j,u_{\alpha_j} \rangle = 0.$$

Assertion (3.15) is evident for $k = 1$. Assume (3.15) holds for $k-1$, and let $x_j \in X(a,U_{\alpha_j})$ $(j = 1,\ldots,k)$ with $x_1 + \ldots + x_k = 0$. Set $x_1' = x_1 + x_2$. Since $u_{\alpha_1} = u_{\alpha_2}$ on $U_{\alpha_1} \cap U_{\alpha_2}$, one obtains

$$\langle x_1,u_{\alpha_1} \rangle + \langle x_2,u_{\alpha_2} \rangle = \langle x_1', u_{\alpha_1}|U_{\alpha_1} \cap U_{\alpha_2} \rangle,$$

and hence the k-case reduces to the (k-1)-case and (3.15) is proved.

We use (3.15) to define a linear functional u on X(a,U). Let $\langle x,u\rangle = \Sigma_j\langle x_j,u_{\alpha_1}\rangle$ where $x = \Sigma_j x_j$ and $x_j \in X(a,U_{\alpha_j})$. To prove that u is continuous, it suffices to show the continuity of each restriction $u|X(a,K)$ for each compact K in U. But for such K, there are $\alpha_1,\dots,\alpha_k$, compact sets K_j in U_{α_j} for $j = 1,\dots,k$ and a constant $M > 0$ such that for each $x \in X(a,K)$ one has

$$x = \Sigma_j x_j, \quad x_j \in X(a,K_j) \quad \text{and} \quad \|x_j\| \leq M\|x\| \quad (j = 1,\dots,k).$$

The continuity of $u_{\alpha_j}|X(a,K_j)$ implies that of $u|X(a,K)$; and since $u|U_\alpha = u_\alpha$ for each α by definition, we have that $\mathcal{F}'$ is a sheaf by [3', Ch. 1, §1.5].

3) Furthermore, $\mathcal{F}'$ can be made into a Frechet sheaf of A-modules. If U is a domain of holomorphy in C^n, then each compact set K in U satisfies

$$\sigma(a',X'/X(a,K)^{\perp}) = \sigma(a,X(a,K)) \subset K^s,$$

where the last inclusion follows from Prop. 5.3.8. Then $K^s \subset U$ by [21'].

By Th. 5.1.39 and Cor. 5.1.40 the action $A(U) \times \mathcal{F}'(U) \to \mathcal{F}'(U)$ given by

$$(3.16) \qquad (f,u) \to f\cdot u = f(a',X'/X(a,K)^{\perp})(u_K/X(a,K)^{\perp})_{U\supset K}$$

defines a separately, and hence jointly, continuous bilinear map between Frechet spaces. Let U, V be domains of holomorphy with $V \subset U$, and let $f \in A(U)$ and $u \in \mathcal{F}'(U)$. Then $(f|V)\cdot(u|V) = (f\cdot u)|V$. A standard argument in the theory of Frechet sheaves shows that there is one, and only one, way to extend (3.16) to a continuous bilinear map $A(U) \times \mathcal{F}'(U) \to \mathcal{F}'(U)$ for each open set U in C^n. Thus $\mathcal{F}'$ becomes a Frechet sheaf of A-modules.

4) In this step we prove that $\mathcal{F}'$ is a soft sheaf. Since $\mathcal{F}'(U) = 0$ for each open set U satisfying $U \cap \sigma(a,X) = \emptyset$, it suffices to show that for bounded open sets U, V, W in C^n with $V \subset V^- \subset W \subset W^- \subset U$ and for each $u \in \mathcal{F}'(U)$, there is a section $v \in \mathcal{F}'(U)$ such that $v|V = u|V$ and supp $v \subset W$. By (3.14) the action of u on $x \in X(a,V)$ is given by $\langle x,u\rangle = \langle x,u_{V^-}\rangle$ (with u defined by (3.13)). By Lemma 5.3.10

$$X(a,C^n \setminus W) + X(a,V^-) = X(a,(C^n \setminus W) \cup V^-),$$

hence the left-hand side is closed since the right-hand side is. It follows that

$$(3.17) \qquad X' = X(a,C^n \setminus W)^\perp + X(a,V^-)^\perp$$

because of the equality $X(a,C^n \setminus W) \cap X(a,V^-) = \{0\}$ and [23', Th. IV.4.8]. Now (3.17) yields the decomposition $u_{V^-} = \tilde{\mathbf{v}} + \tilde{\mathbf{w}}$ with $\tilde{\mathbf{v}} \in X(a,C^n \setminus W)^\perp$ and $\tilde{\mathbf{w}} \in X(a,V^-)^\perp$. Then $v = (\tilde{\mathbf{v}}/X(a,K)^\perp)_{U \supset K}$ is an element in $\mathcal{F}(U)$ with the desired property.

5) Let J be defined by

$$(3.18) \qquad Ju = (u/X(a,K)^\perp)_{C^n \supset K} \qquad (u \in X').$$

Then J is onto. In fact, for each $v \in \mathcal{F}'(\mathbf{C}^n)$ one has $J(v_{\sigma(a,X)}) = v$. But J is also injective. If $Ju = 0$, then (3.18) implies that $u \in X(a,K)^\perp$ for each compact K in $\mathbf{C}^n$. Hence $u = 0$. As an obviously continuous bijection between Frechet spaces, J is a topological isomorphism by the open mapping theorem. By Cor. 5.1.40, J preserves the $A(\mathbf{C}^n)$-module structure, where we regard X' as an $A(\mathbf{C}^n)$-module.This completes the proof.

Theorem 5.3.2 is crucial to the proof of the next theorem.

5.3.14 Theorem. If a' has the LDP, then a has a Frechet soft sheaf model.

Proof. The proof proceeds as in the previous proof.

1) For each open set U in $\mathbf{C}^n$, define the Frechet space F(U) as the projective limit

$$\mathcal{F}(U) = \varprojlim_{U \supset K} X/{}^\perp X'(a',K)$$

with respect to the natural mappings

$$X/{}^\perp X'(a',L) \to X/{}^\perp X'(a',K)$$

where $K \subset L$ are compact in U. By Th. 5.3.2 and the bipolar theorem we may consider the duality system $\langle X/{}^\perp X'(a',K), X'(a',K)\rangle$. Since $\mathcal{F}(U)$

is a reduced projective limit, it follows from [29', Ch. IV, Th. 4.4] that

$$\mathcal{F}(U)_\tau' = \text{'}\varinjlim_{U \supset K} X'(a',K)_\tau,$$

where τ denotes Mackey topology of the duals. Since the inductive limit is formed with respect to the dual spectrum, it can be identified with $X'(a',U)$ together with the inductive locally convex topology of the imbeddings of $X'(a',K)_\tau$ into $X'(a',U)$, where K ranges over all compact sets in U. In the rest of this proof, $X'(a',U)$ will always have this topology. Since $\mathcal{F}(U)$ is barreled, it follows that $\mathcal{F}(U) = (\mathcal{F}(U)_\tau')_\beta' = [X'(a',U)]_\beta'$. The duality between $\mathcal{F}(U)$ and $X'(a',U)$ is given by $\langle u,x\rangle = \langle x_K,u\rangle$, where $u \in X'(a',L)$, L is compact in U and $x = (x_K/{}^\perp X'(a',K))_{U \supset K}$. For open sets $V \subset U$ in C^n the canonical restriction map

$$r_V^U: \mathcal{F}(U) \to \mathcal{F}(V), \qquad x \to x|V = (x_K/{}^\perp X'(a',K))_{V \supset K}$$

is the adjoint of the continuous embedding of $X'(a',V)$ into $X'(a',U)$. Clearly $\mathcal{F}$, together with the restrictions r_V^U, is a presheaf.

2) $\mathcal{F}$ is a sheaf. The proof of this proceeds just as the analogous case in the previous proof up to proving the continuity of the linear functional. Suppose that the linear functional x has been defined on $X'(a',U)$ by the family $\{x_\alpha\}$ with $x_\alpha \in \mathcal{F}(U_\alpha)$ and $U = \cup U_\alpha$; so that $\langle u,x\rangle = \Sigma_j\langle u_j,x_{\alpha_j}\rangle$ where $u = \Sigma u_j$ and $u_j \in X'(a',U_{\alpha_j})$. To prove that x is continuous, it suffices to show that, for each compact K in U, the kernel of the restriction $x|X'(a',K)$ is closed in the weak topology induced on $X'(a',K)$ by the duality between $X/{}^\perp X'(a',K)$ and $X'(a',K)$. Since this topology is the relative one of $\sigma(X',X)$ on $X'(a',K)$, we only have to verify that $\ker(x) \cap X'(a',K)$ is $\sigma(X',X)$-closed. Let $\{u_\gamma\}$ be a net in $\ker(x) \cap X'(a',K)$ converging to u in the $\sigma(X',X)$ topology. By the Krein-Smulyan theorem, we may assume that $\|u_\gamma\| \le 1$ for all γ. As in Th. 5.3.13, let $\{U_\alpha\}$ be an open cover of U. Fix indices $\alpha_1,\dots,\alpha_k$, compact sets $K_j \subset U_{\alpha_j}$ and $M > 0$ such that for each $v \in X'(a',K)$ there are $v_j \in X'(a',K_j)$ with the properties

$$v = v_1 + \dots + v_k, \qquad \|v_j\| \le M\|v\| \quad (j = 1,\dots,k).$$

Then each u_γ has the decomposition

$$u_\gamma = u_{\gamma,1} + \dots + u_{\gamma,k}, \qquad \|u_{\gamma,j}\| \le M\|u_\gamma\|.$$

By Th. 5.3.2 and the Alaoglu-Bourbaki theorem, we may suppose that for each j the net $\{u_{\gamma,j}\}$ converges to $u_j \in X'(a',K_j)$ in the $\sigma(X',X)$ topology. Then $u = u_1 + \dots + u_k$, and

$$\begin{aligned}\langle u,x\rangle &= \Sigma_j \langle u_j, x_{\alpha_j}\rangle = \lim_\gamma \Sigma_j \langle u_{\gamma,j}, x_{\alpha_j}\rangle \\ &= \lim_\gamma \langle u_\gamma, x\rangle = 0.\end{aligned}$$

Hence $u \in \ker(x)$, and by Th. 5.3.2 $u \in X'(a',K)$; this is the desired result.

3) $\mathcal{F}$ can be made into a Frechet sheaf of A-modules. Let U be a domain of holomorphy in $\mathbf{C}^n$. Since the restriction of a' to X'(a',K) is similar to the adjoint of the n-tuple induced by a on $X/{}^\perp X'(a',K)$, one has

$$\sigma(a, X/{}^\perp X'(a',K)) = \sigma(a', X'(a',K)) \subset K^s \subset U.$$

Therefore, the map $A(U) \times \mathcal{F}(U) \to \mathcal{F}(U)$ defined by

$$(f,x) \to f \cdot x = f(a, X/{}^\perp X'(a',K))(x_K/{}^\perp X'(a',K))_{U \supset K}$$

makes $\mathcal{F}(U)$ into a topological A(U)-module. Extending this map to arbitrary open sets in $\mathbf{C}^n$ turns $\mathcal{F}$ into a Frechet sheaf of A-modules.

4) The softness of $\mathcal{F}$ follows in the same way as that of $\mathcal{F}'$. We only have to indicate that the equality $X = {}^\perp X'(a',K) + {}^\perp X'(a',L)$ holds for arbitrary closed disjoint sets K, L in $\mathbf{C}^n$ by [23', Th. IV.4.8] and the fact that X'(a',K), X'(a',L) are $\sigma(X',X)$-closed, hence

$$X'(a', K \cup L) = X'(a',K) + X'(a',L)$$

is $\sigma(X',X)$-closed, and that $X'(a',K) \cap X'(a',L) = \{0\}$.

5) Moreover, the map

$$J: X \to F(\mathbf{C}^n), \qquad Jx = (x_K/{}^\perp X'(a',K))_{\mathbf{C}^n \supset K},$$

is a topological isomorphism of $A(\mathbf{C}^n)$-modules.

In the proof of Th. 5.3.17 we shall use Serre's duality principle that the dual of an exact Koszul complex is also exact [2']. Let ${}^{X}A$ denote the sheaf of analytic X-valued functions on C^n, let $\mathcal{C}^\infty$ be the sheaf of C^∞-functions on R^n and let E be a $\mathcal{C}^\infty$-module. We denote by $E_{p,q}$ the sheaf of differential forms with values in E of type (p,q) with respect to $\sigma = (\sigma_1,\ldots,\sigma_n)$ and $d\bar{z} = (d\bar{z}_1,\ldots,d\bar{z}_n)$.

5.3.15 Proposition. For each domain of holomorphy U in C^n, there is a canonical algebraic isomorphism

$$H_c^p(U,{}^{X}A) \cong \begin{Bmatrix} 0, & \text{if } p \neq n \\ A(U)_\beta' \hat{\otimes} X & \text{if } p = n \end{Bmatrix}. \tag{3.19}$$

Proof. By Lemma 5.1.18 the $\bar{\partial}$-operator yields an exact sequence

$$0 \to A(U) \xrightarrow{i} \mathcal{C}_{0,0}^\infty(U) \xrightarrow{\bar{\partial}} \mathcal{C}_{0,1}^\infty(U) \xrightarrow{\bar{\partial}} \cdots \tag{3.20}$$

$$\cdots \xrightarrow{\bar{\partial}} \mathcal{C}_{0,n}^\infty(U) \to 0.$$

Each space in (3.20) is a nuclear (FM)-space. By Serre's lemma [2'], the strong dual sequence of (3.20) is exact and consists of topological homomorphisms, i.e. open maps onto their ranges. But then the exactness is preserved by topological tensor product with the complete (DF)-space X (Banach spaces are (DF)-spaces):

$$0 \to [\mathcal{C}_{0,n}^\infty(U)]' \hat{\otimes} X \xrightarrow{\bar{\partial}} [\mathcal{C}_{0,1}^\infty(U)]' \hat{\otimes} X \xrightarrow{\bar{\partial}} \cdots$$

$$\cdots \to [\mathcal{C}_{0,0}^\infty(U)]' \hat{\otimes} X \xrightarrow{\bar{\partial}} [A(U)]' \hat{\otimes} X \to 0$$

by [30', Props. 4.2 and 4.5]; here all duals have their strong dual topology. The space $[\mathcal{C}_{0,j}^\infty(U)]' \hat{\otimes} X$ can be identified with the space of sections with compact support over U in the sheaf ${}^{X}\mathcal{D}_{0,n-j}$ of X-valued distributions. Since

$$0 \to {}^X A \xrightarrow{i} {}^X\mathcal{D}_{0,0} \xrightarrow{\bar\partial} {}^X\mathcal{D}_{0,1} \xrightarrow{\bar\partial} \dots \xrightarrow{\bar\partial} {}^X\mathcal{D}_{0,n} \to 0$$

is a resolution of ${}^X A$ by soft sheaves, (3.19) follows from [19', Th. II.4.7.1].

We also need the following lemma, whose proof can be found in [27', Lemma 1.3].

5.3.16 Lemma. Let $\mathcal{F}$ be a sheaf of abelian groups on C^n. If supp($\mathcal{F}$) is compact and $H_c^p(U,\mathcal{F}) = 0$ for $p \geq 1$ and for every domain of holomorphy U in C^n, then $\mathcal{F}$ is a soft sheaf.

5.3.17 Theorem. If the commuting system a and its adjoint a' both have property (β), then a' has a Frechet soft sheaf model.

Proof. It follows from the proof of Prop. 5.3.6 that if a' has SVEP then

$$(3.21) \qquad 0 \to A^0 \to A^1 \to \dots \to A^n \to \mathcal{F} \to 0$$

is a natural resolution of the canonical sheaf model $\mathcal{F}$ of a', where $\mathcal{A}^j$ denotes the sheaf ${}^Y A$ where Y denotes the direct sum of $\binom{n}{j}$ copies of X' and the sequence of sheaf homomorphisms is the one induced by $\delta_{\alpha'} = (z_1 - a_1')\sigma_1 + \dots + (z_n - a_n')\sigma_n$. Let U be a domain of holomorphy. The SVEP implies that $H_c^p(U,A^j) = 0$ for $p = 0,1,\dots,n-1$ and $j = 0,1,\dots,n$. A standard argument shows that the p-th cohomology group of the induced sequence

$$(3.22) \qquad 0 \to H_c^n(U, A^0) \to H_c^n(U, A^1) \to \dots \to H_c^n(U, A^n) \to 0$$

is isomorphic to $H_c^p(U, \mathcal{F})$. In fact, one can split the resolution (3.21) into a system of short exact sequences and study the induced long exact sequence of cohomology groups with compact support. On the other hand, by [19', Th. II.4.7.2] one can show that, up to the isomorphism of Prop. 5.3.15, the sequence (3.22) of cohomology groups coincides with the following sequence

$$0 \to [A(U)]_\beta' \hat{\otimes} X \to [A(U)]_\beta' \hat{\otimes} X^{\binom{n}{1}} \to \dots \to [A(U)]_\beta'^{\binom{n}{n}} \to 0,$$

i.e. with the strong dual sequence of

$$0 \to A(U,X) \xrightarrow{\delta_\alpha} \Lambda^1[\sigma,A(U,X)] \xrightarrow{\delta_\alpha} \dots \xrightarrow{\delta_\alpha} \Lambda^n[\sigma,A(U,X)] \to 0.$$

If a has property (β), then, due to Serre's duality lemma, $H_c^p(U,\mathcal{F}') = 0$ for $p \geq 1$, Lemma 5.3.16 implies that the sheaf $\mathcal{F}$ is soft. Therefore, if a' also has property (β), then the canonical sheaf model $\mathcal{F}$ of a' is a Frechet soft sheaf model for a'. This completes the proof.

We say that a form f is δ_α-closed if it lies in the closure of the range of δ_α.

5.3.18 Proposition. If a has the LDP, then the following statements hold:

(i) if D, U_1, ..., U_k are domains of holomorphy in C^n such that D $\cap(U_1 \cup \dots \cup U_k) = \emptyset$, then every analytic form f on D with values in $X(a,U_1 \cup \dots \cup U_k)$ can be decomposed into $f = f_1 + \dots + f_k$ with $f_j \in X(a,U_j)$ and δ_α-closed for $1 \leq j \leq k$.

(ii) a has property (β).

Proof. (i). If k = 1, then there is nothing to prove. So assume that the assertion holds for k-1. Let f and D, U_1, ..., U_k be given as in (i). Decompose f into a linear sum $f = f_1 + \dots + f_k$ where each f_j is $X(a,U_k)$-valued and defined on D. Since $\delta_\alpha f = 0$, one has $\delta_\alpha f_k = - \delta_\alpha f_1 - \dots - \delta_\alpha f_{k-1}$. By the induction hypothesis, $\delta_\alpha f_k$ can be written as a sum $\delta_\alpha f_k = g_1 + \dots + g_{k-1}$ of δ_α-closed analytic forms g_j with values in $X(a,U_j \cap U_k)$, the last following from the fact that $\delta_\alpha f_k \in X(a,(U_1 \cup \dots \cup U_{k-1}) \cap U_k)$. By Prop. 5.3.8 each $g_j = \delta_\alpha h_j$ for some analytic h_j with values in $X(a,U_j \cap U_k)$. Clearly $\delta_\alpha(f - f_k + h_1 + \dots + h_{k-1}) = 0$ and

$$f - f_k + h_1 + \dots + h_{k-1} = f_1 + \dots + f_{k-1} + h_1 + \dots + h_{k-1} \in X(a,U_1 \cup \dots \cup U_{k-1}).$$

Again, by the inductive hypothesis, there are δ_α-closed $X(a,U_j)$-valued forms ψ_j ($j = 1,\dots,k-1$) such that $f - f_k + h_1 + \dots + h_{k-1} = \psi_1 + \dots + \psi_{k-1}$. Then

$$f = \psi_1 + \dots + \psi_{k-1} + (f_k - h_1 - \dots - h_{k-1})$$

has all properties mentioned in (i).

(ii) Since a has SVEP, by Definition 5.3.12 it suffices to verify the other condition of Def. 5.3.4. Assume that $f \in \Lambda^n[\sigma, A(U,X)]$ is δ_α-closed where U is an open polydisc in $\mathbf{C}^n$. Then there exists a sequence of forms $\{\xi_k\}$ in $\Lambda^{n-1}[\sigma, A(U,X)]$ such that $f = \lim \delta_\alpha \xi_k$. Let G, H be open polydiscs with $G^- \subset H$. By the LDP of a, the following decomposition holds:

$$X = X(a, \mathbf{C}^n \backslash G) + X(a, H^-).$$

Set $Y = X(a, \mathbf{C}^n \backslash G)$ and $Z = X(a, H^-)$. Then

$$\sigma(a, X/Y) = \sigma(a, Z/Y \cap Z) \subset \sigma(a, Z) \cup \sigma(a, Y \cap Z) \subset H^-.$$

Now assume $H^- \subset U$. By Cor. 5.1.50, $\hat{f} = \delta_{\alpha/Y} \hat{\xi}$ for some $\hat{\xi} \in \Lambda^n[\sigma, A(U, X/Y)]$, where $\hat{x}$ denotes the coset $x + Y$ for $x \in X$. Setting $g = f - \delta_\alpha \xi$, we have $g \in \Lambda^n[\sigma, A(U,Y)]$ and

$$g = \lim_{k \to \infty} \delta_\alpha(\xi_k - \xi). \tag{3.23}$$

By part (i) we may choose a suitable decomposition of g so that with the help of Prop. 5.3.8 we can write $g = \delta_\alpha \eta_D$ for some form η_D on every open polydisc D with $D^- \subset U$. Since (β) is a local property by Prop. 5.3.11, it follows that a has property (β).

5.3.19 Theorem. For the commuting system $a = (a_1, \ldots, a_n)$, the following statements are equivalent:

(i) a has a Frechet soft sheaf model;
(ii) a' has a Frechet soft sheaf model;
(iii) a has the local decomposition property;
(iv) a' has the local decomposition property;
(v) a and a' both have property (β).

If one of (i)-(v) holds, then the spectral subspaces of a and a' satisfy

$$X(a,F) = {}^{\perp}X'(a', \mathbf{C}^n \backslash F), \quad X'(a',F) = X(a, \mathbf{C}^n \backslash F)^{\perp}. \tag{3.24}$$

for each closed F in $\mathbf{C}^n$.

Proof. If we prove (i) $\Rightarrow$ (iii) $\Rightarrow$ (ii) $\Rightarrow$ (iv) $\Rightarrow$ (i), then the equivalence of (i)-(v) follows from Th. 5.3.17 and Prop. 5.3.18. But (iii) $\Rightarrow$ (ii) and (iv) $\Rightarrow$ (i) follow from Ths. 5.3.13, 5.3.14, resp. It then remains to prove (i) $\Rightarrow$ (iii) and (ii) $\Rightarrow$ (iv); (3.25) is a result of (3.3), (3.4).

Since implications (i) $\Rightarrow$ (iii) and (ii) $\Rightarrow$ (iv) are essentially the same, we prove only the former. Let $\mathcal{F}$ be a Frechet soft sheaf model for a, and let $J: X \to \mathcal{F}(\mathbf{C}^n)$ be a topological isomorphism of $A(\mathbf{C}^n)$-modules. For each closed F in $\mathbf{C}^n$, the space $X(F) = J^{-1}\Gamma_F(\mathbf{C}^n,\mathcal{F})$ is a closed $A(\mathbf{C}^n)$-submodule of X. Since the $A(\mathbf{C}^n)$-module structure of $\Gamma_F(\mathbf{C}^n,\mathcal{F})$ has a natural extension to an $A(F)$-module structure, the same is true for $X(F)$. Therefore, Th. 5.1.9 shows that for each compact F in $\mathbf{C}^n$ we have $\sigma(a,X(F)) \subset F^s$. If F is a closed polydisc in $\mathbf{C}^n\backslash\sigma(a,X)$, then

$$X(F) = J^{-1}\Gamma_F(\mathbf{C}^n,\mathcal{F}) = J^{-1}\{0\} = \{0\}.$$

Hence $X(F) = X(F \cap \sigma(a,X))$ for all closed F in $\mathbf{C}^n$. Furthermore, we can prove that $X(a,F) = X(F)$ and conditions (i) and (ii) of Def. 5.3.12 are satisfied. Thus a has the LDP.

5.3.20 Lemma. Let a be a commuting system in $L(X)$, and let Y be an a-invariant subspace. If both $a|Y$ and a/Y have property (β), then a also has property (β).

Proof. By Ths. 5.1.1 and 5.1.21, for every domain of holomorphy U in $\mathbf{C}^n$ there is a canonical long exact sequence of cohomology,

$$\text{(3.25)} \qquad \ldots \to H^p(A(U,Y),\delta_\alpha|Y) \to H^p(A(U,X),\delta_\alpha) \to$$

$$H^p(A(U,X/Y),\delta_\alpha/Y) \to H^{p+1}(A(U,Y),\delta_\alpha|Y) \to \ldots,$$

induced by the short exact sequence $0 \to Y \to X \to X/Y \to 0$. If both $a|Y$ and a/Y have SVEP, then a does also by (3.25). Now assume that both $a|Y$ and a/Y have property (β). Let D and U be open polydiscs in $\mathbf{C}^n$ such that $D^- \subset U$ and let $\{\xi_k\}$ be a sequence of forms in $\Lambda^{n-1}[\sigma,A(U,X)]$ such that $\delta_\alpha\xi_k \to 0$. Then $\delta_{\alpha/Y}\hat{\xi}_k \to 0$, where $\hat{\xi}_k$ is the coset determined by ξ_k in $\Lambda^{n-1}[\sigma,A(U,X/Y)]$. By the property ($\beta$) for a/Y, there exists a sequence $\{\hat{\psi}_k\}$ in $\Lambda^{n-1}[\sigma,A(U,X/Y)]$ such that

$$\delta_{\alpha/Y}\hat{\xi}_k = \delta_{\alpha/Y}\hat{\psi}_k \text{ and } \hat{\psi}_k \to 0 \text{ uniformly on } D.$$

Let $\phi_k \in \Lambda^{n-1}[\sigma, A(U,X)]$ represent $\hat{\psi}_k$ with $\phi_k \to 0$. Then $\delta_{\alpha/Y}(\hat{\xi}_k - \hat{\psi}_k) = 0$ implies $\delta_\alpha(\xi_k - \phi_k) \in \Lambda^n[\sigma, A(U,Y)]$. Furthermore, we have $\delta_\alpha(\xi_k - \phi_k) \to 0$. Since $a|Y$ also has property (β), there is a sequence of forms $\{\zeta_k\}$ in $\Lambda^{n-1}[\sigma, A(U,Y)]$ such that $\delta_\alpha(\xi_k - \phi_k) = \delta_\alpha \zeta_k$ and $\zeta_k \to 0$ uniformly on D. If we put $\eta_k = \phi_k + \zeta_k$, then $\delta_\alpha \xi_k = \delta_\alpha \eta_k$ on U and $\eta_k \to 0$ uniformly on D. Since (β) is a local property (Prop. 5.3.11), it follows that a has property (β).

5.3.21 Theorem. The commuting system $a = (a_1,\dots,a_n)$ has LDP if and only if one of the following statements holds:

(i) there exists an a-invariant subspace Y such that both $a|Y$ and a/Y have LDP;

(ii) there exists an a-invariant subspace Y such that both $a|Y$ and $a'|Y^\perp$ have LDP.

Proof. If a has LDP, then (i) holds for $Y = \{0\}$. The equivalence (i) $\Leftrightarrow$ (ii) is evident from duality considerations. Now assume that (i) holds. Then both $a|Y$ and a/Y have property (β), and hence a has property (β) by Lemma 5.3.20. Similarly a' also has property (β).

In the final part of this section, we study the duality theory for σ'-decomposable commuting systems on Banach space. First, some new criteria besides those in Th. 5.2.29 will be proved. The following is a generalization of [32, Th. 5.8].

5.3.22 Theorem. Given the commuting system $a = (a_1,\dots,a_n)$, suppose that for every pair of open polydiscs G, H with $G^- \subset H$ there exists an a-invariant subspace Y such that

(i) $\sigma(a,Y) \subset \mathbf{C}^n \backslash G$,

(ii) $\sigma(a,X/Y) \subset H^-$.

Then a has property (β).

Proof. To prove first that a has SVEP, let U be an open polydisc, choose open polydiscs G, H such that $G^- \subset H \subset H^- \subset U$ and let Y be an a-invariant subspace satisfying (i),(ii). For fixed $p = 1,\dots, n-1$, let $\xi \in \Lambda^p[\sigma, A(U,X)]$ be such that $\delta_\alpha \xi = 0$. In the quotient space, one obtains $\delta_{\alpha/Y}\hat{\xi} = 0$, where $\hat{\xi}$ is the coset determined by ξ. By Th.

5.1.48 there is a form $\zeta^* \in \Lambda^p[\sigma, A(U,X/Y)]$ satisfying $\delta_{\alpha/Y}\zeta^* = \hat{\xi}$. Let ζ be a form such that $\hat{\zeta} = \zeta^*$. Then $\xi - \delta_\alpha\zeta \in Y$ and $\delta_\alpha(\xi - \delta_\alpha\zeta) = 0$ on U. Condition (i) implies that there is a form $\chi \in \Lambda^{p-1}[\sigma, A(G,Y)]$ such that $\delta_\alpha\chi = \xi - \delta_\alpha\zeta$. Therefore $\xi = \delta_\alpha(\chi + \zeta)$. Since G may be taken as arbitrary and SVEP is a local property, a has SVEP by Th. 5.2.4.

Before proving property (β), we make the following claim: if $\{\xi_k\}$ is a sequence of forms in $\Lambda^{n-1}[\sigma, A(U,X)]$ such that

$$\delta_\alpha\xi_k \to 0, \tag{3.26}$$

then for every polydisc G with $G^- \subset U$ there is a sequence $\{\eta_k\}$ in $\Lambda^{n-1}[\sigma, A(G,X)]$ such that $\delta_\alpha\xi_k = \delta_\alpha\eta_k$ and $\eta_k \to 0$. Let H,Y be as in the last paragraph. By (3.26) $\delta_{\alpha/Y}\hat{\xi}_k \to 0$. Since $\sigma(a,X/Y) \subset U$, it follows from Cor. 5.1.50 that there is a form $\zeta_k^* \in \Lambda^{n-1}[\sigma, A(U,X)]$ for each k such that

$$\zeta_k^* \to 0 \text{ and } \delta_{\alpha/Y}\hat{\xi}_k = \delta_{\alpha/Y}\zeta_k^* \text{ on U.} \tag{3.27}$$

Exactness of $0 \to A(U,Y) \to A(U,X) \to A(U,X/Y) \to 0$ and (3.27) imply the existence of a form ζ_k with $\hat{\zeta}_k = \zeta_k^*$ for each k and such that

$$\zeta_k \to 0. \tag{3.28}$$

Now $\delta_{\alpha/Y}\hat{\zeta}_k = \delta_{\alpha/Y}\zeta_k^* = \delta_{\alpha/Y}\hat{\xi}_k$ implies $\delta_\alpha(\xi_k - \zeta_k) \in \Lambda^n[\sigma, A(U,Y)]$, and (3.26), (3.28) imply $\delta_\alpha(\xi_k - \zeta_k) \to 0$. Since $\sigma(a,Y) \subset C^n\backslash G$, for each k there is $\chi_k \in \Lambda^{n-1}[\sigma, A(G,Y)]$ such that $\delta_\alpha\chi_k = \delta_\alpha(\xi_k - \zeta_k)$ on G and $\chi_k \to 0$. Let $\eta_k = \chi_k + \zeta_k$. Then $\delta_\alpha\xi_k = \delta_\alpha\eta_k$ on G and $\eta_k \to 0$.

To complete the proof of the theorem, it suffices to prove that if $\delta_\alpha\xi_k \to 0$, then there exists a sequence $\{\phi_k\}$ in $\Lambda^{n-1}[\sigma, A(U,X)]$ such that $\delta_\alpha\xi_k = \delta_\alpha\phi_k$ for each k and $\phi_k \to 0$ on every compact set in U. Let K be compact in U, and let G be as in the last paragraph with $K \subset G$. The sequence $\{\eta_k\}$ determined above and a having SVEP provide another sequence $\{\psi_k\}$ in $\Lambda^{n-2}[\sigma, A(G,X)]$ such that $\xi_k - \zeta_k = \delta_\alpha\psi_k$ for each k. Using the Taylor expansion, we may write $\psi_k = \psi_k' + \psi_k''$ where the coefficients of ψ_k' are polynomials and $\psi_k'' \to 0$ uniformly on K.

Let $\phi_k = \xi_k - \delta_\alpha \psi_k'$ ($= \eta_k + \delta_\alpha \psi_k''$). Then $\phi_k \in \Lambda^{n-1}[\sigma, A(U,X)]$, $\delta_\alpha \phi_k = \delta_\alpha \xi_k$ and $\phi_k \to 0$ uniformly on K because $\phi_k = \eta_k + \delta_\alpha \psi_k''$. This completes the proof.

5.3.23 Corollary. For the commuting system a, if we use
(i) $\sigma'(a,Y) \subset C^n \backslash G$,
(ii) $\sigma'(a,X/Y) \subset H^-$,
in place of (i), (ii) of Th. 5.3.22, then a has property (β).

Proof. This is immediate from Th. 5.3.22, because $\sigma(a,X) \subset \sigma'(a,X)$ is always true.

5.3.24 Proposition. With the conditions of Cor. 5.3.23,

$$\sigma'(a,X(a,F)) \subset F \tag{3.29}$$

for every closed set F in C^n.

Proof. Let $z \notin F$ be fixed, and let G, H be open polydiscs satisfying $z \in G \subset G^- \subset H$ and $H^- \cap F = \varnothing$. We claim that $X(a,F) \subset Y$ for the Y given in Cor. 5.3.23. Let $x \in X(a,F)$. There exists a form $\xi \in \Lambda^{n-1}[\sigma, A(C^n \backslash F, Y)]$ such that $x\sigma_1 \wedge \dots \wedge \sigma_n = \delta_\alpha \xi$ on $C^n \backslash F$. Passing to the quotient space X/Y, one has $\hat{x}\sigma_1 \wedge \dots \wedge \sigma_n = \delta_{\alpha/Y} \hat{\xi}$ on $C^n \backslash F$. Since $\sigma(a,X/Y) \subset H$ and $H^- \cap F = \varnothing$, it follows from Prop. 5.2.7 that $\hat{x} = 0$, and hence $x \in Y$, equivalently, $X(a,F) \subset Y$. With this inclusion one can prove that X(a,F) is closed. (This latter fact can also be proved directly from property (β)).

It remains to show (3.29). For $z \notin \sigma'(a,Y)$, there exists $b = (b_1, \dots, b_n)$ in the commutant of $a|Y$ such that

$$\sum_{j=1}^{n} (z_j - a_j|Y) b_j = I|Y. \tag{3.30}$$

Because X(a,F) is b-invariant, (3.30) implies

$$\sum_{j=1}^{n} (z_j - a_j|X(a,F))(b_j|X(a,F)) = I|X(a,F).$$

Thus $z \notin \sigma'(a,X(a,F))$, and (3.29) follows.

The next theorem generalizes Th. 1.2.1, (i)-(vi).

5.3.25 Theorem. For the commuting system a, the following statements are equivalent:

(i) a is σ'-decomposable;

(ii) for every pair of open polydiscs G,H with $G^- \subset H$, there exist a-invariant subspaces X_G, X_H such that $X = X_G + X_H$, $\sigma'(a,X_H) \subset H$ and $\sigma'(a,X_G) \subset C^n \backslash G$;

(iii) for every pair of open polydiscs G,H with $G^- \subset H$, there exist a-invariant subspaces Y, Z such that $\sigma'(a,Y) \subset C^n \backslash G$, $\sigma'(a,X/Y) \subset H^-$, $\sigma'(a,Z) \subset H^-$ and $\sigma'(a,X/Z) \subset C^n \backslash G$;

(iv) both a and a' have property (β) and $\sigma'(a,X(a,F)) \subset F$ for every closed set F.

Proof. The proof is organized as (i) $\Rightarrow$ (ii) $\Rightarrow$ (iv) $\Rightarrow$ (i) and (i) $\Rightarrow$ (iii) $\Rightarrow$ (iv).

(i) $\Rightarrow$ (ii). Clear.

(ii) $\Rightarrow$ (iv). Since $\sigma(a,X) \subset \sigma'(a,X)$, one has

$$\sigma(a,X_G) \subset C^n \backslash G, \qquad \sigma(a,X_H) \subset C^n \backslash H$$

. By Remark 5.2.12, a has the SVEP and $X(a,F)$ is closed for closed F. The argument of Th. 5.2.29 implies that $\sigma'(a,X(a,F)) \subset F$. Thus we may set

$$X_G = X(a,\mathbf{C}^n \backslash G) \quad \text{and} \quad X_H = X(a,H^-). \tag{3.31}$$

Then (3.31) implies $\sigma'(a,X_H/X_H \cap X_G) \subset H^-$. Therefore, $\sigma'(a,X/X_G) \subset H^-$ since the coinduced systems of a on $X_H/X_H \cap X_G$ and X/X_G are similar. By Cor. 5.3.23, a has property (β).

As for a', let $Y' = X_H^{\perp}$. Then

$$\sigma'(a',X'/Y') \subset \sigma'(a,X_H) \subset H^-;$$
$$\sigma'(a',Y') \subset \sigma'(a,X/X_H) = \sigma'(a,X_H/X_H \cap X_G) \subset C^n \backslash G.$$

Now a' has property (β) by Cor. 5.3.23.

(iv) $\Rightarrow$ (i) follows from Th. 5.3.19 and the inclusion (3.29).

(i) $\Rightarrow$ (iii). For every pair of open polydiscs G, H with $G^- \subset H$, set $Y = X(a, \mathbf{C}^n \backslash G)$, $Z = X(a, H^-)$. Then Y,Z satisfy the spectral inclusions in (iii).

Since (iii) $\Rightarrow$ (iv) is similar to (ii) $\Rightarrow$ (iv), the proof is complete.

5.3.26 Theorem. If the commuting system a is σ'-decomposable, then a' is also.

Proof. With notations as in Th. 5.3.25(iii), we set $Z' = Y^\perp$ and $Y' = Z^\perp$. Then

$$\sigma'(a',Z') \subset \sigma'(a,X/Y) \subset H^-;$$
$$\sigma'(a',X'/Z') \subset \sigma'(a,Y) \subset \mathbf{C}^n\backslash G;$$
$$\sigma'(a',Y') \subset \sigma'(a,X/Z) \subset \mathbf{C}^n\backslash G;$$
$$\sigma'(a',X'/Y') \subset \sigma'(a,Z) \subset H^-.$$

Hence a' is σ'-decomposable.

Our last corollary is immediate.

5.3.27 Corollary. If X is reflexive, then the commuting system a is σ'-decomposable if and only if a' is σ'-decomposable.

4. Perturbation Theory

This section deals with the perturbation theory for commuting systems of operators on Banach space. Some of these results are generalizations of some theorems in Chapter 2, e.g. Th. 2.2.6.

Let $b = (b_1,\dots,b_m)$ and $a = (a_1,\dots,a_n)$ be commuting systems of operators on some Banach space X. such that (b,a) is a commuting (m+n)-system. We denote the operator

$$(\beta,\alpha) = (z_1 - b_1,\dots, z_m - b_m, z_{m+1} - a_1,\dots, z_{m+n} - a_n).$$

Assume that $D = P \times Q$ where P, Q are open polydiscs in $\mathbf{C}^m$, $\mathbf{C}^n$, resp., and $\sigma(a,X) \subset Q$. Indeterminates will be denoted by $\sigma = (\sigma_1,\dots,\sigma_{m+n})$, $\tau = (\sigma_1,\dots,\sigma_m)$ and

$$\delta_{(\beta,\alpha)} = (z_1-b_1)\sigma_1+\ldots+(z_m-b_m)\sigma_m+(z_{m+1}-a_1)\sigma_{m+1}+\ldots+(z_{m+n}-a_n)\sigma_{m+n}$$

$$\delta_\beta = (z_1-b_1)\sigma_1+\ldots+(z_m-b_m)\sigma_m.$$

5.4.1 Lemma. Under the above assumptions, there are cochain maps

$$\rho: \Lambda^{p+n}[\sigma,A(D,X)] \to \Lambda^p[\tau,A(P,X)],$$

$$i: \Lambda^p[\tau,A(P,X)] \to \Lambda^{p+n}[\sigma,A(D,X)]$$

between the cochains such that

$$I = \rho i \quad \text{and} \quad I - \rho i = \delta_{(\beta,\alpha)}\gamma + \gamma\delta_{(\beta,\alpha)} \tag{4.1}$$

for a suitably chosen map γ:

$$\ldots \overset{\gamma}{\leftarrow} \Lambda^p[\sigma,A(D,X)] \overset{\gamma}{\leftarrow} \Lambda^{p+1}[\sigma,A(D,X)] \overset{\gamma}{\leftarrow} \ldots .$$

Proof. We first consider the case $n = 1$. Define ρ, γ to be zero on forms which do not contain σ_{m+1} and put

$$\rho(f\sigma_I \wedge \sigma_{m+1}) = \frac{1}{2\pi i}\left(\int_\Gamma R(w,a)f(z,w)\,dw\right)\sigma_I; \tag{4.2}$$

$$i(f\sigma_I) = f\sigma_I \wedge \sigma_{m+1}; \tag{4.3}$$

$$\gamma(f\sigma_I \wedge f\sigma_{m+1}) = \frac{(-1)^p}{2\pi i}\left(\int_\Gamma \frac{R(w,a)}{w - z_{m+1}} f(z,w)\,dw\right)\sigma_I \tag{4.4}$$

for $z \in P$, $z_{m+1} \in Q$ and $\sigma_I = \sigma_{j_1} \wedge \ldots \wedge \sigma_{j_p}$ $(1 \le j_p \le m)$. The contour Γ in (4.2),(4.4) must surround $\sigma(a,X)$, $\sigma(a,X) \cup \{z_{m+1}\}$, resp., while $f \in A(D,X)$. By Cauchy's integral formula, both integrals (4.2),(4.4) are independent of the choice of Γ. It is easy to verify that ρ, i, γ satisfy (4.1).

The general case can be obtained by an iteration of 1-dimensional reductions. If, for instance, $I - i_1\rho_1 = \delta_1\gamma_1 + \gamma_1\delta_1$ and the identity on the image space of ρ_1 is of the form $I = i_2\rho_2 + \delta_2\gamma_2 + \gamma_2\delta_2$, then

$$\begin{aligned} I - (i_1 i_2)(\rho_2\rho_1) &= I - i_1\rho_1 + i_1(I - i_2\rho_2)\rho_1 \\ &= \delta_1\gamma_1 + \gamma_1\delta_1 + i_1\delta_2\gamma_2\rho_1 + i_1\gamma_2\delta_2\rho_1 \\ &= \delta_1(\gamma_1 + i_1\gamma_2\rho_1) + (\gamma_1 + i_1\gamma_2\rho_1)\delta_1 \end{aligned}$$

provided ρ_j, i_j $(j=1,2)$ are cochain maps for δ_1, δ_2. Therefore, in the general case, i is given by $if = f \wedge \sigma_{m+1} \wedge ... \wedge \sigma_{m+n}$ and ρ can be defined as the composition of all maps occurring in the 1-dimensional reductions (cf.(4.2)).

5.4.2 Remark. Lemma 5.4.1 states that, under the given assumptions, the cochains induced by $\delta_{(\beta,\alpha)}$,

$$0 \to A(D,X) \xrightarrow{\delta_{(\beta,\alpha)}} \cdots \xrightarrow{\delta_{(\beta,\alpha)}} \Lambda^{m+n}[\sigma,A(D,X)] \to 0$$

and by δ_β

$$0 \to A(D,X) \xrightarrow{\delta_\beta} \cdots \xrightarrow{\delta_\beta} \Lambda^{m}[\sigma,A(D,X)] \to 0$$

are cochain equivalent (see [25', p. 40]).

In what follows we assume that a has the LDP. Let $D \subset C^{m+n}$, $U \subset C^{m+n}$ be open. By an X(a,U)-analytic form of degree p on D we mean a form $f \in \Lambda^p[\sigma,A(D,X)]$ with the property that for some compact $K \subset U$ all coefficients of f have values in X(a,K).

The following lemma is an analogue of Prop. 5.3.18(i).

5.4.3 Lemma. Let a have LDP and let $D = P \times Q$ where P, Q are open polydiscs in $\mathbf{C}^m$, $\mathbf{C}^n$, resp. Assume that $U_1,...,U_k$ are domains of holomorphy in $\mathbf{C}^n$ with $U_j \cap Q = \emptyset$ for $1 \le j \le k$. Then each $\delta_{(\beta,\alpha)}$-closed $X(a,U_1 \cup ... \cup U_k)$-analytic form f on D has a decomposition $f = f_1 + ... + f_k$ into a sum of $\delta_{(\beta,\alpha)}$-closed $X(a,U_j)$-analytic forms f_j on D.

Proof. We prove the lemma by induction on k. For k = 1 there is nothing to prove, hence suppose the lemma holds for k-1. Let D, $U_1,..., U_k$, f be as in the hypothesis. Since a has LDP, f has the decomposition $f = f_1 + ... + f_k$, where f_j is an $X(a,U_j)$-valued analytic form defined on D for $j = 1,...,k$. By the induction hypothesis, the form

$$\delta_{(\beta,\alpha)} f_k = -\delta_{(\beta,\alpha)} f_1 - \dots - \delta_{(\beta,\alpha)} f_{k-1}$$

can be written as

$$\delta_{(\beta,\alpha)} f_k = g_1 + \dots + g_{k-1}$$

of $\delta_{(\beta,\alpha)}$-closed $X(a,U_j \cap U_k)$-valued analytic forms g_j on D. Let $K_j \subset U_j \cap U_k$ be compact such that g_j is $X(a,K_j)$-valued. By the projection property for the Taylor spectrum, one has

$$D \cap \sigma((b,a),X(a,K_j)) \subset P \times (Q \cap K_j^s) \subset P \times (Q \cap U_j) = \emptyset.$$

Therefore, we may write $g_j = \delta_{(\beta,\alpha)} h_j$ for a suitable $X(a,U_j \cap U_k)$-valued analytic form h_j. Again by the induction hypothesis, there are $\delta_{(\beta,\alpha)}$-closed $X(a,U_j)$-valued analytic forms u_j with

$$f - f_k + (h_1 + \dots + h_{k-1}) = u_1 + \dots + u_{k-1}.$$

Then $f = u_1 + \dots + u_{k-1} + (f_k - (h_1 + \dots + h_{k-1}))$ is the required decomposition.

5.4.4 Proposition. Assume that a has LDP and that K is a closed set in the open set U in $\mathbf{C}^n$. If $D = P \times Q$ where P, Q are open polydiscs in $\mathbf{C}^m$, $\mathbf{C}^n$, resp., such that $Q^- \cap K = \emptyset$, then every $\delta_{(\beta,\alpha)}$-closed analytic from f on D with values in X(a,K) can be written as $f = \delta_{(\beta,\alpha)} g$ for a suitable analytic form g on D with values in X(a,U).

Proof. Without loss of generality, we assume that K is compact, for otherwise we replace it with $K \cap \sigma(a,X)$. Let $U, U_1, \dots, U_k$ be open polydiscs in $\mathbf{C}^n$ with the properties

$$K \subset U_1 \cup \dots \cup U_k \subset U, \qquad Q \cap U_j = \emptyset \text{ (all } j).$$

By Lemma 5.4.3, we can write $f = f_1 + \dots + f_k$ where f_j are $\delta_{(\beta,\alpha)}$-closed $X(a,U_j)$-valued forms on D. As in the proof of this lemma, each f_j can be written as $f_j = \delta_{(\beta,\alpha)} g_j$ for suitable $X(a,U_j)$-valued analytic forms on D. Therefore, $f = \delta_{(\beta,\alpha)}(g_1 + \dots + g_k)$.

If a has LDP, then for each open polydisc Q in C^n we define the space $E(Q) = X(a, C^n \backslash Q)$. It is a routine exercise to verify that

$$\sigma(a, X/E(Q)) \subset Q^-. \tag{4.5}$$

5.4.5 Theorem. If a has the LDP, then the following statements are equivalent:

(i) (b,a) has SVEP;

(ii) if P, Q are open polydiscs in C^m, C^n, resp., and f is an analytic form of degree $p \leq m-1$ on P such that $\delta_\beta f$ has values in $E(Q)$, then there are open concentric polydiscs $P_0 \subset P$, $Q_0 \subset Q$ and an analytic form g of degree $p-1$ on P_0 such that $f - \delta_\beta g$ has values in $E(Q_0)$ on P_0.

Proof. Using the notation of Lemma 5.4.1, we assume first that (b,a) has SVEP. Let f have the hypotheses of (ii), and assume that Q_0, Q_1 are open polydiscs in C^n such that $Q_0^- \subset Q_1 \subset Q_1^- \subset Q$. Since $\delta_\beta f$ has values in $E(Q)$, so does the form $\delta_{(\beta,\alpha)}(if) = i(\delta_\beta f)$. By Prop 5.4.4 there is an analytic form h on $D = P \times Q_1$ with values in $E(Q_1)$ such that $\delta_{(\beta,\alpha)}(if) = \delta_{(\beta,\alpha)}h$. Since (b,a) has SVEP, there is an analytic form ϕ on D such that $if - h = \delta_{(\beta,\alpha)}\phi$. Hence $if/E(Q_0) = \delta_{(\beta,\alpha)}(\phi/E(Q_0))$ on $D = P \times Q_1$ (Recall $E(Q_1) \subset E(Q_0)$). The inclusion $\sigma(a, X/E(Q_0)) \subset Q_1$ implies that Lemma 5.4.1 applies to $(b,a)/E(Q_0)$ and yields

$$f/E(Q_0) = \rho i(f/E(Q_0)) = \delta_\beta \rho(\phi/E(Q_0)).$$

Let g be an analytic form on P representing $\rho(\phi/E(Q_0))$. Then $f - \delta_\beta g$ has values in $E(Q_0)$.

Conversely, assume (ii) and that f is an analytic $\delta_{(\beta,\alpha)}$-closed form of degree $p \leq m+n-1$ on the open polydisc $D \subset C^{m+n}$. Let $P = D_1 \times \ldots \times D_m$ and choose an open polydisc Q with Q^- contained in $D_{m+1} \times \ldots \times D_{m+n}$. Applying Lemma 5.4.1 to $(\beta,\alpha)/E(Q)$, one obtains

$$0 = \rho\delta_{(\beta,\alpha)}(f/E(Q)) = \delta_\beta \rho(f/E(Q)).$$

Choose $h \in \rho\,\delta_{(\beta,\alpha)}(f/E(Q))$ on P. By assumption there are open concentric polydiscs $P_0 \subset P$, $Q_0 \subset Q$ and an analytic form ϕ on P such that $h - \delta_\beta\phi$ has values in $E(Q_0)$ on P_0. Since $\sigma(a,X/E(Q_0)) \subset Q_0^-$, we may define $\tilde{\rho}$ with respect to $(b,a)/E(Q_0)$ as described in Lemma 5.4.1. Then it is easy to see that $h/E(Q_0) = \tilde{\rho}(f/E(Q_0))$. Consequently,

$$\begin{aligned}\delta_{(\beta,\alpha)}i(\phi/E(Q_0)) &= i(h/E(Q_0)) = i\tilde{\rho}(f/E(Q_0)) \\ &= f/E(Q_0) - \delta_{(\beta,\alpha)}\gamma(f/E(Q_0)) - \gamma\delta_{(\beta,\alpha)}(f/E(Q_0))\end{aligned}$$

holds on $D_0 = P_0 \times (D_{m+1} \times ... \times D_{m+n})$. The last term is zero, and hence there is an analytic form g_0 on D_0 such that $f - \delta_{(\beta,\alpha)}g_0$ has values in $E(Q_0)$. From Prop. 5.4.4, $f = \delta_{(\beta,\alpha)}g$ holds for a suitable analytic form g defined in a neighborhood of the center of D. Since the SVEP is a local property, the proof is complete.

5.4.6 Remark. We shall use the fact that for a commuting system a with the SVEP an analytic form ξ defined on a domain of holomorphy U in C^n is δ_α-exact on U if and only if it is locally δ_α-exact on U. If ξ is of degree $p < n$, then the conclusion follows from Th. 5.2.4, so it is nontrivial only in case ξ has degree n. In this case, we may first note that $\xi = (\delta_\alpha + \bar{\partial})\eta$ for a suitable C^∞-form η on U by Lemma 5.2.3. Then the application of Lemma 5.2.2 yields the conclusion.

5.4.7 Theorem. Assume that a has the LDP and (b,a) has the SVEP. Then the following are equivalent:

(i) (b,a) has property (β);

(ii) if P, Q are open polydiscs in C^m, C^n, resp., and $\{f_k\}$ is a sequence of forms in $\Lambda^m[\tau,A(P,X)]$ which converges to $f \in \Lambda^m[\tau,A(P,X)]$ such that $f_k/E(Q) = \delta_\beta(g_k/E(Q))$ has a solution $g_k \in \Lambda^{m-1}[\tau,A(P,X)]$ for each $k \in \mathbf{N}$, then there are open concentric polydiscs $P_0 \subset P$, $Q_0 \subset Q$ and a solution $g \in \Lambda^{m-1}[\tau,A(P,X)]$ for $f/E(Q_0) = \delta_\beta(g/E(Q_0))$ on P_0.

Proof. (i) $\Rightarrow$ (ii). For the sequence $\{f_k\}$ in (ii), one has

$$(f_k/E(Q))\wedge\sigma_{m+1}\wedge...\wedge\sigma_{m+n} = \delta_{(\beta,\alpha)}((g_k/E(Q))\wedge\sigma_{m+1}\wedge...\wedge\sigma_{m+n})$$

on $D = P \times Q$. By Prop. 5.4.4 and Remark 5.4.6 there are analytic forms h_k on D such that $f\wedge\sigma_{m+1}\wedge...\wedge\sigma_{m+n} = \delta_{(\beta,\alpha)}h$. Since (b,a) has property (β), it follows that $f_k \wedge \sigma_{m+1} \wedge \ldots \wedge \sigma_{m+n} = \delta_{(\beta,\alpha)} h_k$ for some analytic form h on D. Let Q_0 be an arbitrary open polydisc with $Q_0^- \subset Q$. Then Lemma 5.4.1 applied to $(b,a)/E(Q_0)$ implies

$$f/E(Q_0) = \rho\delta_{(\beta,\alpha)}(h/E(Q_0)) = \delta_\beta\rho(h/E(Q_0)) = \delta_\beta(g/E(Q_0)),$$

where g is a form in the coset $\rho(h/E(Q_0))$.

(ii) $\Rightarrow$ (i). Let D be an open polydisc in C^{m+n} and let $\{g_k\}$ be a sequence of forms in $\Lambda^{m+n-1}[\sigma,A(D,X)]$ such that $f_k = \delta_{(\beta,\alpha)} g_k \to f \in \Lambda^{m+n}[\sigma,A(P,X)]$. Define $P = D_1 \times \ldots \times D_m$ and choose an open polydisc $Q \subset C^n$ with $Q^- \subset D_{m+1} \times \ldots \times D_{m+n}$. Passing to the quotient space $X/E(Q)$ and applying the corresponding cochain map ρ, we obtain

$$\delta_\beta\rho(g_k/E(Q)) = \rho(f_k/E(Q)).$$

Since the cochain maps ρ, i defined in Lemma 5.4.1 are continuous, $\rho(f_k/E(Q)) \to \rho(f/E(Q))$. The canonical embedding

$$q: \Lambda^m[\tau,A(P,X)] \to \Lambda^m[\sigma,A(P,X/E(Q))], \qquad h \to h/E(Q),$$

is a continuous linear surjective map between Frechet spaces. By the open mapping theorem, there are forms h_k, $h \in \Lambda^m[\tau,A(P,X)]$ such that $q(h_k) = \rho(f_k/E(Q))$, $q(h) = \rho(f/E(Q))$ and $h_k \to h$. By (ii) there are open concentric polydiscs $P_0 \subset P$, $Q_0 \subset Q$ and a form $\phi \in \Lambda^{m-1}[\tau,A(P_0,X)]$ with $h/E(Q_0) = \delta_\beta(\phi/E(Q_0))$ on P_0. Exactly as in the proof of Th. 5.4.5, it follows that $f = \delta_{(\beta,\alpha)} g$ for a suitable analytic form g defined in a neighborhood of the center of D. Since (β) is a local property, it follows that (b,a) has property (β).

5.4.8 Remark. If condition (ii) of Th. 5.4.7 is true only for the special case $f_k = x_k \sigma_1 \wedge \ldots \wedge \sigma_m$, $f = x\sigma_1 \wedge \ldots \wedge \sigma_m$, where $x_k \to x$ in X, then one obtains a necessary and sufficient condition for the closedness of the spectral subspaces of (b,a). The proof is almost the same.

Using Ths. 5.3.19 and 5.4.7, we can prove the following criterion for the inheritance of LDP.

5.4.9 Theorem. Assume that a has the LDP. If, for every open polydisc $Q \subset C^n$, the commuting systems induced by b on $X(a,Q^-)$ and on $X/X(a,C^n\backslash Q)$ have LDP, then (b,a) has LDP.

Proof. By Ths. 5.4.5 and 5.4.7, the commuting system (b,a) has property (β). By Th. 5.3.19, we have

$$X'(a',C^n\backslash Q) \subset X(a,Q_0^-)^\perp \subset X'(a',C^n\backslash Q)$$

for every pair of open polydiscs Q_0, Q in $\mathbf{C}^n$ with $Q_0^- \subset Q$. Since the commuting systems $b'/X(a,Q_0^-)^\perp$ and $[b|X(a,Q_0^-)]'$ are similar, $b'/X(a,Q_0^-)^\perp$ has property (β). Ths. 5.4.5 and 5.4.7 can now be used in an obvious way to deduce that (b',a') has property (β). Therefore, (b,a) has the LDP by Th. 5.3.19.

5.4.10 Corollary. Assume that (b,a) satisfies the assumptions of Th. 5.4.9. Then each single operator $f(b,a)$ $f \in A(\sigma(b,a))$ is decomposable. In particular, if a, b are single operators and satisfy the assumptions of Th. 5.4.9, then $a + b$ and ab are decomposable.

Proof. It follows from [6'] that $f(b,a)$ is decomposable if (b,a) has the LDP. Then the corollary follows easily.

In the final part we shall see a similar result for the existence of a strong spectral capacity, defined as follows.

5.4.11 Definition. Assume that $a = (a_1,\dots,a_n)$ is a commuting system. A spectral capacity $\mathcal{E}$ for a is called a <u>strong</u> spectral capacity if it satisfies in addition to Def. 5.2.21(iii) the condition

$$\mathcal{E}(F) = \mathcal{E}(F\cap U_1^-) + \dots + \mathcal{E}(F\cap U_k^-)$$

for every closed F and every open cover $\{U_1,\dots,U_k\}$ of F.

We require the following lemma.

5.4.12 Lemma. Let Y be an invariant subspace for the commuting system a such that a/Y has the SVEP. Then

(i) a has SVEP if and only if $a|Y$ does;

(ii) if a has property (β), then so does $a|Y$;

(iii) for each $y \in Y$ one has $\sigma_{a|Y}(y) = \sigma_a(y)$, or equivalently, $Y(a|Y,F) = X(a,F) \cap Y$ for every closed F in C^n.

Proof. (i) For every domain of holomorphy U in $\mathbf{C}^n$, by Th. 5.1.1 there is a canonical long exact sequence of cohomology,

$$(4.6)\quad 0 \to H^0(A(U,Y),\delta_\alpha|Y) \to H^0(A(U,X),\delta_\alpha) \to H^0(A(U,X/Y),\delta_\alpha/Y)$$

$$\to H^1(A(U,Y),\delta_\alpha|Y) \to \ldots \to H^n(A(U,Y),\delta_\alpha/Y)$$

$$\to H^n(A(U,X),\delta_\alpha) \to H^n(A(U,Y),\delta_\alpha/Y) \to 0,$$

induced by the short exact sequence

$$0 \to Y \to X \to X/Y \to 0.$$

Assertion (ii) follows easily from the fact that all maps in (4.6) are continuous if the cohomology spaces are equipped with their natural quotient topologies. For (iii), let $y \in Y$ and let $f \in \Lambda^{n-1}[\sigma,A(D,X)]$ satisfy $y\sigma_1 \wedge\ldots\wedge \sigma_n = \delta_a f$ on some open polydisc D in $\mathbf{C}^n$. If a/Y has SVEP, then there is an analytic form on D such that $f/Y = \delta_a(g/Y)$. Hence $h = f - \delta_a g$ is a form with values in Y such that $\delta_a h = y\sigma_1 \wedge\ldots\wedge \sigma_n$ on D.

5.4.13 Definition. Let a = be a commuting system. If a has the SVEP, its spectral subspaces are closed and $X = X(a,\overline{U_1}) + \ldots + X(a,\overline{U_k})$ for every finite open cover $\{U_1,\ldots,U_k\}$ of $\mathbf{C}^n$, then a is said to have the <u>global spectral decomposition</u> property (GDP).

5.4.14 Theorem. Assume that (b,a) has SVEP and that its spectral subspaces are closed. If for each closed set F in $\mathbf{C}^{m+n}$, the system $b/X((b,a),F)$ has property (β) and the restrictions $b|X((b,a),F)$, $a|X((b,a),F)$ have the GDP, then (b,a) has a strong spectral capacity.

Proof. Since (b,a) has the SVEP, from [6', Th. 2.1] one has $X((b,a),G \times H) = X(b,G) \cap X(a,H)$ for each pair of closed sets G, H in $\mathbf{C}^m$, $\mathbf{C}^n$ resp. Let $F \subset \mathbf{C}^{m+n}$, $H \subset \mathbf{C}^n$ be closed and let $Y = X((b,a),F)$. We claim

$$(4.7)\qquad Y(a|Y,H), = X((b,a), F\cap(\mathbf{C}^m \times H)).$$

Clearly, the right hand side of (4.7) contains the left hand side. Let Z denote the space on the right hand side. then $Z(a|Z,D) = \{0\}$ for each closed set D in $\mathbf{C}^n\setminus H$. Since $a|Z$ has the GDP, it follows that $Z =$

$Z(a|Z,H)$. Hence

$$(4.8) \qquad \sigma(a,Z) \subset H.$$

In fact, we have first $\cup\{\sigma_{a|Z}(a): x \in Z\} \subset \sigma(a,Z)$. Conversely, if $\zeta \in \cap\{\rho_{a|Z}(x): x \in Z\}$, then $H^n(Z, \zeta - a) = 0$. But $a|Z$ has SVEP, hence one has $H^p(Z, \zeta - a) = 0$ for $0 \le p \le n-1$. Hence $z \notin \sigma(a,Z)$. Thus

$$(4.9) \qquad \cup\{\sigma_{a|Z}(a): x \in Z\} = \sigma(a,Z).$$

Since the left hand side of (4.9) is clearly contained in H, (4.8) holds and implies (4.7). Let $\mathcal{E}_a(H) = Y(a|Y,H)$. Then $\mathcal{E}_a$ is the spectral capacity for $a|Y$. Similarly, $b|Y$ has a spectral capacity $\mathcal{E}_b$ defined by

$$\mathcal{E}_b(G) = Y(b|Y,G) = X((b,a),F\cap(G \times \mathbf{C}^n)).$$

The argument above actually implies that the restrictions $b|Y$ to all spectral subspaces of $a|Y$ have the LDP. It follows from the proof of Th 5.4.9 that $((b,a)|Y)'$ has property (β). Theorem 5.4.7 and Lemma 5.4.12(ii) can be applied to verify that $(b,a)|Y$ has property (β). Just note that $Y/X((b,a),F\cap(\mathbf{C}^m \times H))$ is a subspace of $X/X((b,a),F\cap(G \times \mathbf{C}^n))$ and that the system induced by b on the quotient of these latter two spaces is similar to b/Y, which thus has the SVEP by Lemma 5.4.12.

Since $(b,a)|Y$ and its adjoint both have property (β), $(b,a)|Y$ has the LDP by Th. 5.3.19.

The proof will be complete if we can verify the spectral inclusion $\sigma((b,a),Y) \subset F$. Let $z \in \mathbf{C}^{m+n}\backslash F$ and let $D = P \times Q$ be an open polydisc centered at z with $D^- \subset \mathbf{C}^{m+n}\backslash F$. If P_0 is an open polydisc concentric with P in $\mathbf{C}^m$ and $P_0^- \subset P$, then

$$\begin{aligned} Y &= Y \cap X(b,\mathbf{C}^m\backslash P_0) + Y \cap X(b,P^-) \\ &= Y \cap X(b,\mathbf{C}^m\backslash P_0) + Y \cap X(a,\mathbf{C}^n\backslash Q). \end{aligned}$$

Let X_1, X_2 denote the first, second space, resp., in the last expression. Then

$$\sigma(b,X_1) \subset \mathbf{C}^m\backslash P_0 \quad \text{and} \quad \sigma(a,X_2) \subset \mathbf{C}^n\backslash Q$$

Therefore, $Y = (z - (b,a))Y$, i.e.

$$Y = \Sigma_{j=1}^{m} (z_j - b_j)Y + \Sigma_{k=1}^{n} (z_k - a_k)Y,$$

and hence $H^{m+n}(Y,(z-(b,a))) = 0$. This, together with the fact that $(b,a)|Y$ has SVEP, shows that $z \notin \sigma((b,a),Y)$.

Finally, assume that $\{U_1,\ldots,U_k\}$ is an open cover of F. Then

$$\begin{aligned} Y &= Y((b,a),U_1^-) + \ldots + Y((b,a),U_k^-) \\ &\subset X((b,a),F\cap U_1^-) + \ldots + X((b,a),F\cap U_k^-) \\ &\subset Y. \end{aligned}$$

Hence $\mathcal{E}(F) = X((b,a),F)$ defines a strong spectral capacity $\mathcal{E}$ for (b,a).

5.4.15 Remark. It is not very hard to see that the sufficient conditions given in Th. 5.4.14 for the existence of a strong spectral capacity are also necessary. In fact, if a commuting system has a strong spectral capacity, then each restriction and each quotient modulo one of its spectral subspaces still has a strong spectral capacity and this can be proved as in the one dimensional case. Thus the necessity follows easily from the fact that the decomposability of (b,a) implies that of b.

NOTES AND COMMENTS

CHAPTER I

The main results on equivalent conditions of operators with the spectral decomposition property (Theorem 1.2.1) were proved by S. Wang and I. Erdelyi [95,96] and R. Lange and S. Wang [64,65]. A result like Th. 1.2.1(ix) first appeared in S. W. Huang [49]. The equivalent condition that T is decomposable if and only if both T and T^* have property ($\boldsymbol{\beta}$) is used specifically to prove the main theorem on perturbations of decomposable operators (Theorem 2.2.6).

Equivalent conditions for strongly decomposable operators and operators decomposable relative to the identity (Ths. 1.3.2 and 1.3.5) were proved by R. Lange and S. Wang [66].

Theorems 1.3.11 and 1.4.18 were proved by S. Wang and R. Lange [98]. The motivation for introducing quasi-strong decomposability (called the open-restriction decomposition property in [99]) is that E. Albrecht [6] proved that the operator in Th. 1.4.18 is decomposable but the restriction $T|X(T,[0,1])$ is not decomposable. The question arises whether there exists a decomposable T and a two-dimensionsal closed set F such that $T|X(T,F)$ is not decomposable. Although the answer seems to be unknown, the notion leads naturally to the introduction of the class of operators treated in Th. 1.3.11.

The fact that the class of decomposable operators has a symmetric duality theory was shown by I. Erdelyi and S. Wang in [31]. But Th. 1.4.29, which together with Th. 1.4.28 was proved by S. Wang [94], shows that the class of strongly decomposable operators does not share this symmetry.

Th. 1.4.30 was proved by R. Lange in [55], and Ths. 1.4.32 and 1.4.34 were proved by R. Lange and S. Wang in [65].

Boundedly decomposable operators are due to R. Evans [36], who proved Ths. 1.4.4 and 1.4.5. The basis of the proof of Th. 1.4.4 may be found in [35].

R. Lange introduced the notion of analytically decomposable operators in [55].

We mention finally that the term "superdecomposable" has been used to denote both Classes 4 and 7 (**§4**) ([68] and [83], resp.); but since G. Shulberg's paper [83] appeared first, we use his sense.

The question of M. Radjabalipour whether there is a decomposable Hilbert space operator which is not strongly decomposable was answered affirmatively by J. Eschmeier [12']. The dual property of (**β**) was established by C. K. Fong in his interesting paper [16'].

CHAPTER II

The main theorem on commuting perturbation of operators with the spectral decomposition property (Th. 2.2.6) is new and was proved by I. Erdelyi and S. Wang in [34], where one operator may be unbounded; a similar result with a more complicated proof may be found in [89]. The results on perturbation of ASD-operators (Th. 2.2.13 and Cor. 2.2.16) appeared originally in Lange [55].

Theorem 2.3.1 on noncommuting compact perturbations of certain decomposable operators was proved by Radjabalipour and Radjavi [78]. Our line of proof follows theirs, especially Lemmas 2.3.2-2.3.5 and Th. 2.3.6, up to the juncture Lemma 2.3.9, which allows us to invoke Th. 1.2.1 instead of using the older argument. Cor. 2.3.11 follows from a result in Erdelyi-Wang {32, Th. 17.20].

The first proof that weak contractions are decomposable was given by Jafarian [51], and subsequently Lange [59] gave a much simpler elementary proof of this fact including Cor. 2.3.15. Cors. 2.3.16 and 2.3.17 as well as Th. 2.3.20 first appeared in [61].

Most of the results in Chapter 2.4 concern quasisimilarity of certain operators having spectral decomposition properties. However, Th. 2.4.4 uses some ideas summarized in a paper by Fillmore, Stampfli and Williams [37], but it allows us to apply BDF-theory (Th. 2.3.19) and our results on biquasitriangularity to prove a perturbation of type (B) (e.g. Cor. 2.4.5).

CHAPTER III

The original idea for weak decomposability seems to be due to I. Colojoara and C. Foias [24], who observed that weak contractions satisfy Def.1.4.1(f). A. Jafarian developed this notion in his dissertation [50]. Other contributions were made by Lange [54], [55], and Erdelyi-Lange [30, Ch. 3]. For a time it was unclear whether weakly decomposable operators form a proper generalization of decomposable operators until Albrecht [3] settled this question with Ex. 3.2.11. It is still unknown if the classes of decomposable and weakly decomposable are distinct on reflexive spaces.

Weak decomposability relative to the identity is a natural generalization of Def. 1.3.5. On the other hand, our Ex 3.2.16 (due originally to Albrecht [4]; see also Th. 1.4.14) gives an instance of the violation of the linear order of the previously studied classes as shown in Chap. 1.4. This class was first studied by Wang, Zhong and Lange [101].

The results of Chapter III suggest that weak decomposability relative to the identity is a useful notion. For example, §4 gives a rather wide generalization of the case of A-spectral operators [24, Ch. 3.4]. The material of §3 extends some results of Apostol [13] on algebras of decomposable operators.

The notion of automatic continuity goes back to B. Johnson [52], who proved a version for normal operators. These results are codified and extended by Sinclair in [84].

In the development of the proof of Th.3.5.20, Lemma 3.5.2 is classical [84]. Lemma 3.5.10 is a special case of a result of Albrecht and Neumann [9]. Our proof of Th. 3.5.20 is elementary and complete. Though a variant of earlier proofs for special cases ([84], [68]), the inequality (5.9) gives an explicit crucial step allowing one to infer that θ is continuous on $p(T)X$.

Th. 3.5.20 is a direct generalization of the principal result of Albrecht and Neumann in [9, Th. 4.3]. This extension is not trivial because our Prop. 3.6.8 follows from Th. 3.5.20, but [9, Th 4.3] is not applicable. A similar remark holds for Th. 3.6.9.

The results of §7 are published here for the first time. They show, roughly, that an operator with the automatic continuity property (ACP) has a near converse (to Th. 3.5.20) in the sense that ACP $\Rightarrow$ SVEP (Th. 3.7.8). Cor 3.7.14-Th. 3.7.18 are clearly related to Chapter 4.4 of [24].

Chapter IV

Section 4.2 contains some basic facts from the theory of uniform algebras, which was systematically treated by T. W.

Gamelin in his book [42]. Section 3 is adapted from C. Apostol [14], who made the first successful attempt to extend S. Brown's technique to the case of non-Hilbert space operators.

Section 4 follows Putinar's excellent work [72], in which the author proved that every hyponormal operator must be subscalar of order 2. An immediate consequence of this fact is that every hyponormal operator has property (β) (Cor. 4.4.8).

In §5 we used some work by S. Brown [23] on the existence of invariant subspaces for hyponormal operators with thick spectrum. As pointed out in [23], only the subdecomposability of hyponormal operators is needed to make some variant of the Brown method go through.

After the submission of this manuscript for publication, Albrecht and Chevreau [8] obtained some important results on existence of invariant subspaces for $\mathcal{P}$-operators having property (β) on a "large" part of the spectrum.

CHAPTER V

Most of this chapter is adopted from the book [34'] by F.-H. Vasilescu and the excellent work of J. Eschmeier and M. Putinar. Section 1, the preparation for the chapter, is mainly taken from [34', Chaps. I-III]. The material on homological algebra can be found in [2',25'], while the basic properties about complexes of Frechet and Banach spaces as well as those of commuting systems of operators can be found in J. L. Taylor [30', 31', 32'].

Most results in §2 were proved by S. Frunza, J. Eschmeier, and M. Putinar [7'-15',17']. We omit the details on sources on each item. Theorems 5.2.28 and 5.2.30 first appeared in G. Liu [24'].

In §3, Ths. 5.3.1-5.3.3 are adopted from J. Eschmeier [9'], and the important Th. 5.3.9 is due to Eschmeier and Putinar [14']. Theorems 5.3.20 and 5.3.23 seem to be new and were proved by the authors. Th. 5.3.24 appears in [9'], but the proof given here is new.

The material of §4 was taken completely from [6'].

In the development of multivariate operator theory, sheaf theory seems to be quite important; the reader can easily find its basic facts in [3'].

BIBLIOGRAPHY

The following references include a few items not used in the text (marked (*)), nor are they exhaustive. It should, however, provide the interested reader with enough background for further exploration of the topics discussed in the book.

1. E. Albrecht, An example of a non-regular generalized scalar operator, Rev. Roum. Math. Pures Appl. 18 (1973), 983-985.

2. ________, An example of a $C^\infty(\mathbb{C})$-decomposable operator which is not $C^\infty(\mathbb{C})$-spectral, Rev. Roum. Math. Pures Appl. 19(1974), 131-139.

3. ________, An example of a weakly decomposable operator which is not decomposable, Rev. Roum. Math. Pures Appl. 20 (1975) 855-861.

4. ________, Generalized spectral operators, in Functional analysis: surveys and recent results, Proc. Paderborn Conf. Functional Analysis, Math Studies 27, North-Holland, Amsterdam (1977), 259-277.

5. ________, On some classes of generalized spectral operators, Archiv Math. (Basel) 30(1978), 297-303.

6. ________, On two questions of I. Colojoara and C. Foias, Manuscripta Math. 25(1978), 1-15.

7. ________, On decomposable operators, Integral Equations Operator Theory 2(1979), 1-10.

8. E. Albrecht, B. Chevreau, Invariant subspaces for ℓ^p-operators having Bishop's condition (β) on a large part of the spectrum, J. Operator Theory 18(1987), 339-372.

9. E. Albrecht, M. Neumann, Automatische stetigheitseigenschaften einer klassen linearer operatoren, Math. Ann. 240(1979), 251-280.

10. ________, Automatic continuity of generalized local linear operators, Manuscripta Math. 32(1980), 263-294.

11. C. Apostol, Restrictions and quotients of decomposable operators, Rev. Roum. Math. Pures Appl. 13(1968), 607-610.

12. ________, Spectral decomposition and functional calculus, Rev. Roum. Math. Pures Appl.13(1968),1481-1528.

13. ________, Decomposable multiplication operators, Rev Roum. Math. Pures Appl.17(1972), 323-333.

14. ________, The spectral flavor of Scott Brown's techniques, J. Operator Theory 6(1980), 3-12.

15. C. Apostol, C. Foias, D. Voiculescu, Some results on nonquasitriangular operators. IV, Rev. Roum. Math. Pures Appl. 18(1973), 387-514

16. I. Bacalu, On restrictions and quotients of decomposable operators, Rev. Roum. Math. Pures Appl. 18(1973), 809-813.

17*. ________, S-decomposable operators in Banach space, Rev. Roum. Math. Pures Appl.20(1975), 1101-1107

18. ________, Some properties of decomposable operators, Rev. Roum. Math. Pures Appl. 21 (1976), 177-194.

19. C. A. Berger, B. I. Shaw, Intertwining, analytic structure and the trace norm estimate, in Proc. Conf. Operator Theory, Lecture Notes in Math. 345, Springer, New York, 1973.

20. E. Bishop, A duality theorem for an arbitrary operator, Pacific J. Math. 9(1959), 379-397.

21. L. G. Brown, R. G. Douglas, P. Fillmore, Unitary equivalence module the compact operators and extensions of C*-algeras, in Proc. Conf. Operator Theory, Math Lecture Notes 345, Springer, 1973.

22. S. W. Brown, Some invariant subspaces for subnormal operators, Integral Equations Operator Theory 1(1978), 310-333.

23. ________, Hyponormal operators with thick spectrum have invariant subspaces, Ann. Math. 125(1987), 93-103.

24. I. Colojoara, C. Foias, Theory of generalized spectral operators, Gordon & Breach, New York, 1968.

25. H. R. Dowson, Spectral theory of linear operators, Academic Press, London, 1978.

26.* N. Dunford, Spectral operators, Pacific J. Math 4(1954), 321-354.

27. N. Dunford, J. T. Schwartz, Linear operators, Part I, Wiley, New York, 1962.

28. ________, Linear operators, Part II, Wiley, New York, 1967.

29. ________, Linear operators, Part III, Wiley, New York, 1971.

30. I. Erdelyi, R. Lange, Spectral decomposition on Banach spaces, Lecture Notes in Math 623, Springer, New York, 1977.

31. I. Erdelyi, S. Wang, A spectral duality theorem for closed operators, Pacific J. Math. 114(1984), 73-93.

32. ________, A local spectral theory for closed operators, Cambridge Univ. Press, Cambridge, 1985.

33. ________, Equivalent conditions to the spectral decomposition property for closed operators, submitted.

34. ________, The spectral decomposition property of the sum and product of two commuting operators, Tohoku Math. J. 41(1989), 657-672..

35. R. Evans, Embedding C(K) in B(X), Math. Scand. 48(1981), 119-136.

36. ________, Boundedly decomposable operators and the continuous functional calculus, Rev. Roum. Math. Pures Appl. 28(1983) 465-473.

37. P. A. Fillmore, J. G. Stampfli, J. P. Williams, On the essentially

numerical range, the essential spectrum and a problem of Halmos, Acta Sci. Math. (Szeged) 33(1972), 179-192.

38. J. Finch, The single-valued extension property on a Banach space, Pacific J. Math 58(1975), 61-69.

39 C. Foias, Spectral maximal spaces and decomposable operators in Banach spaces, Archiv Math. (Basel) 14(1963), 341-349.

40. S. Frunza, A duality theorem for decomposable operators, Rev. Roum. Math. Pures Appl. 16(1971), 1055-1058.

41. ________, The single-valued extension property for coinduced operators, Rev. Roum. Math. Pures Appl. 18(1973), 1061-1065.

42. T. W. Gamelin, Uniform algebras, Chelsea, New York, 1984.

43. T. W. Gamelin, J. Garnett, Pointwise bounded approximation and Dirichlet algebras, J. Funct. Analysis 8(1971), 360-404.

44. I. Gohberg, M. G. Krein, Introduction to Theory of nonself-adjoint operators, AMS Translations 18, Amer. Math. Soc., Providence, 1969.

45. P. R. Halmos, A Hilbert space problem book, Van Nostrand, New York, 1967.

46. D. Herrero, Indecomposable compact perturbations of the bilateral shift, Proc. Amer. Math. Soc. 62(1977), 254-258.

47. ________, Approximation of operators, Pitman, San Francisco, 1983.

48. L. Hørmander, Linear partial differential equations, Springer-Verlag, Berlin, 1963.

49. S. Huang, On operator-characterization of decomposable operators, Southwestern Normal College Research Rep. 2(1985), 11-18.

50. A. Jafarian, Weak and quasi-decomposable operators, Rev. Roum.

Math. Pures Appl. 22(1977), 195-212.

51. ________, Weak contractions of Sz.-Nagy and Foias are decomposable, Rev. Roum. Math. Pures Appl. 22(1977). 489-497.

52. B. E. Johnson, Continuity of linear operators commuting with continuous linear operators, Trans. Amer. Math. Soc. 128(1967), 88-102.

53. B. E. Johnson, A. M. Sinclair, Continuity of linear operators commuting with continuous linear operators, II, Trans. Amer. Math. Soc.146(1969), 533-540.

54. R. Lange, Roots of almost decomposable operators, J. Math Analysis Appl. 49 (1975), 721-724.

55. ________, Analytically decomposable operators, Trans. Amer. Math. Soc. 244 (1978), 225-240.

56. ________, Strongly analytic subspaces, in Operator Theory and Functional Analysis, Research Notes Math. 38, Pitman, San Francisco, 1979,16-30.

57. ________, A purely analytic criterion for a decomposable operator, Glasgow Math. J. 21(1980), 69-70.

58. ________, On generalization of decomposability, Glasgow Math. 22(1981), 77-81.

59. ________, On weak contractions, Bull. London Math. Soc. 13(1981), 69-72.

60. ________, Duality and asymptotic spectral decomposition, Pacific J. Math. 121 (1985), 93-108.

61. ________, Commuting boundedly decomposable perturbations, J. Math Analysis and Appl. 145(1990), 555-561.

62. ________, Biquasitriangularity and spectral continuity, Glasgow Math. J. 26(1985), 177-180.

63.* R. Lange, S. Wang, Cohyponormal operators with single-valued extension property, International J. Math. and Math. Sciences

9(1986), 659-663.

64. ________, New criteria for a decomposable operator, Illinois J. Math. 31 (1987), 438-445.

65. ________, Strongly analytic subspaces in spectral decomposition, Glasgow Math. J. 30(1988), 249-257.

66. ________, Universal notions characterizing spectral decompositions, Glasgow Math. J., to appear

67. K. B. Laursen, Algebraic spectral subspaces and automatic continuity, Czechoslovak Math. J. 38(113)(1988), 157-172

68. K. B. Laursen, M. Newman, Decomposable operators and automatic continuity, J. Operator Th. 15(1986), 33-51.

69. Yu. I. Ljubic, V. I. Macaev, On operators with decomposable spectrum, Mat. Sbornik 56(98)(1962), 433-468 (Russian).

70. B. Nagy, Operators with spectral decomposable property are decomposable, Studia Sci. Math. Hungar. 13(1978), 429-432.

71. B. Sz.-Nagy, C. Foias, Harmonic analysis of operators on Hilbert space, North-Holland, Amsterdam, 1970.

72. M. Putinar, Hyponormal operators are subscalar, J. Operator Th. 12(1984), 385-395.

73.* ________, Hyponormal operators and eigendistributions, in Operator Theory, Advances and Applications, Birkhauser, Basel, 1986.

74.* M. Radjabalipour, On decomposability of compact perturbations of operators, Proc. Amer. Math. Soc. 53(1975), 159-164.

75. ________, Some decomposable subnormal operators, Rev. Roum. Math. Pures Appl. 22(1977)341-345.

76. ________, Equivalence of decomposable and 2-decomposable operators, Pacific J. Math, 77(1978) 243-247.

77. ________, Decomposable operators. Bull. Iranian Math Soc. 9(1978),1-49.

78. M. Radjabalipour, H. Radjavi, On decomposability of perturbations of normal operators, Canadian J. Math. 27(75), 725-735.

79. ________, On invariant subspaces of compact perturbations of operators, Rev. Roum. Math. Pures Appl. 21(1976), 147-160.

80. H. Radjavi, P. Rosenthal, Invariant subspaces, Springer, New York, 1973.

81.* C. E. Rickart, General theory of Banach algebras, van Nostrand, Princeton, 1960.

82. A. Shields, A note on invariant subspaces, Mich. Math. J. 17(1970), 231-233.

83. G. W. Shulberg, Decomposable restrictions and extensions, J. Math. Analysis 83 (1981), 144-158.

84. A. M. Sinclair, Automatic continuity of linear operators, Cambridge U. Press, Cambridge, 1976.

85.* J. Snader, Strongly analytic subspaces and strongly decomposable operators, Pacific J. Math. 115(1984), 193-202.

86. J. G. Stampfli, Hyponormal operators, Pacific J. Math. 12(1962), 145-158.

87. ________, Quasisimilarity of operators, Proc. Royal Irish Acad. 81A, No. 1(1981), 109-119.

88. J. G. Stampfli, B. L. Wadhwa, An asymmetric Putnam-Fuglede theorem for dominant operators, Indiana U. Math. J. 25(1976), 359-365.

89. S. L. Sun, Some problems for decomposable operators, Doctoral dissertation, Jilin Univ., 1985 (Chinese).

90.* P. Vrbova, On the continuity of linear transformations

commuting with generalized scalar operator Casopis pro Pestovani mathmatiky 97(1972), 142-150.

91. ________, Structure of maximal spectral spaces of generalized scalar operators, Czech. Math. J. 23(1973) 493-496.

92. S. Wang, Local resolvents and operators decomposable with respect to the identity, Acta Math. Sinica 26(1983), 113-162.

93. ________, Theory of spectral decomposition with respect to the identity for closed operators, Chinese Math. Ann. Ser. B 6(1985), 269-279.

94. ________, A characterization of strongly decomposable operators and their duality theorem, Acta Math. Sinica 29(1986), 145-155>

95. S. Wang, I. Erdelyi, A duality theorem for unbounded closed operators, C. R. Rep. Acad. Sci. Canada 6(1983), 105-110.

96. ________, A spectral duality theorem for closed operators II, Chinese Math. Ann. Ser. A 6(1985), 143-151.

97. ________, A spectral duality theorem for closed operators III, Sci. Sinica, Issue A, No. 6 (1985), 25-31.

98. ________, Introduction to the theory of SDP operators, Nanjing Univ. J. Math. Issue (1985), 152-160.

99. S. Wang, R. Lange, Open-restriction decomposition property, J. Math. Analysis and Appl., to appear.

100. S. Wang, D. Sun, Strongly spectral decomposition properties, Nanjing Univ. J. of Sci. 21(1985), 223-232.

101. S. Wang, Y. Zhong, R. Lange, Automatic continuity for weakly decomposable operators, submitted.

SUPPLEMENTARY BIBLIOGRAPHY

1'. E. Albrecht, Spectral decompositions for systems of commuting operators, Proc. Roy. Irish Acad. Sec. A, 81(1981), 81-98.

2'. C. Banica, O. Stanasila, Methodes algebriques dan la theorie des espaces complexes, Paris, 1977

3'. G. E. Bredon, Sheaf theory, McGraw-Hill, Princeton, 1967.

4'. A. D. Dash, Joint spectra, Studia Math. 45 (1973), 225-237.

5'. J. Eschmeier, Spektralzerlegungen und funktional kalkule fur vertauschende tupel stetiger und abgeschlossener operatoren in Banach raumen, Schr. Math. Inst. Munster, (2) no. 20, 1981.

6' ________, Local properties of Taylor's analytic functional calculus, Invent. Math 68 (1982), 103-116.

7' ________, On two notions of the local spectrum for several commuting operators, Michigan Math. J. 30(1983), 245-248.

8' ________, Equivalence of decomposability and 2-decomposability for several commuting operators, Math. Ann. 262(1983), 305-312.

9' ________, Are commuting systems of decomposable operators decomposable?, J. Operator Theory 12(1984), 213-219.

10'. ________, Some remarks concerning the duality problem for decomposable systems of commuting operators, Operator Theory: Advances and Applications, Proceedings of the 8th Conf. on Operator Theory, Timisoara and Herculane, Birkhauser, Basel, 1984.

11'. ________, Spectral decompositions and decomposable multipliers, Manuscripta Math. 51(1985), 146-154.

12'. ________, A decomposable Hilbert space operator which is not strongly decomposable, Integral Equations and Operator Theory 11(1988), 161-172.

13' ________,Analytic spectral mapping theorems for joint spectra, Operator Theory: Adv. Appl. 24, Birkhauser, Basel,

1987, 167-181.

14'. J. Eschmeier, M. Putinar, Spectral theory and sheaf theory, III, J. fur Reine Angw. Math. 354, (1984), 150-163.

15'. ________, Invariant subspaces for subscalar operators, Arch Math. (Basel) 52(1989), 562-570.

16'. C. K. Fong, Decomposability into spectral manifolds and Bishop's property (β), Northeast Math. J. 5(1989), 391-394.

17'. S. Frunza, The Taylor spectrum and spectral decompositions, J. Funct. Analysis, 19 (1975), 390-431.

18'. ________, A characterization for the spectral capacity of a finite system of operators, Czech. Math. J. 27 (102) (1977), 356-362.

19'. R. Godement, Topologie algebrique et theorie des faisceaux, Paris, 1958.

20'. A. Groethendieck, Produits tensoriels topologiques et espaces nucleaires, Mem. Amer. Math. Soc. No. 16, Providence, 1955.

21'. R. C. Gunning, H. Rossi, Analytic functions of several complex variables, Prentice-Hall, New York, 1966.

22'. L. Hørmander, An introduction to complex analysis in several variables, Van Nostrand, Princeton, 1966.

23'. T. Kato, Perturbations theory for linear operators, Springer-Verlag, New York, 1966.

24'. G. Liu, Spectral decomposition for several commuting multiplication operators, Chinese Ann. Math. Ser. A, 6(2985) 169-178.

25'. S. MacLane, Homology, Springer-Verlag, New York, 1963.

26'. M. Putinar, Spectral theory and sheaf theory I, Operator Theory: Advances and Applications VII, Proceeding of the 7th Conf. on Operator Theory, Timisoara and Herculane, Birkhauser, Basel, 1983.

27'. ________, Spectral theory and sheaf theory II, Math. Zeitschrift 192 (1986), 473-490.

28'. ________, Spectral theory and sheaf theory IV, Proceedings of a Symposium in Pure Mathematics 51, Part 2 (1990), 273-293.

29'. H. H. Schaefer, Topological vector spaces, Van Nostrand, New York, 1966.

30'. J. L. Taylor, Homology and cohomology for topological algebras, Adv. in Math. 9(1972), 137-182.

31'. ________, A joint spectrum for several commuting operators, J. Funct. Analysis 6(1970), 172-191.

32'. ________, The analytic functional calculus for several commuting operators, Acta Math. 125 (1970), 1-38.

33'. F. Treves, Topological vector spaces, distributions and kernels, Academic Press, London, 1967.

34'. F.-H. Vasilescu, Analytic functional calculus and spectral decomposition, Reidel, Dordrecht, 1982.

INDEX

Recent Titles in This Series

(Continued from the front of this publication)

98 **Kenneth Appel and Wolfgang Haken,** Every planar map is four colorable, 1989
97 **J. E. Marsden, P. S. Krishnaprasad, and J. C. Simo, Editors,** Dynamics and control of multibody systems, 1989
96 **Mark Mahowald and Stewart Priddy, Editors,** Algebraic topology, 1989
95 **Joel V. Brawley and George E. Schnibben,** Infinite algebraic extensions of finite fields, 1989
94 **R. Daniel Mauldin, R. M. Shortt, and Cesar E. Silva, Editors,** Measure and measurable dynamics, 1989
93 **M. Isaacs, A. Lichtman, D. Passman, S. Sehgal, N. J. A. Sloane, and H. Zassenhaus, Editors,** Representation theory, group rings, and coding theory, 1989
92 **John W. Gray and Andre Scedrov, Editors,** Categories in computer science and logic, 1989
91 **David Colella, Editor,** Commutative harmonic analysis, 1989
90 **Richard Randell, Editor,** Singularities, 1989
89 **R. Bruce Richter, Editor,** Graphs and algorithms, 1989
88 **R. Fossum, W. Haboush, M. Hochster, and V. Lakshmibai, Editors,** Invariant theory, 1989
87 **Laszlo Fuchs, Rüdiger Göbel, and Phillip Schultz, Editors,** Abelian group theory, 1989
86 **J. Ritter, Editor,** Representation theory and number theory in connection with the local Langlands conjecture, 1989
85 **Bor-Luh Lin, Editor,** Banach space theory, 1989
84 **Stevo Todorcevic,** Partition problems in topology, 1989
83 **Michael R. Stein and R. Keith Dennis, Editors,** Algebraic K-theory and algebraic number theory, 1989
82 **Alexander J. Hahn, Donald G. James, and Zhe-xian Wan, Editors,** Classical groups and related topics, 1989
81 **Kenneth R. Meyer and Donald G. Saari, Editors,** Hamiltonian dynamical systems, 1988
80 **N. U. Prabhu, Editor,** Statistical inference from stochastic processes, 1988
79 **Ernst Kunz and Rolf Waldi,** Regular differential forms, 1988
78 **Joan S. Birman and Anatoly Libgober, Editors,** Braids, 1988
77 **Wang Yuan, Yang Chung-chun, and Pan Chengbiao, Editors,** Number theory and its applications in China, 1988
76 **David C. Hobby and Ralph McKenzie,** The structure of finite algebras, 1988
75 **Frank M. Cholewinski,** The finite calculus associated with Bessel functions, 1988
74 **William M. Goldman and Andy R. Magid, Editors,** Geometry of group representations, 1988
73 **Rick Durrett and Mark A. Pinsky, Editors,** Geometry of random motion, 1988
72 **R. F. Brown, Editor,** Fixed point theory and its applications, 1988
71 **James A. Isenberg, Editor,** Mathematics and general relativity, 1988
70 **Jerome Kaminker, Kenneth C. Millett, and Claude Schochet, Editors,** Index theory of elliptic operators, foliations, and operator algebras, 1988
69 **Walter A. Carnielli and Luiz Paulo de Alcantara, Editors,** Methods and applications of mathematical logic, 1988
68 **Mladen Luksic, Clyde Martin, and William Shadwick, Editors,** Differential geometry: The interface between pure and applied mathematics, 1987
67 **Kenneth A. Ribet, Editor,** Current trends in arithmetical algebraic geometry, 1987
66 **Narian Gupta,** Free group rings, 1987

(See the AMS catalogue for earlier titles)